Vortex Dynamics Theories and Applications

Edited by Zambri Harun

Published in London, United Kingdom

IntechOpen

Supporting open minds since 2005

Vortex Dynamics Theories and Applications
http://dx.doi.org/10.5772/intechopen.87912
Edited by Zambri Harun

Contributors
Martin Abrahamyan, Zambri Harun, Tajul Ariffin Norizan, Wan Hanna Melini Wan Mohtar, Luz Estella Torres Molina, Luis F. Carrion, Sara Morales, Masaaki Mori, Shigeru Yoshikawa, Evgeny Nikolaev, Maria Nikolaeva, Lunji Song, Olumayowa Kajero, Lokman Abdulkareem, Barry Azzopardi, Mukhtar Abdulkadir, Bilel Ben Amira, Ammar Meriem, Ahmed Kaffel, Zied Driss, Mohamed Salah Abid, Benjamin Oyegbile, Guven Akdogan, Brian Oyegbile

First published in London, United Kingdom, 2020 by IntechOpen
IntechOpen is the global imprint of INTECHOPEN LIMITED, registered in England and Wales, registration number: 11086078, 5 Princes Gate Court, London, SW7 2QJ, United Kingdom
Printed in Croatia

British Library Cataloguing-in-Publication Data
A catalogue record for this book is available from the British Library

Additional hard and PDF copies can be obtained from orders@intechopen.com

Vortex Dynamics Theories and Applications
Edited by Zambri Harun
p. cm.
Print ISBN 978-1-83962-616-6
Online ISBN 978-1-83962-617-3
eBook (PDF) ISBN 978-1-83962-618-0

For EU product safety concerns:
IN TECH d.o.o., Prolaz Marije Krucifikse Kozulić 3, 51000 Rijeka, Croatia, info@intechopen.com or visit our website at intechopen.com.

Meet the editor

Dr. Zambri Harun is a senior fellow at the Faculty of Engineering and Built Environment, National University Malaysia (UKM). He graduated with a BSc and MEng both from Rensselaer Polytechnic Institute, New York, US. His experience in the industry stretches over almost a decade; being an engineer at Motorola Malaysia and in a civil project. His PhD was on turbulence flow from the University of Melbourne, Australia. His research covers the wind tunnel and atmospheric turbulence as well as building and tunnel flow management.

Contents

Preface

This book discusses the latest research and findings on the complexity of flow; vortex dynamics. The first section contains vortex dynamics and turbulence theories. Traditional approaches of characterising vortices and length scales such as eddies, Batchelors and Kolmogorov length scales are discussed. Non-conventional vortex theories go beyond just delta and earth landscape formations, including also the decisive role of vortices in the process of planet formation and vortex sound theory on acoustic energy-generation used today in flue instruments. The second section contains mathematical modelling and simulations for aerodynamic bodies such as turbine and helicopter blades. The third section is on applications. Vortex characteristics are important in many aspects of our lives, from blood circulation in the arteries to the high-speed jet. Flow control and manipulation of vortices have been used in engineering applications. Here, flow-influencing devices are inserted into pump inlets and in a tank and the results show that these devices are effective in controlling flow behaviours. On a larger scale, a case study on urban heat island (UHI) study is also discussed. Larger scale vortex theories can be used to predict temperature patterns. Daily temperature, surface albedo, evapotranspiration, and anthropogenic heating are not the only parameters that must be used to explain the UHI phenomenon in a central business district (CBD) region, wind flows, which contain vortex theories, must also be considered for an adequate UHI explanation.

Zambri Harun
The National University of Malaysia,
Malaysia

Section 1

Vortex Dynamics and Turbulence Theories

Chapter 1

The Role of Micro Vortex in the Environmental and Biological Processes

Benjamin Oyegbile, Brian Oyegbile and Guven Akdogan

Abstract

This work presents a short review of the theoretical developments in the application of vortex dynamics to the processing of environmental and biological systems. The mechanisms of complex fluid-particle interaction in vortex dominated and non-vortex dominated flows are briefly discussed from theoretical and practical perspectives. Micro vortex propagation, characteristics and their various applications in environmental process engineering are briefly discussed. Several existing and potential applications of vortex dynamics in turbulent flows are highlighted and as well as the knowledge gaps in the current understanding of turbulence phenomenon with respect to its applications in the processing of solid-liquid suspension and biological systems.

Keywords: hydrodynamics, turbulence, eddies, vortex, aggregation

1. Introduction

Hydrodynamic-mediated interactions often occur in many technical and natural environmental processes. In the case of turbulent flows, this leads to the formation of turbulent eddies of various scales and sizes [1, 2]. These energy-carrying eddies often interact with particles and biological materials on various temporal and spatial scales. Eddy-particle interactions often play a crucial role in these processes and it is largely the dominant driver of mass and momentum transfer. In studying the dynamics of such complex interplay of forces, a good knowledge of the vortex dynamics and its influence on the fluid and particle dynamics is highly indispensable [3, 4].

Turbulent mixing, particle dispersion, and bioreactions have been topics of intense and sustained interest in many scientific inquiries [5, 6]. The role of mixing and turbulence-driven particle dispersion in many fluid-particle processes is well-understood owing an abundant body of knowledge from many scientific interrogations. However, turbulence as a phenomenon is still poorly understood due to its complex nature. Since mixing and chemical reactions are impacted by the presence of turbulence, it is therefore extremely important to understand the different scales of turbulence in mixing applications.

As mentioned earlier, turbulence has been shown to lead to the formation of eddies on different scales [7, 8]. The spatial degree of mixing such as the macro, meso and micro mixing are governed by the different scales of turbulence [5]. Mixing especially at high Reynolds number is often characterized by irregular, rotational, and dissipative motion containing vorticities of different energy spectra or eddy sizes [3]. It is therefore imperative to carry out qualitative and quantitative assessment of mixing efficiency in many of the practical applications involving mixing and dispersion. A number of techniques are available for quantifying mixing performance in a wide range of applications. One widely used parameter for quantifying the mixing performance is the coefficient of variation proposed by Alloca and Streiff [6]. This approach relies on the statistical analysis of the spatio-temporal homogeneity of the particle dispersion in mixing applications [9]. Several other techniques are also available for quantifying the degree of mixing in bioreactor systems. In terms of the different phenomena responsible for fluid-particle mixing, advective, turbulent, and diffusive transport depicted in **Figure 1** are the dominant ones.

In turbulent mixing, energy transfer occurs on different eddy scales or energy spectrum. These turbulent eddies consist of the large energy carrying eddies at the inertia sub-range to the smallest ones at the dissipation sub-range as shown in **Figure 2**. The important scales of energy spectrums with respect to the different mixing regimes (i.e. micro, meso, and macro mixing) are the Kolmogorov, Batchelor, and Taylor length scales (Eqs. (1)–(3)). The Kolmogorov length scale of turbulence is used as a convenient reference point for comparison of different scales of mixing. A detailed description of the different time and length scales in turbulent flows is beyond the scope of this work. Further discussion on the subject matter can be found in the following reference texts [11, 12]. Therefore, getting the desired mixing regime is highly imperative for enhanced mass and momentum transfer.

In fluid-particle mixing there exist three distinct mixing regimes in most practical mixing applications namely: micro, meso, and macro mixing. In typical mixing conditions, the dividing line between micro and macro scale is between 100 and 1000 μm, respectively [13]. Consequently, it is often necessary to tailor the mixing performance to the physical, chemical, and biological processes in the target reactor systems.

Macromixing is largely driven by the largest scale of motion in the fluid or the integral length scale. Meso mixing on the other hand involves mixing on a smaller scale than the bulk circulation, but larger than the micro mixing, while micro mixing refers to the mixing on the smallest scale of fluid motion or molecular level. The largest eddies in turbulent dispersion which represents the macro scale of turbulence, are produced by the stirrer or the agitator head and contains most of the fluid energy [11, 14]. Turbulent flow can be viewed as an eddy continuum, with their sizes ranging from the dimension of the turbulence generating device to the Kolmogorov length scale. In between the energy-containing and energy-dissipating

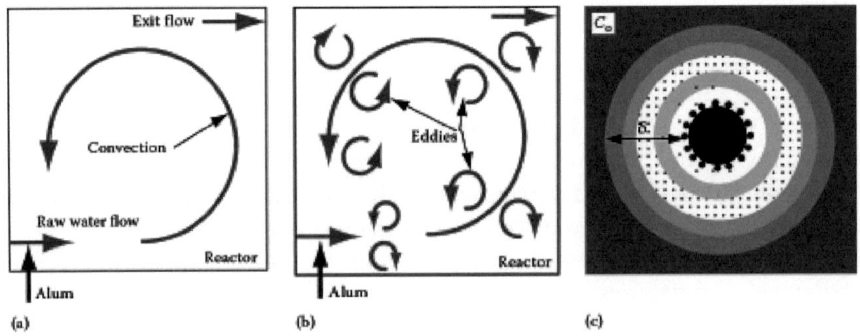

Figure 1.
Schematic representation of different mixing and transport mechanisms (a) advection (b) turbulence (c) diffusion (reproduced from [10] with permissions © 2011 Taylor & Francis).

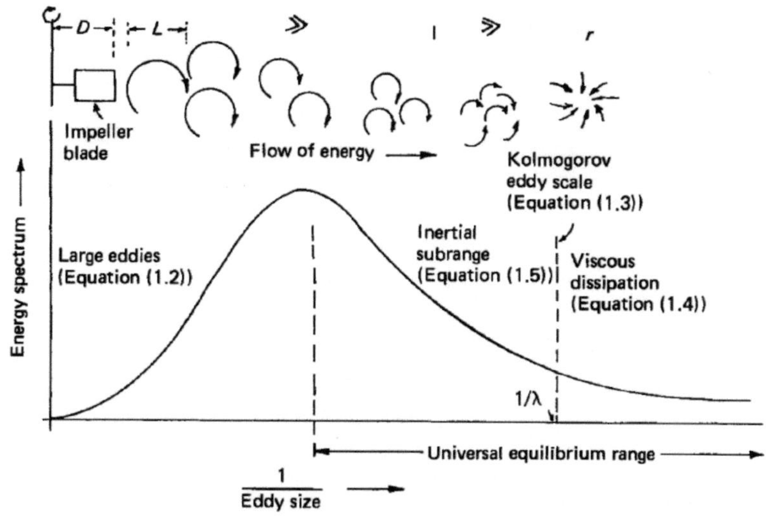

Figure 2.
Schematic view of energy spectrum in turbulent mixing (reproduced from [15] with permissions © 2007 IWA publishing).

eddies, there exists many eddies of other scales smaller than the integral scale that continually transfer the kinetic energy of the fluid through the other length scales. The Batchelor and Taylor scales given in Eqs. (2) and (3) are the examples of other important length scales of fluid motion.

The Taylor scale is an intermediate length scale in the viscous subrange that is representative of the energy transfer from large to small scales, but not a dissipation scale and does not represent any distinct group of eddies. Batchelor scale on the other hand is a limiting length scale where the rate of molecular diffusion is equal to the rate of dissipation of turbulent kinetic energy of the fluid.

$$\lambda_0 = \left[\frac{v^2}{\varepsilon}\right]^{\frac{1}{4}} \qquad (1)$$

$$\lambda_B = \left[\frac{D^2 v}{\varepsilon}\right]^{\frac{1}{4}} \qquad (2)$$

$$\lambda_T = \frac{u\sqrt{15}}{\sqrt{\frac{\varepsilon}{v}}} \qquad (3)$$

2. Colloidal stability and interaction forces

Colloidal materials in environmental and biological systems consist of small particles with very large surface area. Their typical sizes are in the range of 0.001–10 μm as shown in **Table 1**. The stability of these colloidal materials when dispersed in fluid can be explained by their tendency to acquire electrostatic charges by adsorbing ions from their surroundings. Forces mediating particle-particle interactions can broadly be classified into the following categories depending

Particle sizes, mm	Classification	Examples	Total surface area, $m^2\ cm^{-3}$	Time required to settle 100 mm if SG = 2.65
$10–10^{-1}$	Coarse dispersion	Sand, mineral substances, precipitated and flocculated substances, silt, microplankton	$6 \times 10^{-4}–6 \times 10^{-2}$	0.1–13 s
$10^{-2}–10^{-4}$	Fine particle dispersion	Mineral substances, precipitated and flocculated substances, silt, bacteria, plankton, and other micro organisms	0.6–60	11 min–2 years
$10^{-5}–10^{-6}$	Colloidal dispersion	Mineral substances, hydrolyzed and precipitated products, macromolecules, biopolymers, viruses	6×10^3	20 years
$<10^{-6}$	Solution	Inorganic simple and complex ions, molecules and polymeric species, polyelectrolytes, organic molecules and undissociated solutes	—	—

Table 1.
Classification of particles in dispersion [16].

on the chosen approach: contact forces due to the particle-particle collisions and non-contact forces due to molecular interactions at contact or interface. A brief description of these forces is hereby presented in the following sections.

2.1 Molecular dynamics approach

The molecular dynamics description of particle-particle interaction forces shown in **Figure 3** is based on the molecular interactions through interfacial forces on the surface of the interacting particles [17]. Interfacial forces are generally assumed to act on a length scales smaller than the particle size and interactions are only possible when particles are in close proximity or during collisions. The interactions of particles in suspension depend on these surface forces which consists of the long-range ionic electrostatic repulsive forces and the short-range London-van der Waals attractive forces. Classical DLVO or colloid stability theory provides a quantitative description of the forces experienced by particles in close proximity by considering such interactions forces to be additive [18].

The magnitude of the London-van der Waals attractive force between two charged particles and the electrostatic repulsive force due to the electric double layer can be derived from their corresponding interaction potential energies [17]. Eqs. (4)

and (5) represent the magnitude of these forces while Eqs. (6) and (7) represent their respective potential energies. In addition to these surface forces, a number of other interfacial interactions such as the hydration effects, hydrophobic attraction, steric repulsion, and polymer bridging have been observed to mediate particle-particle interactions [19]. Additional forces due to the fluid-particle interactions must also be considered to fully resolve all the forces experienced by particles in suspension and this is briefly discussed in Section 3.

$$F_{vdw} = \frac{dU_{vdw}}{dR_{ij}} \tag{4}$$

$$F_{elec} = \frac{dU_{elec}}{dR_{ij}} \tag{5}$$

$$U_{vdw} = \frac{A_H}{6}\left[\frac{2R_i^2}{R_{ij}^2 - 4R_i^2} + \frac{2R_i^2}{R_{ij}^2} + ln\left\{1 - \frac{4R_i^2}{R_{ij}^2}\right\}\right] \tag{6}$$

$$U_{elec} = 2\pi\varepsilon R_i \Psi_o^2 \ln\left[1 + \exp\left(-K\left(R_{ij} - 2R_i\right)\right)\right] \tag{7}$$

where R_{ij} is the distance between two interacting particles (center-to-center), A_H is the Hamaker constant, ε is the permittivity of the medium, ψ_0 is the surface potential of the particles, K is the reciprocal of the Debye length.

2.2 Micromechanical approach

In contrast to the molecular dynamics approach, the micromechanical description of particle-particle interaction relies on the geometric analysis of finite number

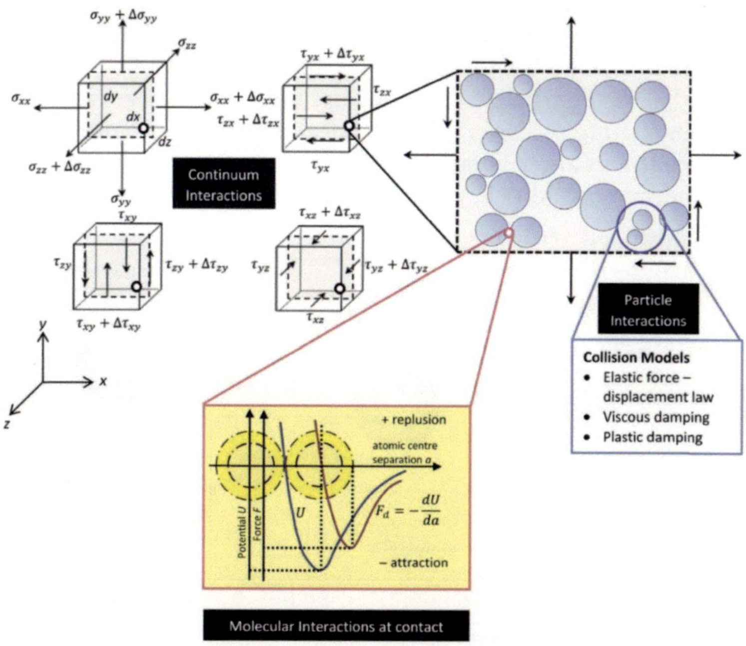

Figure 3.
Classification of different phenomena fluid-particle interactions (reproduced from with [17] permissions © 2019 Elsevier).

of discrete sub-elements as shown in **Figure 3**. All particle-particle interactions within this context are described by contact forces in the normal and tangential directions, while considering the elastic force-displacement, inelastic deformations or plastic dislocations, solid friction, and viscous damping [17]. Depending on the simplicity of these interactions, a soft or hard sphere description can be given. In the hard sphere model, only elastic force-displacement is allowed. Soft sphere model on the other hand allows for most of the interactions that are possible when two particles are in direct contact.

When all these contact forces are fully resolved, the behavior of the particles upon collisions or impact on a wall such as their translational and rotational velocities can be predicted with a high degree of accuracy. A detailed description of the micromechanical theory of particle collisions and its importance in the determination of particle trajectory in dispersed suspension is beyond the scope of this communication and is available elsewhere [20]. In addition to the contact forces, body forces such as gravity and buoyancy and surface forces due to the fluid are some of the other important forces acting on the particles and their quantification is highly indispensable in resolving the dynamics of particles in suspension [21]. Some of these additional forces are discussed in the next section.

3. Particle-vortex interactions in turbulence

In order to initiate meaningful interactions through the interfacial forces, particles must be brought in close proximity. This can be achieved through the mechanisms of Brownian motion, differential settling, or turbulent dispersion as shown in **Figure 4**. The probability of particle collisions and the frequency of such collisions also depend on the trajectory of the particle motion. Thomas et al. [22] identified two types of particle trajectories leading to particle collisions namely: curvilinear and rectilinear particle trajectories. The particles in suspension under the influence of turbulence will experience fluctuating fluid motion with the particle being transported by the fluid eddies that exists within the flow vortex [3].

Consequently, small particles suspended in fluid exist in an environment of small energy-dissipating eddies in most practical flow devices. Under such conditions, particle collisions are facilitated by eddy size similar to that of the colliding particles. Furthermore, studies have shown that in addition to the fluid properties,

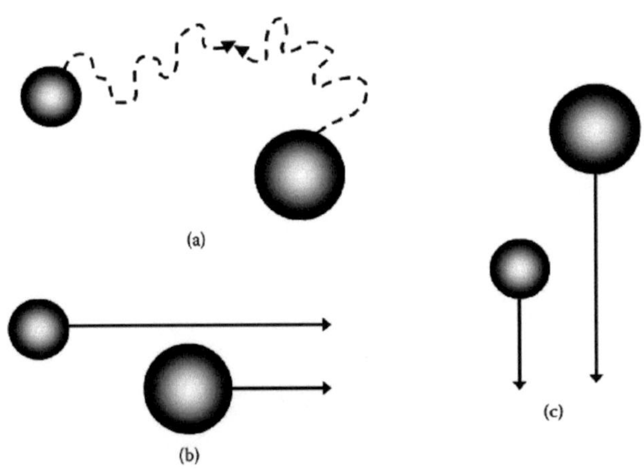

Figure 4.
Mechanisms of particle transport in fluid leading to collisions (a) Brownian motion (b) fluid turbulence (c) differential settling (reproduced from [19] with permissions © 2006 CRC press).

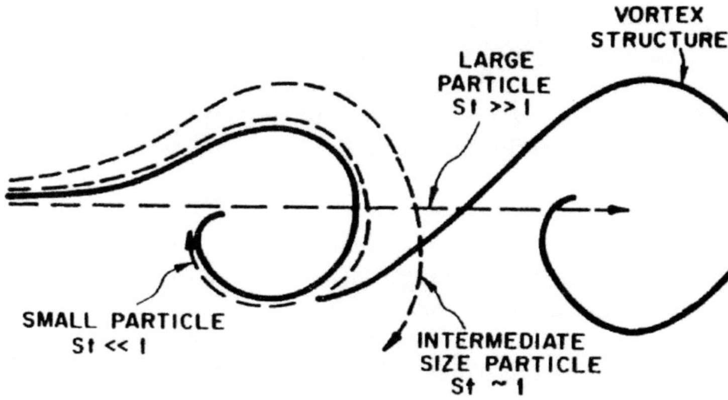

Figure 5.
Vortex-Particle interactions in turbulent flow (reproduced from [1] with permissions © 1995 springer).

the particle properties such as size, density, and porosity also play important role in particle-vortex interactions [1, 23]. Smaller particles with lower density than the flow fluid have been shown to be fully entrained within the vortex in the case of a vortical flow produced by the interaction of two flow streams of unequal velocity. The particles will be in dynamic equilibrium with the carrier fluid and will faithfully follow the flow streamlines. On the other hand, large particles will be unaffected by the vortex due to their large inertia, while the intermediate particles will be driven from the center of the vortex to the periphery as shown in **Figure 5**. The determining factor in particle-vortex interactions is the ratio of the particle relaxation time to that of the fluid, which is given by the Stokes number (Eq. (8)). Depending on the flow scenario, the particles in suspension will experience additional forces such as the drag, lift, pressure gradient, virtual mass, basset, and viscous stress forces due to the fluid-particle interactions [4, 24]. Taken together, all these forces will ultimately determine the trajectory, dynamics, and fate of particles in suspension. The trajectory equation is given in Eq. (9).

$$\mathrm{St} = \frac{\tau_A}{\tau_F} = \frac{\varrho_p d^2 U}{18 \mu L} \tag{8}$$

where is τ_A the particle relaxation time, τ_F is the time associated with fluid motion (fluid time), ϱ_p is the particle density, d is the particle diameter, L is the length scale associated with the vortex while U is the flow velocity.

$$m \frac{d\vec{v}}{dt} = \vec{F}_{mol} + \vec{F}_{mic} + \vec{F}_{f-p} \tag{9}$$

4. Particle dynamics and aggregate disruption

Turbulence is the main driver of particle interactions in many practical applications. Consequently, the particle dynamics in terms of the particle collisions, coalescence/aggregation, growth, and breakage is primarily controlled by the fluid turbulence. The aggregate stability under the influence of hydrodynamic force has been suggested to be a function of the binding or cohesive force F_B and the hydrodynamic breaking force F_H. While the binding force is determined by the aggregate structure and physicochemical attributes discussed Section 2, flow turbulence is the principal factor in the case of the hydrodynamic force. Therefore, the dynamics of particle behavior depends on an interplay of collision-induced particle aggregation and cohesive force and the rate of aggregate breakage due to the hydrodynamic stress.

The aggregate cohesive strength τ is a function of the physicochemical and biological conditions as well as the aggregate properties, while the hydrodynamic stress σ depends on the design of the aggregation unit and the prevailing process conditions. Several empirical and theoretical formulations are available for predicting the aggregate cohesive force and the maximum hydrodynamic breaking force. The global hydrodynamic stress σ due to the shearing action of the fluid motion on the aggregate as well as the overall cohesive strength τ of the aggregate resisting the hydrodynamic loading assuming a uniform shape and constant porosity can be expressed mathematically in Eqs. (10) and (11) [25, 26]. An equilibrium of particle dynamics is reached at the steady-state condition. In this state, a continued particle or micro flocs/aggregate attachment to the larger flocs/aggregate is prevented, or the breakup kinetics is equal to the turbulence-induced collision rates.

In assessing aggregate strength and resistance to hydrodynamic-induced breakup, two common approaches are normally followed namely: limiting growth and limiting strength. The former relies on the determination of the maximum floc size before rupture, while the latter is based on the micromechanical analysis of aggregate strength. Many empirical and theoretical formulations based on the mentioned approaches are available in literature (Eqs. (12) and (13)). Liu et al. [27] presented the yield stress approach for calculating maximum aggregate tensile strength τy at which breakage is likely to occur in the inertia range of turbulence (Eq. (12)). Similarly, Attia [28] presented a model for predicting the critical fluid velocity above which there will be aggregate disruption by estimating floc yield stress σy resulting from the dynamic pressure acting on the floc (Eq. (13)).

$$\sigma = \mu G = \mu\sqrt{\frac{\varepsilon}{v}} \tag{10}$$

$$\tau = \frac{(1-p)F_B}{p d_p^{\,2}} \tag{11}$$

$$\tau_y = \frac{F_B}{d_p^{\,2}}\left(\frac{\lambda_0}{d_F}\right)^k \tag{12}$$

$$\sigma_y = \frac{1}{2}\rho_f v^2 \tag{13}$$

5. Case studies in environmental and biological systems

This section presents a short review of the recent studies on the applications of turbulence or hydrodynamics in the environmental and biological systems (**Table 2**). Water purification, irrigation, water quality assessment, sludge dewatering, bio-flocculation, and bio-clogging are some of the technical areas of application identified.

Technical application	Flow regime	Study type	References
Biofouling/biofilms	Turbulent	Experimental/CFD	[25–30]
Water disinfection/irrigation	Turbulent	Experimental/CFD	[31–35]
Water self-purification	Turbulent	Experimental	[43–46]
Solid-liquid separation	Turbulent	Experimental	[2, 36–38]
Food and paper processing	Turbulent	Experimental	[39–42]

Table 2.
Selected studies on computational and experimental studies of turbulence applications in environmental and biological processes.

It should be noted that improvements in the performance of the engineered processes (e.g., stirred tanks, shear reactors and photobioreactors etc.) in the identified areas of applications continue to shape the research focus in the field of environmental process engineering [29]. In this respect, studies have been conducted to determine how to accurately quantify the impact of hydrodynamic characteristics on the infectivity of bacteriophage MS2, a norovirus surrogate. Several studies also involved the development of bioreactors for testing the effect of hydrodynamic characteristics on microalgae and human enteric viruses [29–33]. The results obtained from the studies indicated that the hydrodynamic cavitation could trigger the inactivation waterborne viruses to levels defined in water quality directives. This was reportedly due to OH-radicals that form an AOP during the cavitation process and high shear forces inside the cavitation structure. Also, flow structures in a hydrodynamic filter have been numerically investigated [34]. In this study, tangential component of velocity was defined, and the three-dimensional pattern of the flow current/streamlines was obtained using their two-dimensional projection in the meridian cross-section of the filter, which allows one to discover the vortex structures. It was concluded that the optimal flow regime can be implemented by selecting the optimal correlation between the flow of liquid regime to be processed and the rotation frequency of the filtering baffle in the hydrodynamic filter. The remaining sub-sections describe how hydrodynamics, turbulence, and vortex dynamics are applied to achieve the desired process efficiency in other identified areas of applications—water purification, sludge dewatering, food processing, and self-purification of the water bodies.

5.1 Hydrodynamics in water purification process

In fluid engineering problems, research has consistently focused on identifying parameters that improve engineered processes including water purification and inactivation of pathogens [35]. While the conventional technique of disinfection by chlorination has been employed to kill pathogenic microorganisms in raw water, recent studies have shown that chlorine reacts with organic compounds in water and generates disinfection byproducts (DBPs), such as trihalomethanes (THMs), haloacetic acids (HAAs), etc. As a result, turbulence-induced inactivation has been studied as an alternative approach.

The effect of hydrodynamic parameters such as orifice size, orifice number and orifice layout of multi-orifice plate, cavitation number, cavitation time and orifice velocity on the microbial population have been investigated to determine how the desired process efficiency can be achieved [36]. Experimental results have shown that cavitation effects increased with decrease in orifice size and increase in orifice number, cavitation time and orifice velocity. Flow hydrodynamics and pipe material have also been shown to influence biofilm development in drinking water distribution systems (DWDS). Furthermore, biofilm development was inhibited at higher flows indicating shear forces imposed by the flow conditions were above the critical levels for biofilm attachment. Experimental data from these studies were used to characterize the hydrodynamic behavior for numerical simulation and validation [37, 38]. Low-cost pipe-based pathogen reduction system was also demonstrated by Thomas et al. Their approach has a huge potential for application in developing countries due to its simple design [39].

5.2 Hydrodynamics in solid-liquid separation

In the design of process reactors, it is often necessary to tailor the separation technique to the dynamics and characteristics of the waste slurry that is being treated. Hence, several studies have been conducted to determine the influence of hydrodynamics on sludge processing. The optimum dosage values, which were obtained when flocculation performance was assessed based on surrogate indicators such as sludge volume index and supernatant turbidity, confirmed polymer bridging as the primary flocculation mechanism. Specific apparatus construction and reactor geometry were found to help maintain sludge suspension in a metastable state that is crucial for the formation of pellet-like compact agglomerates with better dewaterability properties [2, 40]. Similarly, sludge disintegration, using rotor-stator type hydrodynamic cavitation reactor (HCR), has been experimentally investigated [41]. To determine the effects of cavitation (including thermal energy) and shear stress on sludge disintegration, the performance of the HCR with and without the dimples and temperature control was analyzed. The results indicated that when dimples were present and there was no temperature control, the reduction of sludge particles increased by 50–80%. Further, the disintegration performance increased with the rotational speed and was minimized at the highest inlet pressure. Several other studies leveraging on the hydrodynamics of the process reactor for fluid-particle separation are available in literature [42, 43]. Many of these lab-scale studies have demonstrated the feasibility of turbulence-induced fluid-particle separation.

5.3 Hydrodynamics in food and paper processing

Hydrodynamic cavitation (HC) is a process in which high energy is released in a flowing liquid upon bubble implosion due to decrease and subsequent increase in local pressure. In food and beverage industries, hydrodynamic cavitators can be utilized for the purpose of extraction, emulsification, sterilization, disinfection, and homogenization [44]. HC, which can effectively induce sonochemistry by mechanical means, creates extraordinarily high of pressures of ~1000 bar, local hotspots with ~5000 K, and high oxidation (hydroxyl radicals) in room environment, without introducing new chemicals. For possible industrial application, the efficiency of HC has been studied by comparing the chemical oxygen demand (COD) removal efficiency of a Venturi device to that of an orifice plate. A sucrose solution and an effluent from a sucrose-based soft drink industry were treated. Results showed that the Venturi device recorded 90% COD removal efficiency after treatment period of three minutes. On the other hand, the orifice plate recorded 90% COD removal efficiency after 9 minutes [44–46]. Developing high-performance HCRs and revealing the corresponding disinfection mechanisms constitute the most crucial issues today [45].

Refining of cellulose pulp is a critical step in obtaining high quality paper characteristics, however, this process is slow and costly especially for refining longer conifer fibers which are the preferred source for high quality paper production and give the paper its strength. Recently, hydrodynamic cavitation was applied to the refinement conifer rich pulp samples [47].

5.4 Hydrodynamics in self-purification of water bodies

The self-purification ability of water bodies is related to the prevailing hydrodynamic conditions. Coupled hydrodynamic and water quality models have been used to investigate the spatial and temporal water quality variations of the water bodies. Using an Acoustic Doppler Velocimeter, the efficiency of aeration plug-flow device (APFD) in terms of water flow and dissolved oxygen (DO) have been determined experimentally [48–50]. Recent findings have shown that discharges from several rivers flowing into the New York/New Jersey (NY/NJ) harbor interact and interfere with one another. Such interactions can improve or inhibit water and contaminant flushing from the harbor. In Poyang Lake, three-dimensional velocity at various locations as well as the velocity distribution and turbulence characteristics were assessed, and plug-flow characteristics were analyzed. The two patterns of velocity and turbulence in horizontal sections observed are (1) near the aeration plug-flow device (APFD), the water flow was intensively pushed downstream and simultaneously recirculated; (2) farther away, the reflux area gradually decreased, and the velocity and turbulence distribution were more or less uniform. At the interfaces between two immiscible fluids – water and alkane of small carbon number, the amphiphilic PEO chain diffuses laterally, experiencing hydrodynamic drags from both phases. The absolute values of interfacial diffusion coefficients demonstrate a bigger contribution from the hydrodynamics from the water phase, which may be attributed to a stronger attraction between water and the PEO molecules [51].

5.5 Vortex patterns in fluid-particle reactors

The flow vortex patterns in many fluid-particle reactors are quite diverse and their analysis and characterization can provide an insight into the fluid and particle dynamics within the reactor. One important feature in many of these flow types is the presence of rotation or swirling or a combination of both resulting in anisotropic turbulence. A few of the dominant vortex patterns will be discussed in this section. It is also worth mentioning that the flow pattern is a function of the reactor geometry, stirrer or agitator, baffles, and the operating conditions. The focus will be on the common reactor geometries and flow regimes that are typically encountered in many practical applications.

A qualitative and quantitative analysis of the flow pattern in a processing reactor may be performed experimentally using an optical diagnostic imaging technique such as PIV or PDA or by numerical modeling. The selected approach in a given scenario will depend on the required detail of the flow field as well as the available resources. While the later provide a flexible option for the investigation of fluid flow problems, the former often complements or used in the validation of numerical model. A detailed description of specific techniques is beyond the scope of this submission. A few of the vortex patterns in some of these reactors are discussed below.

5.5.1 Rotatory flow vortex pattern (e.g., tubular reactor)

This type of flow pattern is commonly found in rotating tubular reactors with an enclosed flow induced by a stirrer. Previous studies have shown that this flow pattern has a profound influence on the performance of fluid-particle reactors, and this type of flow pattern can be found in many technical applications [29, 39, 52]. For instance, in terms of the stirrer-vessel configuration, the flocculation performance is significantly influenced by the impeller type and its speed. The axial impeller has been found to promote floc formation over a range of impeller speeds as it produces a more homogeneous distribution of local velocity gradients in comparison to the radial impeller. Also, high velocity gradients occur in the region around the impeller, which might significantly hamper the high rate activated sludge (HRAS) flocculation process [53, 54].

5.5.2 Oscillatory flow vortex pattern (e.g., oscillatory reactor)

An oscillator is a device whose action causes intermittent velocity gradients over time, space, and direction, and can be used for colloid removal based on physical flow manipulation. An oscillatory flow pattern is commonly found in reactors with an enclosed flow induced by an oscillator. Results from previous studies have shown that gentle oscillation can promote simultaneous flocculation and particle agglomeration over a relatively short periods of time [42]. This technique, whose implementation can result in reductions of the reactor sizes and process times, has a strong potential to improve conventional separation processes. It has also been observed that higher oscillating frequency promotes faster upward vertical velocities, resulting in different sedimentation patterns and removal efficiencies generated by the different oscillation frequencies [43].

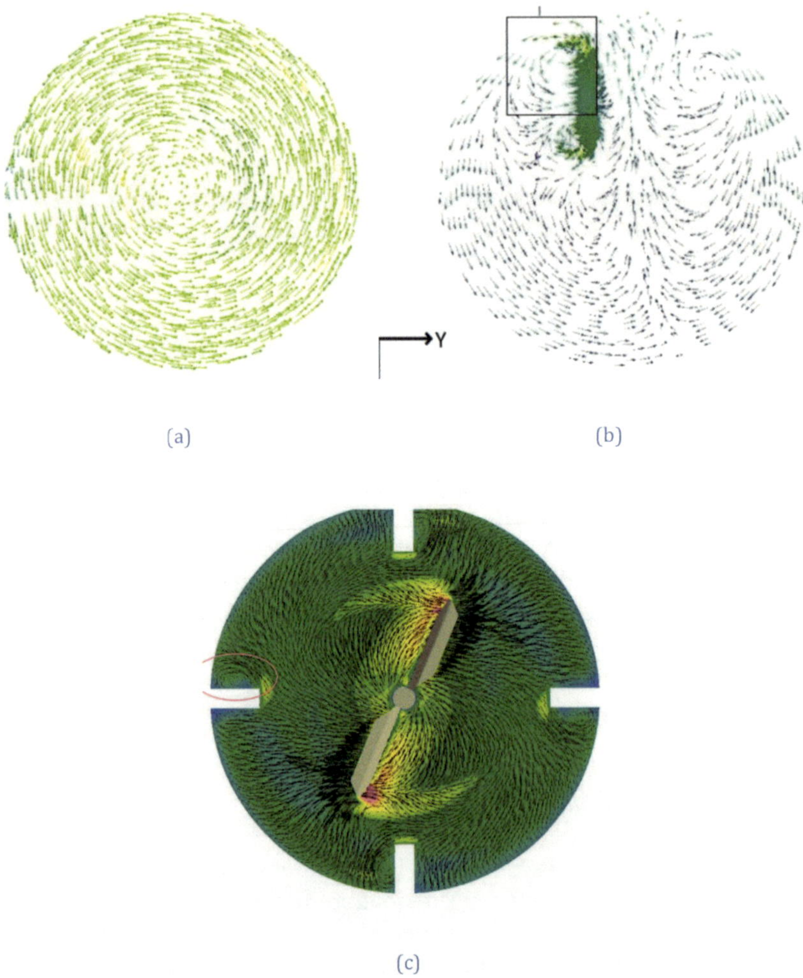

Figure 6.
Typical cross-sections of vortex patterns in fluid-particle reactors (a) rotatory-type flow field (b) oscillatory-type flow field (c) swirling-type flow field.

5.5.3 Swirling flow vortex pattern (e.g., stirred tank, hydrocyclone)

A swirling type of flow pattern is commonly found in hydrocyclone [55–57] and counter-rotating reactors with an enclosed flow induced by a rotational swirling effect [58, 59]. In addition to hydrocyclone, this type of flow field is also common in rotating reactors with baffles such as mixing tank, in which case, the flow is a combination of rotational and swirl dominated flow as shown in **Figure 6c**. In the case of a counter-rotating vortex reactor, it has also been found that the swirl ratio and micromixing time of the flow increases as the vortex reactor (MIVR) is scaled up, indicating a flow with stronger swirl yet less mixing efficiency [60]. In order to promote mixing and enhance the floc formation process, some baffles should be installed to break water flow co-rotation with the impeller [54]. In modifying the flow pattern to suit a particular condition, it is advisable to perform the optimization of the impeller shape for a particular vessel geometry [61, 62].

6. Conclusions and future perspectives

This brief communication is a summary of the main concepts involved the fluid-particle research and technical applications. It is by no means an exhaustive contribution and readers interested in the details of the subject-matter are advised to consult other scientific information available on the subject-matter. The fluid process engineering is a field of active research and the there is an ever-increasing scope for the application of hydrodynamics, turbulence, and vortex dynamics. In addition to the identified areas of application, there are several emerging areas of application. As the turbulence research advance with better computing power and algorithms, it is hoped that there will be limitless scope for the application of vortex dynamics in fluid process engineering.

Conflict of interest

The authors declare no conflict of interest.

Nomenclature

APFD	aeration plug-flow device
COD	chemical oxygen demand
DBPs	disinfection byproducts
DO	dissolved oxygen
DWDS	drinking water distribution systems
HAAs	haloacetic acids
HC	hydrodynamic cavitation
HCR	hydrodynamic cavitation reactor
THMs	trihalomethanes

Author details

Benjamin Oyegbile[1*], Brian Oyegbile[2] and Guven Akdogan[1]

1 Department of Process Engineering, Faculty of Engineering, Stellenbosch University, Stellenbosch, South Africa

2 College of Engineering, University of Georgia, Athens, GA, USA

*Address all correspondence to: hollander196@yahoo.com

References

[1] Crowe CT, Troutt TR, Chung JN. In: Green SI, editor. Fluid Vortices. Dordrecht: Springer; 1995. pp. 829-861. DOI: 10.1007/978-94-011-0249-0_19

[2] Oyegbile BA. Optimization of Micro Processes in Fine Particle Agglomeration by Pelleting Flocculation. Leiden: CRC Press; 2016. DOI: 10.1201/9781315671871

[3] Oyegbile B, Ay P, Narra S. Flocculation kinetics and hydrodynamic interactions in natural and engineered flow systems: A review. Environmental Engineering Research. 2016;**21**:1-14. DOI: 10.4491/eer.2015.086

[4] Oyegbile B, Akdogan G, Karimi M. Modelling the dynamics of granular particle interactions in a vortex reactor using a coupled DPM-KTGF model. South African Journal of Chemical Engineering. 2020;**34**:31-46. DOI: 10.1016/j.sajce.2020.05.008

[5] Morchain J. Bioreactor Modeling. Oxford: Elsevier; 2017. pp. 29-84. DOI: 10.1016/b978-1-78548-116-1.50002-1

[6] Paul EL, Atiemo-Obeng VA, Kresta SM, editors. Handbook of Industrial Mixing: Science and Practice. Hoboken, NJ: John Wiley & Sons; 2004. DOI: 10.1002/0471451452

[7] Harun Z, Lotfy ER. In: Barille R, editor. Turbulence and Related Phenomena. London: InTech Open; 2019. DOI: 10.5772/intechopen.76854

[8] Yeoh GH, Cheung CP, Tu J. Multiphase Flow Analysis Using Population Balance Modeling: Bubbles, Drops and Particles. Waltham, MA: Butterworth-Heinemann; 2014

[9] Peker SM, Helvacı ŞŞ, Yener HB, İkizler B, Alparslan A, editors. Solid-Liquid Two Phase Flow. Oxford: Elsevier; 2008. DOI: 10.1016/b978-0-444-52237-5.x5001-2

[10] Hendricks D. Fundamentals of Water Treatment Unit Processes. Boca Raton, FL: CRC Press; 2011. DOI: 10.1201/9781439895092

[11] Zlokarnik M. Stirring: Theory and Practice. Weinheim: Wiley-VCH; 2008

[12] Kockmann N. Transport Phenomena in Micro Process Engineering. Heidelberg: Springer; 2008

[13] Oldshue JY, Trussell RR, Trussell R. Design of Impellers for mixing. In: Amirtharajah A, Clark MM, Trussell R, editors. Mixing in Coagulation and Flocculation. Denver, CO: American Water Works Association; 1991. pp. 309-342

[14] Logan BE. Environmental Transport Processes. Hoboken, NJ: John Wiley & Sons; 2012

[15] Bache DH, Gregory R. Flocs in Water Treatment. London: IWA Publishing; 2007. DOI: 10.2166/9781780402000

[16] Bratby J. Coagulation and Flocculation in Water and Wastewater Treatment. London: IWA Publishing; 2016

[17] Yeoh GH, Tu J. Computational Techniques for Multiphase Flows. 2nd ed. Oxford: Elsevier; 2019

[18] Taboada-Serrano P, Chin C-J, Yiacoumi S, Tsouris C. Modeling aggregation of colloidal particles. Current Opinion in Colloid & Interface Science. 2005;**10**:123-132. DOI: 10.1016/j.cocis.2005.07.003

[19] Gregory J. Particles in Water: Properties and Processes. Boca Raton, FL: CRC Press; 2006. DOI: 10.1201/9780203508459

[20] Norouzi HR, Zarghami R, Sotudeh-Gharebagh R, Mostoufi N.

Coupled CFD-DEM Modeling: Formulation, Implementation and Application to Multiphase Flows. Chichester: John Wiley & Sons; 2016. DOI: 10.1002/9781119005315

[21] Crowe CT, Schwarzkopf JD, Sommerfeld M, Tsuji Y. Multiphase Flows with Droplets and Particles. Boca Raton, FL: CRC Press; 2011. DOI: 10.1201/b11103

[22] Thomas DN, Judd SJ, Fawcett N. Flocculation modelling: A review. Water Research. 1999;**33**:1579-1592. DOI: 10.1016/s0043-1354(98)00392-3

[23] Marshall JS. Effect of particle collisions on the expulsion of heavy particles from a vortex core. Physics of Fluids. 2006;**18**:113301. DOI: 10.1063/1.2370427

[24] Sun R, Xiao H, Sun H. Investigating the settling dynamics of cohesive silt particles with particle-resolving simulations. Advances in Water Resources. 2018;**111**:406-422. DOI: 10.1016/j.advwatres.2017.11.012

[25] Coufort C, Bouyer D, Liné A. Flocculation related to local hydrodynamics in a Taylor–Couette reactor and in a jar. Chemical Engineering Science. 2005;**60**:2179-2192. DOI: 10.1016/j.ces.2004.10.038

[26] Shamlou PA, Titchener-Hooker N. Processing of Solid–Liquid Suspensions. Oxford: Butterworth-Heinemann; 1993. pp. 1-25. DOI: 10.1016/b978-0-7506-1134-3.50005-3

[27] Liu SX, Glasgow LA. Aggregate disintegration in turbulent jets. Water, Air, and Soil Pollution. 1997;**95**:257-275. DOI: 10.1007/bf02406169

[28] Attia YA, Laskowski JS, Ralston J. Flocculation. In: Colloid Chemistry in Mineral Processing. Amsterdam: Elsevier; 1992. pp. 277-308

[29] Zhang Q, Wu X, Xue S, Liang K, Cong W. Study of hydrodynamic characteristics in tubular photobioreactors. Bioprocess and Biosystems Engineering. 2012;**36**:143-150. DOI: 10.1007/s00449-012-0769-2

[30] Kosel J, Gutiérrez-Aguirre I, Rački N, Dreo T, Ravnikar M, Dular M. Efficient inactivation of MS-2 virus in water by hydrodynamic cavitation. Water Research. 2017;**124**:465-471. DOI: 10.1016/j.watres.2017.07.077

[31] Zhang QH, Wu X, Xue SZ, Wang ZH, Yan CH, Cong W. Hydrodynamic characteristics and microalgae cultivation in a novel flat-plate photobioreactor. Biotechnology Progress. 2012;**29**:127-134. DOI: 10.1002/btpr.1641

[32] Yan H, Guan C, Jia Y, Huang X, Yang W. Mixing characteristics, cell trajectories, pressure loss and shear stress of tubular photobioreactor with inserted self-rotating helical rotors: Mixing characteristics, cell trajectories, pressure loss and shear stress. Journal of Chemical Technology and Biotechnology. 2017;**93**:1261-1269. DOI: 10.1002/jctb.5484

[33] Muller-Feuga A, Pruvost J, Guédes RL, Déan LL, Legentilhomme P, Legrand J. Swirling flow implementation in a photobioreactor for batch and continuous cultures of porphyridium cruentum. Biotechnology and Bioengineering. 2003;**84**:544-551. DOI: 10.1002/bit.10818

[34] Devisilov VA, Sharai EY. Numerical study of the flow structure in a hydrodynamic filter. Theoretical Foundations of Chemical Engineering. 2016;**50**:209-216. DOI: 10.1134/s0040579516020044

[35] Baranova MP. Hydrodynamic cavitation unit for improving the performance of irrigation water

during plant growing. IOP Conference Series: Earth and Environmental Science. 2019;**315**:062018. DOI: 10.1088/1755-1315/315/6/062018

[36] Kudryashova T, Polyakov S, Tarasov N. A novel parallel algorithm for 3D modelling electromagnetic purification of water. MATEC Web of Conferences. 2018;**210**:04027. DOI: 10.1051/matecconf/201821004027

[37] Dong Z, Zhao W. Killing rate of colony count by hydrodynamic cavitation due to square multi-orifice plates. IOP Conference Series: Earth and Environmental Science. 2018;**121**:022004. DOI: 10.1088/1755-1315/121/2/022004

[38] Cowle MW, Webster G, Babatunde AO, Bockelmann-Evans BN, Weightman AJ. Impact of flow hydrodynamics and pipe material properties on biofilm development within drinking water systems. Environmental Technology. 2019:1-13. DOI: 10.1080/09593330.2019.1619844

[39] Thomas SF, Rooks P, Rudin F, Cagney N, Balabani S, Atkinson S, et al. Swirl flow bioreactor containing dendritic copper-containing alginate beads: A potential rapid method for the eradication of Escherichia coli from waste water streams. Journal of Water Process Engineering. 2015;**5**:6-14. DOI: 10.1016/j.jwpe.2014.10.010

[40] Oyegbile B, Ay P, Narra S. Optimization of physicochemical process for pre-treatment of fine suspension by flocculation prior to dewatering. Desalination and Water Treatment. 2016;**57**:2726-2736. DOI: 10.1080/19443994.2015.1043591

[41] Kim H, Koo B, Sun X, Yoon JY. Investigation of sludge disintegration using rotor-stator type hydrodynamic cavitation reactor. Separation and Purification Technology. 2020;**240**:116636. DOI: 10.1016/j.seppur.2020.116636

[42] Halfi E, Brenner A, Katoshevski D. Separation of colloidal minerals from water by oscillating flows and grouping. Separation and Purification Technology. 2019;**210**:981-987. DOI: 10.1016/j.seppur.2018.08.054

[43] Halfi E, Arad A, Brenner A, Katoshevski D. Development of an oscillation-based technology for the removal of colloidal particles from water: CFD modeling and experiments. Engineering Applications of Computational Fluid. 2020;**14**:622-641. DOI: 10.1080/19942060.2020.1748114

[44] Asaithambi N, Singha P, Dwivedi M, Singh SK. Hydrodynamic cavitation and its application in food and beverage industry: A review. Journal of Food Process Engineering. 2019;**42**. DOI: 10.1111/jfpe.13144

[45] Sun X, Liu J, Ji L, Wang G, Zhao S, Yoon JY, et al. A review on hydrodynamic cavitation disinfection: The current state of knowledge. Science of the Total Environment. 2020;**737**:139606. DOI: 10.1016/j.scitotenv.2020.139606

[46] Alves PHL, de Silva PSL, Ferreira DC, de Gonçalves JCSI. COD removal from sucrose solution using hydrodynamic cavitation and hydrogen peroxide: A comparison between Venturi device and orifice plate. RBRH. 2019;**24**. DOI: 10.1590/2318-0331.241920180147

[47] Kosel J, Šinkovec A, Dular M. A novel rotation generator of hydrodynamic cavitation for the fibrillation of long conifer fibers in paper production. Ultrasonics Sonochemistry. 2019;**59**:104721. DOI: 10.1016/j.ultsonch.2019.104721

[48] Li B, Yang G, Wan R, Li H. Hydrodynamic and water quality

modeling of a large floodplain Lake (Poyang lake) in China. Environemental Science and Pollution Research. 2018;**25**:35084-35098. DOI: 10.1007/s11356-018-3387-y

[49] Li Y, Feng H, Zhang H, Sun J, Yuan D, Guo L, et al. Hydrodynamics and water circulation in the New York/New Jersey Harbor: A study from the perspective of water age. Journal of Marine Systems. 2019;**199**:103219. DOI: 10.1016/j.jmarsys.2019.103219

[50] Li X, Huang M, Wang R. Numerical simulation of Donghu Lake hydrodynamics and water quality based on remote sensing and MIKE 21. ISPRS International Journal of Geo-Information. 2020;**9**:94. DOI: 10.3390/ijgi9020094

[51] Li Z, Yang J, Hollingsworth JV, Zhao J. Lateral diffusion of single polymer molecules at interfaces between water and oil. RSC Advances. 2020;**10**:16565-16569. DOI: 10.1039/d0ra02630a

[52] Visscher F, van der, Schaaf J, Nijhuis TA, Schouten JC. Rotating reactors – A review. Chemical Engineering Research and Design. 2013;**91**:1923-1940. DOI: 10.1016/j.cherd.2013.07.021

[53] Bridgeman J, Jefferson B, Parsons SA. Computational fluid dynamics modelling of flocculation in water treatment: A review. Engineering Applications of Computational Fluid. 2014;**3**:220-241. DOI: 10.1080/19942060.2009.11015267

[54] Balemans S, Vlaeminck SE, Torfs E, Hartog L, Zaharova L, Rehman U, et al. The impact of local hydrodynamics on high-rate activated sludge flocculation in laboratory and full-scale reactors. PRO. 2020;**8**:131. DOI: 10.3390/pr8020131

[55] Mousavian SM, Najafi AF. Influence of geometry on separation efficiency in a hydrocyclone. Archive of Applied Mechanics. 2008;**79**:1033-1050. DOI: 10.1007/s00419-008-0268-8

[56] Dai GQ, Li JM, Chen WM. Numerical prediction of the liquid flow within a hydrocyclone. Chemical Engineering Journal. 1999;**74**:217-223. DOI: 10.1016/s1385-8947(99)00044-3

[57] Liu Y, Cheng Q, Zhang B, Tian F. Three-phase hydrocyclone separator – A review. Chemical Engineering Research and Design. 2015;**100**:554-560. DOI: 10.1016/j.cherd.2015.04.026

[58] Derksen J. Confined and agitated swirling flows with applications in chemical engineering. Flow, Turbulence and Combustion. 2002;**69**:3-33. DOI: 10.1023/a:1022419316418

[59] Jakirlic S, Hanjalic K, Tropea C. Modeling rotating and swirling turbulent flows: A perpetual challenge. AIAA Journal. 2002;**40**:1984-1996. DOI: 10.2514/2.1560

[60] Liu Z, Ramezani M, Fox RO, Hill JC, Olsen MG. Flow characteristics in a scaled-up multi-inlet vortex nanoprecipitation reactor. Industrial and Engineering Chemistry Research. 2015;**54**:4512-4525. DOI: 10.1021/ie5041836

[61] Jaszczur M, Młynarczykowska A, Demurtas L. Effect of impeller design on power characteristics and Newtonian fluids mixing efficiency in a mechanically agitated vessel at low Reynolds numbers. Energies. 2020;**13**:640. DOI: 10.3390/en13030640

[62] Escamilla-Ruíz IA, Sierra-Espinosa FZ, García JC, Valera-Medina A, Carrillo F. Experimental data and numerical predictions of a single-phase flow in a batch square stirred tank reactor with a rotating cylinder agitator. Heat and Mass Transfer. 2017;**53**:2933-2949. DOI: 10.1007/s00231-017-2030-7

Chapter 2

Vortices in Rotating and Gravitating Gas Disk and in a Protoplanetary Disk

Martin G. Abrahamyan

Abstract

Nonlinear equations describing dynamics of 2D vortices are very important in the physics of the ocean and the atmosphere and in plasma physics and Astrophysics. Here linear and nonlinear 2D vortex perturbations of gravitating and light gaseous disks are examined in the geostrophic and post-geostrophic approximations. In the frame of geostrophic approximation, it is shown that the vortex with positive velocity circulation is characterized by low pressure with negative excess mass density of substance. Vortex with negative circulation has higher pressure and is a relatively tight formation with the positive excess mass density. In the post-geostrophic approximation, structures of the isolated monopole and dipole vortex (modons) solutions of these equations are studied. Two types of mass distributions in dipole vortices are found. The first type of modon is characterized by an asymmetrically positioned single circular densification and one rarefaction. The second type is characterized by two asymmetrically positioned densifications and two rarefactions, where the second densification-rarefaction pair is crescent shaped. The constant density contours of a dipole vortex in a light gas disk coincide with the streamlines of the vortex; in a self-gravitating disk, the constant density contours in the vortex do not coincide with streamlines. Possible manifestations of monopole and dipole vortices in astrophysical objects are discussed. Vortices play decisive role in the process of planet formation. Gas in a protoplanetary disk practically moves on sub-Keplerian speeds. Rigid particles, under the action of a head wind drags, lose the angular momentum and energy. As a result, the ~10 cm to meter-sized particles drift to the central star for hundreds of years. Long-lived vortical structures in gas disk are a possible way to concentrate the ~10 cm to meter sized particles and to grow up them in planetesimal. Here the effect of anticyclonic Burgers vortex on formation of planetesimals in a protoplanetary dusty disc in local approach is also considered. It is shown that the Burgers vortex with homogeneously rotating kernel and a converging radial stream of substance can effectively accumulate in its nuclear area the meter-sized rigid particles of total mass $\sim 10^{28}$ g for characteristic time $\sim 10^{6}$ year.

Keywords: anticyclone, Burgers vortex, dipole, gravitating disk, monopole, planetesimals, protoplanetary disk, vortex

1. Introduction

Nonlinear equations describing dynamics of 2D vortices are important in the physics of the ocean and the atmosphere, in plasma physics, and in astrophysics. The same type of nonlinear equations describes these vortical structures. In fluid dynamics, Hasegawa-Mima equation is well-known [1].

$$\frac{\partial}{\partial t}(1-\Delta)\psi - v_0\frac{\partial\psi}{\partial y} - (\boldsymbol{e}_z \times \nabla\psi)\nabla\Delta\psi = 0, \tag{1}$$

which describes the nonlinear Rossby waves in the atmosphere [2] and drift nonlinear waves in plasma [3]. Here $\psi(x,y,z)$ is a stream function: $v = e_z \times \nabla\psi$. In plasma physics ψ is the electric potential, and constant v_0 is defined by equilibrium density gradient.

The exact solution of the equation, describing a stationary solitary dipole vortex (modon) drifting along the y-axis on rotating shallow water, was obtained in [4]. The same type of solutions later received a large number of similar equations [5–10].

Nonlinear vortex disturbances of uniformly rotating gravitating gaseous disk were considered in [9]. For short-scale (much smaller than the Jeans wavelength: $\lambda \ll \lambda_J$) and long-scale $\lambda \gg \lambda_J$ perturbations, nonlinear equation turns into Eq. (1).

IR, submillimeter, and centimeter radiation of protoplanetary disk analyses shows that vortices serve as incubators for the growth of dust particles and formation of planetesimals [11–14]. The initial stage of growth probably proceeds through the nucleation of submicron-sized dust grains from the primordial nebula, which then forms the monomers of fractal dust aggregates up to ~1 mm to ~10 cm for characteristic time of an order of 10^3 years [15, 16]. The best astrophysical evidence for grain growth to specified sizes is the detection of 3.5 cm dust emission from the face-on disk of radius 225 AU round classical T Tauri star TW Hya [17]. When the planetesimals reached a size of about 1 km, they began to attract other smaller bodies due to their gravity.

In models of protoplanetary disks, gas practically moves on sub-Keplerian speeds. Rigid particles, under the action of a head wind drag, lose the angular momentum and energy. As a result, the ~10 cm to meter-sized particles drift to the central star for hundreds of years, that is, much less than the lifetime of a disk which makes several millions of years [18, 19].

Long-lived vortical structures in gas disk are a possible way to concentrate the ~10 cm to meter-sized particles and to grow up them in planetesimal. Similar effect of vortices on the Earth was observed in special laboratories and also in the ocean [20].

In some areas of the stratified protoplanetary disks, the current has a 2D turbulent character. An attractive feature of such hydrodynamic current consists in the fact that in it, through a background of small whirlpools, long-living vortices will spontaneously be formed without requirement of special initial conditions [21–23]. In laboratory experiments [24, 25], formation of Burgers vortex, which will be considered here, is often observed in 2D turbulent flows. Anticyclonic vortices in a protoplanetary disk merge with each other and amplify, while cyclonic ones are destroyed by a shear flow [26].

In cylindrical system of coordinates (r, θ, z), the Burgers vortex is defined as

$$v_r = -Ar, v_\theta = \omega {r_0}^2\left[1 - \exp\left(-r^2/r_0^2\right)\right]/r, \quad v_z = 2Az. \tag{2}$$

This is a vortex with a converging stream of substance to its center with gradient $-A$, ω and r_0 as the circulation and the size of a trunk of a vortex. Rotation of a

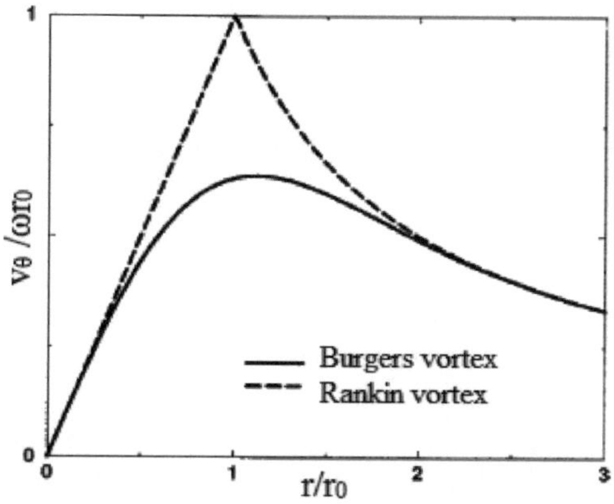

Figure 1.
Rotational velocity profiles of Burgers and Rankin vortices.

vortex trunk is almost uniform, then falls down under hyperbolic law, and at distance $r_{eff} = 4.5r_0$ (effective radius of vortex) makes 1/3 of the maximum value of v_θ (**Figure 1**). The asymptotic behavior of Burgers vortex in small and big distances from the vortex center represents the Rankin vortex [27, 28].

2. Magnitude of some parameters of circumstellar disks

A typical circumstellar disk is a few hundred AU (astronomical unit, 1 AU = 1.5 10^{13} cm) in size. It is mainly composed of hydrogen and helium gas. We consider a vortex in such axially symmetrical viscous accretion disk with effective temperature T and gas density ρ, of almost Keplerian rotation.

The sound speed in gas is estimated by

$$c_s = \sqrt{\gamma kT/m_H} \approx (\gamma T/100K)^{1/2} km/s, \tag{3}$$

where $\gamma = 1.4$ is the gas adiabatic index, k is Boltzmann constant, m_H is hydrogen atom mass.

In a vertical direction, the gas is in hydrostatic balance with a characteristic scale height:

$$H \sim \frac{c_s}{\Omega} \approx 0.03\left(\frac{T}{100K}\right)^{1/2}\left(\frac{M_\odot}{M}\right)^{1/2}\left(\frac{R}{AU}\right)^{3/2} AU. \tag{4}$$

The thickness-to-radius ratio (aspect ratio) is usually $\sim 1/10$ and increases slowly with radius, R. The superficial density of the gas in a disk can be estimated as $\Sigma \approx 2H\rho$.

In "α-model" of accretion disk [29], the expense of gas occurs with a speed $dm/dt = 3\pi\nu\Sigma$, where ν is the kinematic viscosity of gas, $\nu = \alpha c_s H$.

The dynamic time scale of a disk is

$$\tau \sim \frac{1}{\Omega} \approx \frac{1}{5}\left(\frac{M_\odot}{M}\right)^{1/2}\left(\frac{R}{AU}\right)^{3/2} yr \tag{5}$$

For Keplerian disk, radial momentum equation solution yields to a difference between the speeds of rigid particles and surrounding gas [30]. In a thin gas disk ($c_s \ll \Omega R$), rigid particles drift relative to gas with a speed

$$\frac{\Delta v}{c_s} \sim \frac{c_s}{\Omega R} \approx 0.03 \left(\frac{T}{100K}\right)^{1/2} \left(\frac{M_\odot}{M}\right)^{1/2} \left(\frac{R}{AU}\right)^{1/2}. \tag{6}$$

At $c_s \sim 1 km/s$, typical drift speed is of order 30 m/s. The characteristic scale of drift time [18, 19] almost by two orders surpasses the dynamic time: $\tau_d \sim r/\Delta v \sim$ (R/A.E.) 10^2yr.

For a characteristic time $\tau_s \sim \Sigma/\alpha\Omega\rho*$ [31], where ρ^* is the mass density of a particle, dust settled on a midplane of a disk. The characteristic time between collisions of rigid particles among themselves is estimated as $\tau_{col} \sim D\rho*/\Sigma*\Omega$, where D is the diameter of a particle and Σ^* is the superficial density of rigid particles in a disk which is more than by two orders less than a disk Σ.

Viscous dissipation and orbital shear limit the sizes of a vortex. Viscous dissipation destroys vortices of sizes less than the viscous scale [32]:

$$L_\nu = \frac{\alpha c_s H}{v_\theta} \approx 0.003 \left(\frac{\alpha}{0.01}\right) \left(\frac{0.1c_s}{v_\theta}\right) \left(\frac{M_\odot}{M}\right)^{1/2} \left(\frac{R}{AU}\right)^{3/2} AU, \tag{7}$$

where v_θ is the rotational speed of a vortex.

The Keplerian shear flow forbids the formation of circular structures with the sizes larger than the shear length scale:

$$L_{shear} = \sqrt{v_\theta \left|\frac{d\Omega}{dR}\right|^{-1}} \approx 0.05 \left(\frac{v_\theta}{0.1c_s}\right)^{1/2} \left(\frac{M_\odot}{M}\right)^{1/4} \left(\frac{R}{AU}\right)^{5/4} AU. \tag{8}$$

The vortices, whose sizes surpass L_{shear}, are extended in an azimuthal direction that allows them to survive longer. In [33] we have shown the possibility of formation in a disk extended in an azimuthal direction three-axis ellipsoidal vortex, with a linear field of circulation, similar to Riemann S ellipsoids [34]. However, in a disk round the central star of solar mass, at distance 30 AU, the vortex of characteristic speed of rotation, $0.01c_s$, can be circular and have the size of an order of ~1 AU.

In a gas disk, drag force on rigid particles from gas is exposed, which, depending on size of a particle, is expressed either by Stokes or Epstein's formula (see, e.g., [25]).

Here our main results obtained by investigations of the linear and nonlinear perturbation equations of differentially rotating gravitating gaseous disk in geostrophic and post-geostrophic approximations are presented [35], as well as the results on formation of planetesimals by Burgers vortex in a protoplanetary disk [26].

3. Model and basic equations for linear perturbations

Consider at first a gravitating pure gas disk of mass density $\rho(r)$, which rotates with angular velocity $\Omega(r)$ around the z-axis. Explore 2D perturbations in plane of the disk, ignoring its vertical structure. Present any characteristic functions of the disk as $f_0(r) + f(r, \varphi, t)$, where $f_0(r)$ describes the equilibrium state and $f(r, \varphi, t)$ is a small but finite perturbation.

We will consider isentropic perturbations (S = constant) and therefore enthalpy H(S, P) = H(P),

where P is the pressure and

$$dH = dP/\rho = c_s^2 d\rho/\rho, \tag{9}$$

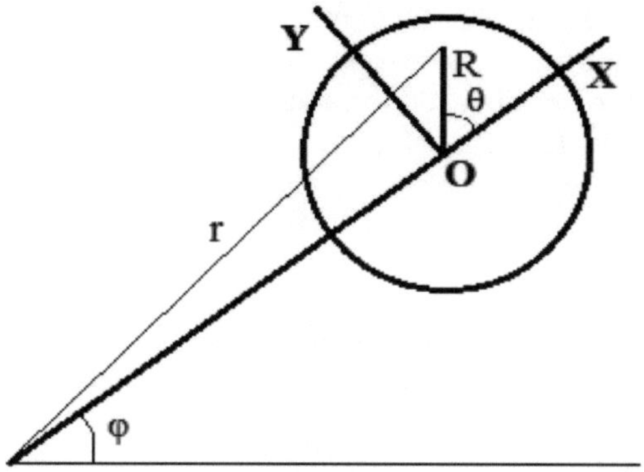

Figure 2.
The local frame of reference 1.

where c_s is the sound speed. Evidently, Eq. (9) is the equation of state of disk substance.

Perturbations of the disk in a rotating with angular velocity $\Omega_0 \equiv \Omega(r_0)$ cylindrical coordinate system (**Figure 2**) are described by 2D hydrodynamic equations[1]:

$$d\mathbf{v}/dt + 2\Omega_0\mathbf{e}_z \times \mathbf{v} + \mathbf{e}_\varphi v_r r\Omega' + \nabla\Phi = 0, \tag{10}$$

$$d\rho/dt + \rho\nabla(\mathbf{v}) = 0, \tag{11}$$

where the velocity was presented in the form of

$$\mathbf{V}_0 \equiv \mathbf{e}_\varphi(\Omega - \Omega_0)r,$$

Φ is the sum of perturbations of gravitational potential U and enthalpy

$$\Phi \equiv U + H, \tag{12}$$

$$d/dt = \partial/\partial t + V_0\partial/r\partial\varphi + \mathbf{v}\nabla; \tag{13}$$

and the Poisson equation is

$$\Delta U = 4\pi G\rho, \tag{14}$$

In Eq. (10) we have used the radial equilibrium condition for the disk:

$$\Omega^2 r = d\Phi_0/dr. \tag{15}$$

Taking into account Eq. (9), the continuity Eq. (11) can be written as

$$dH/dt + {c_s}^2\nabla\mathbf{v} = 0. \tag{16}$$

Taking operator **curl** on Eq. (10) and then by combining the equation of continuity, after simple transformation, we obtain

$$d/dt\{[\mathrm{curl}_z\mathbf{v} + 2\Omega + r\Omega']/\rho\} = 0. \tag{17}$$

[1] Here and below the bar indicates the differentiation of equilibrium parameters of the disk on the radial coordinate r.

The expression in the curly brackets in this equation is a generalized vortencity. The equation shows that for 2D isentropic perturbations, generalized vortencity is conserved along the current lines. So for stationary perturbations, generalized vortencity is an arbitrary function of ψ:

$$(\mathrm{curl}_z \mathbf{v} + 2\Omega + r\Omega')/\rho = B(\psi). \quad (18)$$

In a uniformly rotating ($\Omega = \mathrm{const.}$) gravitating disk, no drifting stationary vortex solution can be obtained without specifying the function $B(\psi)$, because Eq. (18) can be represented as a Jacobean $J\{\psi, (\mathrm{curl}_z \mathbf{v} + 2\Omega)/\rho\} = 0$, which satisfies the arbitrary circularly symmetric vortex disturbance around point O.

4. Vortices in the geostrophic approximation

This approach assumes that Coriolis forces and gravity balance the pressure gradient in the disk.

Then from the equation of motion (17), we get the perturbed geostrophic velocity:

$$\mathbf{v}_G = (1/2\Omega_0)\mathbf{e}_z \times \nabla\Phi. \quad (19)$$

Using the last, Eq. (18) takes the form

$$(1/2\Omega_0\rho)\,\Delta\Phi + (2\Omega + r\Omega_0')/\rho = B(\psi). \quad (20)$$

In further analysis of this topic, we will introduce the local Cartesian coordinate system (X,Y) such that (**Figure 2**)

$$\partial/\partial x = \partial/\partial r, \partial/\partial y = \partial/r\partial\varphi, \quad (21)$$

and will explore the vortical perturbations around a point O in a linear approximation.

The stream function $\psi(x, y)$ of perturbed velocity (19) is expressed through perturbations Φ by the following formula:

$$\psi = \Phi/2\Omega_0. \quad (22)$$

Imagine around a point O function $\rho(x)$ and $\Omega(x)$ in the form of

$$\rho(x) = \rho_0(r_0) + x\rho_0' + \rho(x, y); \Omega(x) = \Omega_0 + x\Omega_0'. \quad (23)$$

In this case $V_0 \approx r_0\Omega_0' x$.

Perturbations of density and enthalpy (9) in linear approach are connected by the following formula:

$$\rho(x, y) = \rho_0 H(x, y)/c_s^2, \quad (24)$$

Then Eq. (13) with an accuracy to a constant term will be in the form of

$$B(\psi) = \rho_0^{-1}\{\Delta\psi - \kappa_0^2 k_R^2 H/2\Omega_0\Omega^2) + x\beta\}, \quad (25)$$

where $k_R = \Omega_0/c_s$ is the Rossby wave number, $\kappa_0{}^2 = 2\Omega_0(2\Omega_0 + r_0\Omega_0')$ is the square of the epicyclical frequency, and

$$\beta \equiv 3\Omega_0' - \kappa_0{}^2\rho_0'/2\Omega_0\rho_0. \tag{26}$$

If to take the relationship of density perturbations with perturbations of gravitational potential using Poisson equation

$$\rho(x, y) = \Delta U(x, y)/4\pi(G),$$

instead of Eq. (25), we obtain the equation

$$B(\psi) = \rho_0{}^{-1}\{\Delta\psi - \kappa_0{}^2\Delta U/2\Omega_0\,\omega_J{}^2 + x\beta\}, \tag{27}$$

where $\omega_J{}^2 \equiv 4\pi G\rho_0$ is the square of the Jeans frequency.

The order of magnitude of $|H/U|$ can be estimated using the definition $c_s{}^2 = (dP_0/d\rho_0)$ and $|\Delta\Phi| = k^2\Phi$.

where k is wavenumber of perturbations

$$|H/U| \approx k^2 c_s{}^2/\omega_J{}^2 = k^2/k_J{}^2, \tag{28}$$

where $k_J = \omega_J/c_s$ is the Jeans wavenumber. Eq. (28) shows that the case $|H| \gg |U|$ describes the small-scale disturbances: matching $k^2/k_J{}^2 \gg 1$, or $\lambda \ll \lambda_J$. In this case $\psi = H/2\Omega_0 \equiv h$, and Eq. (25) takes the form

$$B(h) = \rho_0^{-1}\{\Delta h - (\kappa_0^2 k_R^2/\Omega_0^2)h + xb\}. \tag{29}$$

Limit $|H| \ll |U|$ corresponds to the large-scale perturbations $-\lambda \gg \lambda_J$. Then $\psi = U/2\Omega_0 \equiv \phi$, and Eq. (27) turns into

$$B(\phi) = \rho_0^{-1}\left\{\left(1 - \kappa_0^2/\omega_J^2\right)\Delta\varphi + xb\right\}. \tag{30}$$

By selection of function B, we can explore the stationary vortex solutions of Eqs. (29) and (30).

Let's take a look at the simplest case of uniformly rotating disk of homogeneous density, $\beta = 0,\ \kappa_0^2 = 4\Omega_0^2$, when the vortencity is constant and is equal to $\Gamma/\rho_0\pi a^2$ where Γ is the velocity circulation. We assume that the velocity circulation Γ differs from zero only in a circle of radius $a(\ll r_0)$ around point O. Using now the polar coordinates $(R, \theta) : x = R\cos\theta, y = R\sin\theta$ (**Figure 2**), the equation for disturbances (29) can be written as

$$\Delta h - 4k_R^2 = \begin{cases} \Gamma/\pi a^2, & R \le a \\ 0, & R \ge a \end{cases} \tag{31}$$

which gives a circularly symmetric solution for relative perturbed density of mass

$$\frac{\rho}{\rho_0} = -\frac{\Gamma}{\pi a s c_s}\begin{cases} 1 - \dfrac{K_1(s)I_0(sR/a)}{K_1(s)I_0(s) + I_1(s)K_0(s)}, & R \le a \\[2ex] \dfrac{I_1(s)K_0(sR/a)}{K_1(s)I_0(s) + I_1(s)K_0(s)}, & R \ge a \end{cases} \tag{32}$$

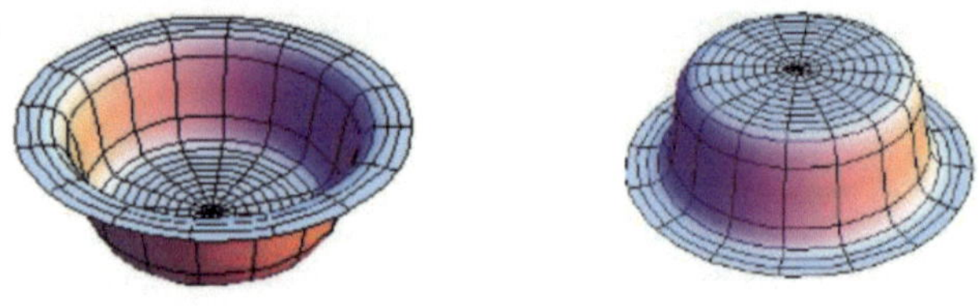

Figure 3.
3D image of relative density perturbations of whirlwind in the range $0 \leq R/a < 1.4$ *for* $2ak_R = 20$.

where $s \equiv 2a\,k_R$. This is a monopole vortex with the following perturbed velocity field:

$$v_\theta = \frac{\Gamma}{\pi a s} \begin{cases} \dfrac{K_1(s) I_1(sR/a)}{K_1(s) I_0(s) + I_1(s) K_0(s)}, & R \leq a \\ \dfrac{I_1(s) K_1(sR/a)}{K_1(s) I_0(s) + I_1(s) K_0(s)}, & R \geq a. \end{cases} \tag{33}$$

Note that the vortices with positive and negative velocity circulation Γ have different properties. Whirlwind with positive circulation is characterized by low pressure, with negative excess mass density of substance. Vortex with negative circulation has a higher pressure and relatively tight formation with the positive excess mass density.

To illustrate these results, we will take into account the fact that the Rossby wavenumber usually is of the order of the inverse thickness of the disk. Considering that the size of the vortex a as the disk thickness order, we will get for the Bessel function argument $2ak_R \approx 12$. **Figure 2** shows 3D image relative density perturbations in monopole whirlwind occupying the region $0 \leq R/a \leq 1.35$, for the value of the argument $s = 12$. This vortex is a retrograde-circulating rarefaction around the center O condensation in the case of $\Gamma < 0$ and prograde-circulating rarefaction in the case of $\Gamma > 0$ (**Figure 3**). The decrease of density in the area $R > a$ of larger vortex is steeper. If the size of the vortex tends to zero, we get a simple classic case of point vortex.

For long-scale perturbations (33), the Rankin vortex velocity profile is given [27, 28]:

$$v = \frac{\Gamma}{2\pi\gamma a} \begin{cases} R/a, & R \leq a \\ a/R, & R \geq a. \end{cases} \tag{34}$$

where $\gamma = 1 - \Omega^2/\pi G \rho_0$.

5. Vortices in the post-geostrophic approximation

In this section, we will get nonlinear perturbation equation, taking into account the inertia term in the equation of motion (10) for homogeneously rotating disk. The cross product of Eq. (10) with e_z: $e_z \times$ Eq. (10), gives

$$\mathbf{v} = \mathbf{v}_G + \mathbf{v}_I, \tag{35}$$

where the first term is geostrophic speed (19) and the second is

$$\mathbf{v}_\mathrm{I} \equiv (1/2\Omega_0)\mathbf{e}_z \times d\mathbf{v}/dt. \tag{36}$$

Substituting Eq. (35) to Eq. (36) and taking approximation $d/dt \ll \Omega$ (slowly varying perturbations), dropping the term $\mathbf{v}_\mathrm{I}\nabla$ in the expression (13) for d/dt, we get.

$$\mathbf{v}_\mathrm{I} = \left(1/4\Omega_0{}^2\right) \mathbf{e_z} \times d[\mathbf{e}_z \times \nabla\Phi]dt. \tag{37}$$

With the use of Eqs. (19) and (34), we find

$$\nabla\mathbf{v}_\mathrm{G} = 0, \tag{38}$$

$$\nabla\mathbf{v}_\mathrm{I} = -\left(1/4\Omega_0{}^2\right)L\Delta\Phi. \tag{39}$$

where

$$L \equiv \partial/\partial t + (1/2\Omega_0)(\nabla\Phi \times \nabla)_z. \tag{40}$$

The continuity equation now takes the form

$$d\left(\rho_+\rho_0\right)/dt + \left(\rho_+\rho_0\right)\nabla\mathbf{v}_\mathrm{I} = 0, \tag{41}$$

or, using Eqs. (14), (20), and (21)

$$L\rho - \left(\rho_0/4\Omega_0{}^2\right)L\Delta\Phi. \tag{42}$$

Here we have served the terms that are of second order in perturbed amplitude and neglected terms of highest order.

Using the Poisson equation, we get from Eq. (42) the basic nonlinear equation

$$L\Delta U - ½\alpha\, L\Delta\Phi + \beta\partial\Phi/r\partial\varphi = 0, \tag{43}$$

where

$$\alpha \equiv \omega_J{}^2/2\Omega_0{}^2; \alpha' \equiv d\alpha/dr; \beta = \alpha'\Omega_0; \omega_J{}^2 \equiv 4\pi G\rho_0 \tag{44}$$

In view of the assessment (28), for short-scale perturbations ($\lambda \ll \lambda_J$), Eq. (42) takes the form

$$\left(\frac{\partial}{\partial t} + \frac{2}{2\Omega_0}(\nabla H \times \nabla)_z\right)\Delta H + \frac{2\beta}{\alpha}\frac{1}{r}\frac{\partial H}{\partial\varphi} = 0. \tag{45}$$

On the limit $|H| \ll |U|$ that corresponds to large-scale disturbances: $\lambda \gg \lambda_J$, Eq. (42) turns into [9].

$$\left(\frac{\partial}{\partial t} + \frac{2}{2\Omega_0}(\nabla U \times \nabla)_z\right)\Delta U + \frac{2\beta}{\alpha - 2}\frac{1}{r}\frac{\partial H}{\partial\varphi} = 0. \tag{46}$$

Eqs. (45) and (46) have the same structure differing only by their coefficients, and are Hasegawa-Mima type (see Eq. (1)).

6. A solitary dipole vortex

In a Cartesian coordinate system (X,Y) (**Figure 2**), we will look for stationary solutions of Eq. (45) (and (46)) in a small neighborhood of the guiding center O with a radius of $a \ll r_0$ in the form of a vortex drifting in y-direction at a constant speed u. Introducing the wave variable

$$\eta = \text{y-ut} \quad (47)$$

Eq. (46) can be rewritten in the form

$$\{\partial/\partial\eta - A(\nabla U \times \nabla)_z\}\delta U = \Lambda\partial U/\partial\eta, \quad (48)$$

or in the form of the Jacobean

$$J(U\text{-}x/A, \Delta U + \Lambda x/A) = 0, \quad (49)$$

where

$$(A)^{-1} = 2u\Omega, \Lambda = -4\Omega_0{}^2 A(\ln|\alpha - 2|)'. \quad (50)$$

On basis of Eq. (49)

$$\Delta U + \Lambda x/A = F(U - x/A), \quad (51)$$

where F is an arbitrary function. As we are interested in the restricted solutions, then in the limit of large values η, solution U should vanish for arbitrary values x; therefore

$$F(-x/A) = -\Lambda x/(A). \quad (52)$$

We will assume that the function F (51) in the equation is linear not only for large η but across the whole plane (x, η). In general, F can be represented as $\propto (U - x/A)$. Introducing polar coordinates $R, \theta : x = R\cos\theta, \eta = R\sin\theta$, we can write Eq. (49) in the form

$$(\Delta + k^2)U = A^{-1}(k^2 - \Lambda)R\cos\theta, \quad R \le a, \quad (53)$$

$$(\Delta - p^2)U = 0, \quad R \ge a, \quad (54)$$

where k and p are real constants. Soon the sense of splitting the (R, θ) plane into two parts will be obvious. Eq. (54) turns out to be uniform, because for a restricted solution, we have $U \to 0$ for large R. This condition implies

$$p^2 = -\Lambda. \quad (55)$$

Eqs. (53) and (54) have the following stationary solution [9]:

$$U(R,\theta) = \Omega u a \begin{cases} \left[\left(1 - \dfrac{s^2}{g^2}\right)\dfrac{R}{a} + \dfrac{s^2}{g^2}\dfrac{J_1(gR/a)}{J_1(g)}\right]\cos\theta, & R \le a \\ -\dfrac{K_1(sR/a)}{K_1(s)}\cos\theta & R \ge a \end{cases} \quad (56)$$

where J_1 and K_1 are Bessel and Macdonald functions, respectively, and $g = ka$ and $s = pa$ are connected by "dispersion equation" which is transcendental

$$(J)_1(g)(K)_3(s) + (J)_3(g)(K)_1(s) = 0. \quad (57)$$

For long-scale perturbations

$$s^2 = \left(2\Omega(a)^2/u\right)\left(\ln|\alpha - 2|\right)', \tag{58}$$

while for small-scale disturbances

$$U \to H = c_s{}^2\rho|\rho_0, \text{ and } s^2 = \left(2\Omega^2 a/u\right)(\ln\alpha)'. \tag{59}$$

From Eqs. (56) and (19), we get the velocity field of a vortex in the form

$$v_R = -\frac{1}{2\Omega}\frac{\partial\Phi}{R\partial\theta} = u\begin{cases}\left[1 - \frac{s^2}{g^2}\left(1 - \frac{aJ_1(gR/a)}{RJ_1(g)}\right)\right]\sin\theta, & R \le a \\ \frac{aK_1'(sR/a)}{RK_1(s)}\sin\theta, & R \ge a\end{cases} \tag{60}$$

$$v_\theta = \frac{1}{2\Omega}\frac{\partial\Phi}{\partial R} = u\begin{cases}\left[1 - \frac{s^2}{g^2}\left(1 - g\frac{J_1'(gR/a)}{J_1(g)}\right)\right]\cos\theta, & R \le a \\ \frac{sK_1{}'(sR/a)}{K_1(s)}\cos\theta, & R \ge a\end{cases} \tag{61}$$

Moreover, the condition (57) is derived from the requirements of continuity (61) on the circle $R = a$.

The current lines are determined by $dR/v_r = Rd\theta/v_\theta$ that gives

$$const. = \begin{cases}\left[\left(1 - \frac{s^2}{g^2}\right)\frac{R}{a} + \frac{s^2}{g^2}\frac{J_1(gR/a)}{J_1(g)}\right]\sin 2\theta, & R \le a \\ \frac{K_1(sR/a)}{K_1(s)}\sin 2\theta, & R \ge a.\end{cases} \tag{62}$$

Figure 4 shows the current lines of drifting solitary dipole vortex, the appropriate formula (62).

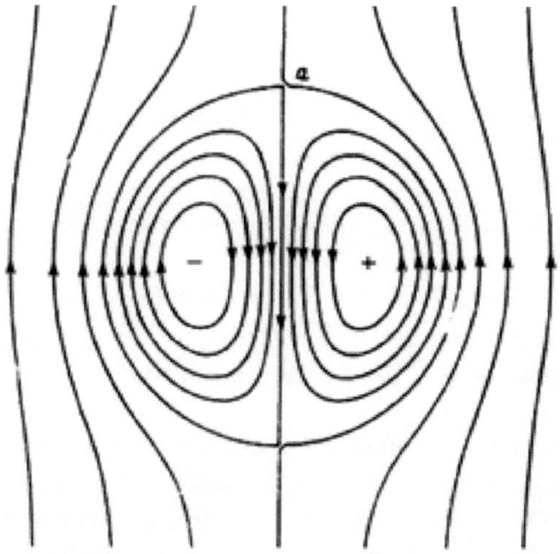

Figure 4.
The stream lines of solitary dipole vortex [9].

7. The contours of constant density

As shown in Section 3, $\Phi/2\Omega$ is the current function of perturbed speed, ψ in the case of short-scale disturbances $\Phi = H$. Then we have the $\psi = H(\rho)/2\Omega$; the constant values of the contours of constant density ρ match the lines of the current $\psi =$ constant: in short-scale modon, substance flows along the lines of constant density.

In the long-scale limit, $\Phi = U$ and $\psi = U/2\Omega$. Therefore, in this case the current lines coincide with equipotentials of the gravitational field, not with the contours of constant density. The last can be found using the Poisson equation:

$$\Delta\psi = 2\pi G\rho/\Omega. \qquad (63)$$

Since equipotentials U = constant, generally speaking, do not coincide with the contours of constant density, it follows that the stream lines ψ = constant do not coincide with the contours of constant density.

Define the contours of constant density of modon. Relative density perturbations in the short-scale range are expressed by the following formula:

$$\sigma = \rho/\rho_0 = \beta_{sw} H/au\Omega, \text{where } \beta_{sw} \equiv au\Omega/c_s^{\,2}. \qquad (64)$$

The relative perturbed density in long-scale range turns out to be in the form

$$\sigma = \frac{1}{\omega_J^2}\Delta\psi = \beta_{lw}\begin{cases} \dfrac{J_1(gR/a)}{J_1(g)}\cos\theta, & R \le a \\ \dfrac{K_1(sR/a)}{K_1(s)}\cos\theta, & R \ge a \end{cases} \qquad (65)$$

where

$$\beta_{lw} \equiv -u\Omega s^2/a\omega_J^{\,2} = -(\ln|\alpha - 2|)''/\alpha. \qquad (66)$$

To illustrate, consider a logarithmic model of the disk, describing in equilibrium by the following functions of potential, mass density, and angular velocity:

$$U_0(r) = ½v_0^{\,2}\ln(R_c^{\,2} + r^2), \rho_0(r) = v_0^{\,2}R^2/2\pi G(R_c^{\,2} + r^2), \quad \Omega^2 = v_0^{\,2}/(R_c^{\,2} + r^2), \qquad (67)$$

where R_c and v_0 are constants, rotation of a disk in the small area $(a \ll R_c)$ can be considered as uniform, and β_{lw} for this model turns out to be equal to

$$\beta_{lw} \approx 8ar/3R_{(c)}^{\;2} \approx 8a(R_c + R\cos\theta)/3R_c^{\,2}, \qquad (68)$$

where we used the relation $r^2 \approx R_c^{\,2} + 2RR_c\cos\theta$, by placing the center O in $R = R_c$.

For illustrations of a perturbed density distribution in dipole vortex (65), we used the following solutions of "dispersion equation" (57): (g,s) = (4.0, 1.52); (4.2, 2.90); (4.5, 6.0); (4.7, 10.0).

The curves in **Figure 5** show perturbed density as a function of dimensionless distance *R/a* from the guiding center O in the short-scale (curves increasing towards the center) and in the long-scale (curves descending towards the center) limit.

The density distribution is antisymmetrical to the guiding center. Depending on the choice of the dispersion curve (57) range, there are two types of mass distribution in dipole vortex. One is antisymmetrically located almost round condensation,

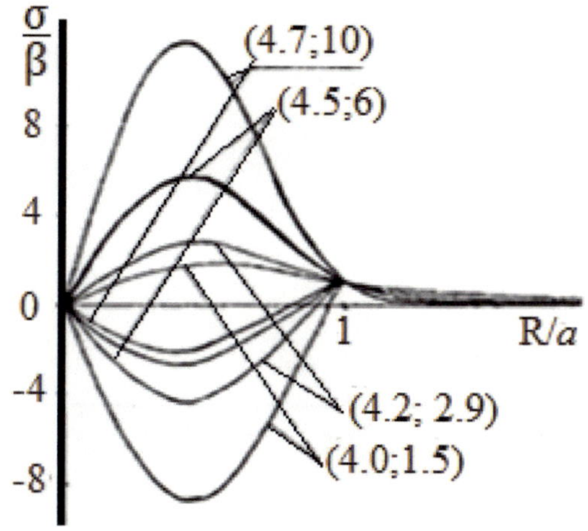

Figure 5.
Dependence of relative perturbed density from the dimensionless distance R/a in short-scale and long-scale modons.

and one rarefaction (**Figure 6**) characterizes the first type. The second type is characterized by the two antisymmetrically located condensations and two rarefactions, and second condensation-rarefaction pair has sickle-form (**Figure 7**). For small values g and s, the short-scale modon is of the second type, with distinctive two condensations (see **Figure 8**). In the middle part of the dispersion curve, the short-scale and long-scale modons have roughly the same structure. They have one antisymmetrical located prominent pair of condensation-rarefaction and another weak pair of sickle forms. For large values of g and s, the short-scale modon is the first type and has the character of a cyclone-anticyclone couple; the long-scale one is the second type and is characterized by a nearly round and sickle-shaped condensations. In laboratory experiments, the solitary dipole vortices on shallow water,

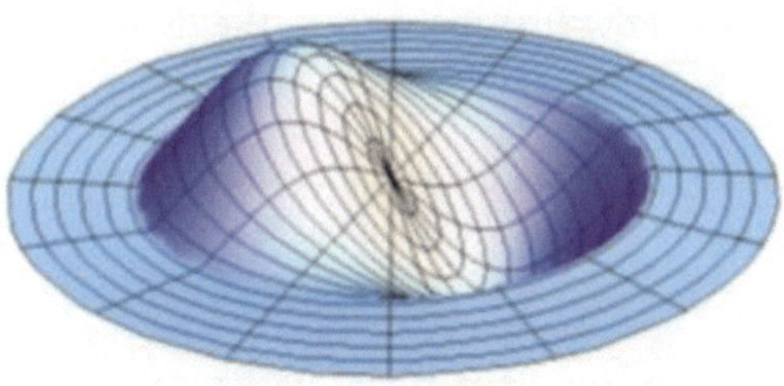

Figure 6.
3D image of density distribution in the first type modon.

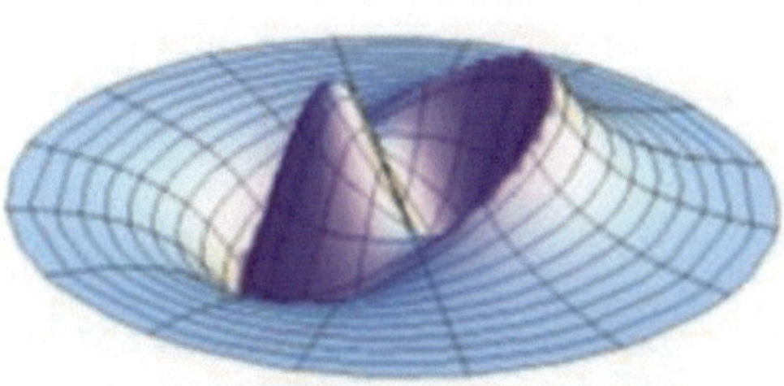

Figure 7.
3D image of density distribution in the second type modon.

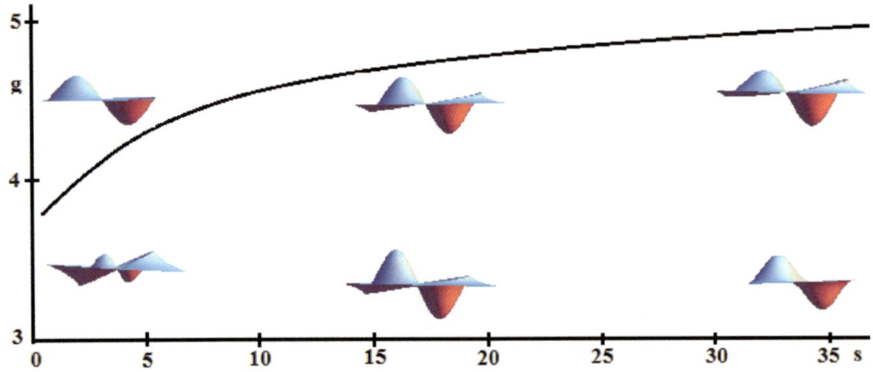

Figure 8.
"Dispersive curve"—the solution of Eq. (57). In the three ranges of the curve, 3D images described the relative perturbed density of modon in short-scale (lower row) and long-scale (upper row) limits. The blue color indicates condensations, and red color indicates rarefactions.

obviously, are the short-scale modons of the first type with the asymmetry between high- and low-pressure centers.

Let's estimate the masses of condensations in long-scale modon:

$$m_1 = \frac{2\pi h\rho_0}{J_1(g)}\beta_{lw}\int_{\pi/2}^{3\pi/2}\cos\theta d\theta\int_0^{x_1} J_1(gx)xdx = \frac{4\pi a^2 h\rho_0}{J_1(g)}\beta_{lw}J_0(gx_1)H_1(gx_1), \tag{69}$$

where $x = R/(a)$, h is the thickness of the gas disk, $H_1(gx)$ is the Struve function of the first order, and x_1 is the root of equation $J_1(gx_1) = 0$. Similarly

$$m_2 = \frac{2\pi a^2 h\rho_0}{J_1(g)}\beta_{lw}\int_{-\pi/2}^{\pi/2}\cos\theta d\theta\int_{x_1}^{1} J_1(gx)xdx + \frac{2\pi a^2 h\rho_0}{K_1(s)}\beta_{lw}\int_{-\pi/2}^{\pi/2}\cos\theta d\theta\int_1^{\infty} K_1(sx)xdx. \tag{70}$$

Numerical estimations show that the ratio of the masses of condensations in the long-scale modon, depending on values of parameters g and s, varies in the range $m_1/m_2 \sim 2 - 30$.

Now we will focus our attention on a role of vortices for the formation of planetesimals in a protoplanetary light dusty disk.

8. The Burgers vortex in local frame of reference

Let's use the local approach, choosing frame of reference, rotating with a disk with angular speed Ω_0 at distance R_0 round the central star of mass M. In this approach, assuming the effective radius of a vortex is much smaller than R_0, we will choose the Cartesian system of coordinates with center O (**Figure 9**), directing the y-axis to a star and the x-axis in direction of Keplerian flow of gas. We will present the disk rotation as

$$\Omega(R) \propto R^{-q}. \tag{71}$$

In case when only the gravitation of the central star operates, rotation will be Keplerian with q = 3/2, and for homogeneously rotating disk, $q = 2$, i.e.$2 \geq q \geq 3/2$.

The substance stream in chosen frame of reference, has X component of speed -$\mathbf{i}q\Omega_0 y$, centrifugal force is compensated by gravitation of the central star at distance

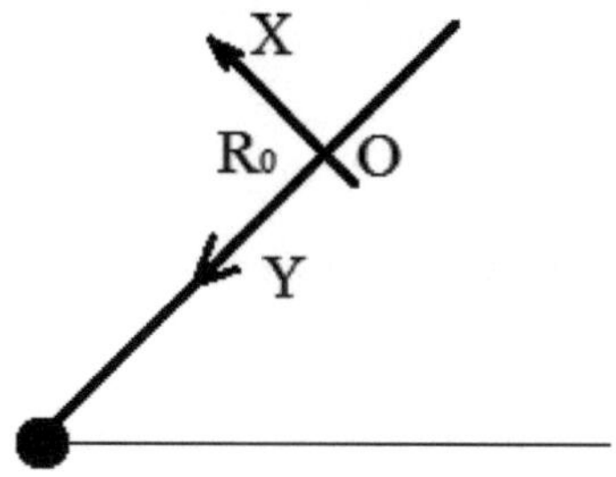

Figure 9.
The local frame of reference 2.

R_0, in other points their sum gives the tidal force $\mathbf{j}3\Omega_0{}^2y$. The vertical component of gravitation $-\Omega_0{}^2z$ is returning force along the z-axis.

In the local approach, the equation stationary isentropic shear flow of dusk viscous substance is described by Navier–Stokes and continuity equations:

$$(\mathbf{v}\nabla)\mathbf{v} = \mathbf{j}3\Omega_0{}^2y - \mathbf{k}\Omega_0{}^2z - 2\Omega_0 \times \mathbf{v} - \nabla h + \nu\Delta\mathbf{v} \quad (72)$$

$$\nabla(\rho\mathbf{v}) = 0, \quad (73)$$

where h is specific enthalpy $\left(h = \int \rho^{-1}dp\right)$ and **i**, **j**, and **k** are Cartesian unit vectors. The first term in the right-hand side of Eq. (72) is tidal acceleration, the second term is vertical gravitation, the third is acceleration of Coriolis, and the last is a viscous stress.

In the Cartesian coordinate system, the Burgers vortex (2) will be presented in the form

$$\begin{aligned} v_x &= -Ax - \omega r_0{}^2 y\left[1 - \exp\left(-r^2/r_0{}^2\right)\right]/r^2, \\ v_y &= -Ay + \omega r_0{}^2 x\left[1 - \exp\left(-r^2/r^2\right)\right]/r^2, \\ v_z &= 2Az, \end{aligned} \quad (74)$$

where $r^2 \equiv x^2 + y^2$.

9. Motion of rigid particles in Burgers vortex

Let us study the two-dimensional dynamics of dust rigid particles in a Burgers vortex. We will neglect the influence of rigid particles on dynamics of gas and the interaction of rigid particles among themselves.

As we consider centimeter- to meter-sized particles, then D considerably surpass the mean free path of gas molecules; therefore, the friction of rigid particles with gas will be described by Stokes drag force:

$$f = \beta(\mathbf{v} - \mathbf{u}), \text{where } \beta \equiv 18\ \rho\nu/\rho * D^2, \quad (75)$$

u = (dX/dt, dY/dt) is velocity of a particle, and X and Y are particle coordinates.

In a dimensionless form, the equation of motion of particles in the accepted approach looks like

$$du_x/dt = 2u_y + \gamma\left(v_x|_{r=(X,Y)} - u_x\right) - \partial h/\partial x\big|_{r=(X,Y)}, \quad (76)$$

$$du_y/dt = 3y - 2u_x + \gamma\left(v_y|_{r=(X,Y)} - u_y\right) - \partial h/\partial y\big|_{r=(X,Y),} \tag{77}$$

where γ is a dimensionless parameter

$$\gamma = \beta/\Omega_0 = 18\rho\nu/\rho * D^2\Omega_0. \tag{78}$$

In Eqs. (76) and (77) a characteristic length is accepted: the size of a trunk of a vortex r_0, for characteristic time and speed $-1/\Omega_0$ and $\Omega_0 r_0$, respectively.
In the vortex trunk area $(r^2/{r_0}^2 < 1)$, the profile of rotation has uniform character:

$$v_x = -Ax - \omega y + O\left(r^2/{r_0}^2\right), v_y = -Ay + \omega x + O\left(r^2/{r_0}^2\right). \tag{79}$$

where A and ω are measured in unit Ω_0. Therefore

$$\partial h/\partial x = -\left(A^2 - \omega^2 - 2\omega\right)x - 2\,A(\omega + 1)y, \tag{80}$$

$$\partial h/\partial y = 2A(\omega + 1)x - \left(3 + A^2 - \omega^2 - 2\omega\right)y. \tag{81}$$

With the use of Eqs. (76)–(81), we receive the equations of motion of rigid particles in the field of a vortex trunk:

$$\begin{Bmatrix} \dot{X} \\ \dot{Y} \\ \dot{u}_x \\ \dot{u}_y \end{Bmatrix} = \begin{Bmatrix} 0 & 0 & 1 & 0 \\ 0 & 0 & 0 & 1 \\ a & b & -\gamma & 2 \\ -b & a & -2 & -\gamma \end{Bmatrix} \begin{Bmatrix} X \\ Y \\ u_x \\ u_y \end{Bmatrix}, \tag{82}$$

where

$$a = A(A - \gamma) - (\omega + 1)2 + 1; b = 2A(\omega + 1) - \gamma\omega. \tag{83}$$

From Eq. (82) it follows that the equilibrium position of rigid particles in a vortex trunk is its center $X = Y = 0$, where $u_x = u_y = 0$ and $\dot{u}_x = \dot{u}_y = 0$. Particles gradually come nearer to the center of the vortex by helicoidal trajectories.
For establishing the stability of this position of balance, it is necessary to require real parts of eigenvalues of a matrix in Eq. (82) to be zero or negative.
Eigenvalues are complex:

$$\Lambda_{1,2,3,4} = -\gamma/2 \mp i \pm \sqrt{\left[a - 1 + \gamma^2/4 \pm i(b - \gamma)\right]},.$$

which gives stability condition

$$(b - \gamma)^2 + \gamma^2(a - 1) \leq 0, \tag{84}$$

Taking into account Eq. (83), Eq. (84) leads to stability criterion $\gamma > A$, which for viscosity, ν, in a dimensional form, gives

$$\nu > \rho * AD^2/18\rho \tag{85}$$

Hence, the unique position of balance for rigid particles in a Burgers vortex is its center where all particles captured by a vortex will gather during the characteristic time:

$$\tau \sim \omega r_{eff}/A\sqrt{\beta\nu}. \tag{86}$$

The mass of the rigid particles captured by a vortex during this time is in the order

$$M_p \approx \pi {r_{eff}}^2 \Sigma *, \tag{87}$$

which forms a planetesimal.

10. The thickness of disk in nuclear area of Burgers vortex

So far we have considered the behavior of a whirlwind in a disk plane. However the whirlwind of Burgers is in 3D formation. We will discuss now a question on a thickness of a disk in the area where the Burgers vortex is located. For this purpose we will address a z-projection of the Navier–Stokes Eq. (72). Integrating this equation taking into account the formula for speed v_z, we will receive dependence enthalpy from the z coordinate:

$$h\,(z) = {c_{s0}}^2 - \left(4A^2 + {\Omega_0}^2\right)z^2/2,$$

where c_{s0} is a sound speed at the vortex center (enthalpy, $h_0 = {c_{s0}}^2$, at the center of vortex is estimated by Clapeyron equation) and Ω_0 is an angular speed of rotation of local frame of reference. Whence we obtain half thickness of a disk at the kernel area of a whirlwind:

$$z_0 = c_{s0}\left(2A^2 + {\Omega_0}^2/2\right)^{-1/2}. \tag{88}$$

The question arises whether the disk thickness in area of vortex localization changed. On radius of R_0 the half-thickness of Keplerian disk from (4) is of order $z_K \cong c_{s0}/2\Omega_0$. Therefore the relative thickening

$$\frac{\Delta z}{Z_K} \equiv \frac{Z_0}{Z_K} - 1 = \frac{2\sqrt{2}}{\sqrt{\left(1 + 4A^2/\Omega_0^2\right)}} - 1, \tag{89}$$

is positive if $A < 1.3\Omega_0$. This condition is carried out in all areas of a typical protoplanetary disk. Therefore, the disk in the area of localization of a whirlwind of Burgers is thicker.

11. Discussion and conclusion

First let's pay attention to the nontrivial structure of monopoly and dipole vortices in a rotating and gravitating pure gas disk. Monopole vortices (33) with mass distribution (32) are localized formations and can have positive and negative velocity circulation, and $\Gamma.\Gamma > 0$vortex, characterized by low pressure, has negative excess mass density of substance, in contrast of $\Gamma < 0$ vortex of higher pressure, with the positive excess mass density (see **Figure 3**).

More interesting are properties of solitary dipole vortex - modon (60), (61) with mass distribution (65) in short-scale and long-scale limits. There exist two types of mass distribution in dipole vortex. Anti-symmetrically located one almost round condensation and one rarefaction (**Figure 6**) characterizes the first type. The second type is characterized by the anti-symmetrical located two condensations and two rarefactions, and second condensation-rarefaction pair has sickle-form

Figure 10.
The isodense picture of the galaxy Markaryan 266 with two nuclei, rotating in the opposite direction [36].

(**Figure 7**). Circulation of substance in different parts of modon occurs in opposite direction (**Figure 4**)!

Now it is difficult to judge about a way of evolution of these structures, for example, whether monopole vortices lead to the formation of planets in circumstellar disks, or the formation of stars or clouds in the galactic gas disk? Or, if it could transformed the dipole vortices to well-known double objects, such as double stars, double nuclei in galaxies (as Mrk 266 [36], **Figure 10**), as well as in giant molecular clouds, or a planet with a companion in circumstellar disk, or not?

As for dusty protoplanetary disks, long-lived anticyclonic vortical structures can capture the ~10 cm to meter-sized particles and grow up them into planetesimals. Let's estimate an order of magnitudes of time (86), and mass (87) for planetesimal formation by Burgers vortex for a model of a disk of radius 30 AU and mass 0.5 $M_\odot$ round a star of solar mass: $M \approx M_\odot$. Taking $R_0 = 20$ AU we will obtain estimations $\Omega_0 \approx 8 \cdot 10^{-9} s^{-1}$ and $\Sigma \sim 1600\, g/cm^2$. For a typical protoplanetary disk at considering distance the vertical scale height is of order $H \approx 10^8$km and sound speed $c_s \approx H\Omega_0 \approx 0.8$ km/s.

Let the maximum rotation speed of a vortex be ~10 m/s at distance $r_0 \approx 10^{10}$m from its center, and converging speed of a stream be $v_r = A \cdot r_0 \approx 5$ m/s. Then we will have

$$\omega \approx 10^{-9} s^{-1}, A \approx 5 \cdot 10^{-10} s^{-1}.$$

The condition (85) is carried out with a large supply for protoplanetary disks. The molecular viscosity of gas, estimated by the formula $\nu \sim \lambda c_s$, in which λ is the mean free path of molecules, c_s is the speed of a sound, does not play an appreciable role in processes of a protoplanetary disk. For this reason, the "α-disk" model [29] is used, in which turbulent viscosity is represented by the expression $\nu \sim \alpha c_s H \approx \alpha H^2 \Omega_0$. The dimensionless parameter α is constant value of an order $\alpha \sim 10^{-2}$. The scale of viscous length thus makes $L_\nu \approx 10^6$km, so Burgers vortex of big sizes cannot be destroyed by viscosity. Keplerian shear length makes $L_{shear} \approx 6 \cdot 10^9$km. Hence, vortices with the sizes $r_{eff} < L_{shear}$ can have circular form.

Taking $\rho * / \rho \approx 10^{10}$ in a midplane of a disk, using in (87) and (86) also the average value for viscosity from stability condition (85), we will receive the estimations:

$$M_p \approx 10^{28} g; \tau \sim 3 \cdot 10^6 (m/D) yrs.$$

Therefore, during an order of $\sim 10^6$ year, for meter-sized rigid particles, in the vortex trunk the mass amount comparable with mass of Venus accumulates.

Finally, note that the disk in the Burgers vortex localization area is thicker.

Author details

Martin G. Abrahamyan
Yerevan Haybusak University, Armenia

*Address all correspondence to: haybusaksci@gmail.com

References

[1] Hasegawa A, Mima K. Pseudo-three-dimensional turbulence in magnetized non-uniform plasma. Physics of Fluids. 1978;**21**:87

[2] Hasegawa A, Maclennan CG, Kodama Y. Nonlinear behavior and turbulence spectra of drift waves and Rossby waves. Physics of Fluids. 1979; **22**:212

[3] Meiss JD, Horton W. Solitary drift waves in the presence of magnetic shear. Physics of Fluids. 1983;**26**:990

[4] Larichev E, Reznik GM. On the 2D solitary Rossby waves. USSR AS Reports. 1976;**231**:1077

[5] Pavlenko V, Petviashvili V. Solitary waves in a plasma and in atmosphere. Plasma Physics. 1983;**9**:603

[6] Mikhailovskii AB, Aburdzhaniya GD, Onishchenko OG, Churikov AP. Alfven vortex solution in homogeneous magnetized plasma. Journal of Experimental and Theoretical Physics. 1984;**59**:1198

[7] Horton W, Liu J, Meiss JD, Sedlak JE. Physics of Fluids. 1986;**29**:1004

[8] Nycander J, Pavlenko VP, Stenflo I. Physics of Fluids. 1987;**30**:1367

[9] Dolotin V, Fridman AM. Generation of an observable turbulence spectrum and solitary dipole vortices in rotating gravitating systems. Journal of Experimental and Theoretical Physics. 1991;**72**(1):1

[10] Adams FC, Watkins R. Vortices in circumstellar disks. The Astrophysical Journal. 1995;**451**:314

[11] Heng K, Kenyon SJ. arXiv: 1005.1660v3 [astro-ph. EP]; 2010

[12] Youdin AN. From grains to planetesimals. European Astronomical Society (EAS) Publication Series. 2010; **41**:187

[13] Armitage PJ. Lecture notes on the formation and early evolution of planetary systems. arXiv: astro-ph/ 0701485v2; 2007

[14] Armitage PJ. Astrophysics of Planet Formation. UK: Cambridge University Press; 2010

[15] Blum J, Wurm G. The growth mechanisms of macroscopic bodies in protoplanetary disks. Annual Review of Astronomy and Astrophysics. 2008;**46**:21

[16] Zsom A, Ormel CW, Guttler C, Blum J, Dullemond CP. arXiv: 1001.0488v1; 2010

[17] Wilner DJ, D'Alessio P, Calvet N, Claussen MJ, Hartmann L. Toward planetesimals in the disk around TW Hydrae: 3.5 centimeter dust emission. The Astrophysical Journal. 2005;**626**: L109

[18] Adachi I, Hayashi C, Nakazawa K. The gas drag effect on the elliptical motion of a solid body in the primordial solar nebula. Progress in Theoretical Physics. 1976;**56**:1756

[19] Weidenschilling SJ. Formation and evolution of exoplanets. Monthly Notices of the Royal Astronomical Society. 1977;**180**:57

[20] Paterson HL, Feng M, Waite AM, Gomis D, Beckley LE, Holliday D, et al. Physical and chemical signatures of a developing anticyclonic vortexes. Journal of Geophysical Research. 2008; **113**(C7):C07049

[21] Carnevale GF, McWilliams JC, Pomeau Y, Weiss JB. Evolution of

vortex statistics in two-dimensional turbulence. Physical Review Letters. 1991;**66**:2735

[22] Weiss JB, McWilliams JC. Temporal scaling behavior of decaying two-dimensional turbulence. Physics of Fluids A. 1993;**5**:3

[23] Tabeling P. Two-dimensional turbulence: A physicist approach. Physics Reports. 2002;**362**:1

[24] Inaba S, Barge P, Daniel E, Guillard H. A two-phase code for protoplanetary disks. Astronomy & Astrophysics. 2005; **431**:365

[25] Inaba S, Barge P. Dusty vortices in protoplanetary disks. The Astrophysical Journal. 2006;**649**:415

[26] Abrahamyan MG. The vortex of Burgers in protoplanetary disk. Astronomical Society of the Pacific. 2017;**511**:254-264

[27] Rankine WJM. On the thermal energy of molecular vortices. Philosophical Magazine. 1870;**39**:211

[28] Abrahamyan MG, Matveenko LI. Initial phase of protostar formation. Astrophysics. 2012;**55**:443

[29] Shakura NI, Sunyaev RA. Black holes in binary systems. Astronomy & Astrophysics. 1973;**24**:337

[30] Pringle JE. Astrophysics of planet formation. Annual Review of Astronomy and Astrophysics. 1981; **19**:137

[31] Chiang EI, Goldreich P. Formation and evolution of exoplanets. The Astrophysical Journal. 1997;**490**:368

[32] Godon P, Livio M. The formation and role of vortices in protoplanetary disks. The Astrophysical Journal. 2000; **537**:396

[33] Abrahamyan MG. Anticyclonic vortex in a protoplanetary disk. Astrophysics. 2016;**59**:309

[34] Chandrasekhar S. Ellipsoidal Figures of Equilibria. New Haven and London: Yale University Press; 1969

[35] Abrahamyan MG. Vortices in rotating gravitating gas disks. Astrophysics. 2015;**58**:105

[36] Petrosyan AP, Sahakyan KA, Khachikyan EY. Spectroscopic investigation of double nucleus galaxy Markarian 266. Astrophysics. 1980;**16**:621

Chapter 3

Vortices on Sound Generation and Dissipation in Musical Flue Instruments

Shigeru Yoshikawa

Abstract

Musical flue instruments such as the pipe organ and flute mainly consist of the acoustic pipe resonance and the jet impinging against the pipe edge. The edge tone is used to be considered as the energy source coupling to the pipe resonance. However, jet-drive models describing the complex jet/pipe interaction were proposed in the late 1960s. Such models were more developed and then improved to the discrete-vortex model and vortex-layer model by introducing fluid-dynamical viewpoint, particularly vortex sound theory on acoustic energy generation and dissipation. Generally, the discrete-vortex model is well applied to thick jets, while the jet-drive model and the vortex-layer model are valid to thin jets used in most flue instruments. The acoustically induced vortex (acoustic vortex) is observed near the amplitude saturation with the aid of flow visualization and is regarded as the final sound dissipation agent. On the other hand, vortex layers consisting of very small vortices along both sides of the jet are visualized by the phase-locked PIV and considered to generate the acceleration unbalance between both vortex layers that induces the jet wavy motion coupled with the pipe resonance. Vortices from the jet visualized by direct numerical simulations are briefly discussed.

Keywords: edge tone, pipe tone, jet wave, jet-drive model, discrete-vortex model, vortex-layer model, vortex sound theory, flow visualization, acoustically induced vortex, PIV, direct numerical simulation

1. Introduction

Musical wind instruments have a mechanism converting the direct energy of the fluid flow into the alternative energy of the sound. Such a system is called the self-sustained oscillation system. The fluid flow that drives the instruments may be regarded as the aerodynamical sound source or aeroacoustical source. Wind instruments are a very extensive subject of research over the vibration theory, acoustics, and fluid dynamics. The interaction between the resonance of the instrument [called generically *flue instruments* such as an organ pipe, flute, and recorder in this chapter (see **Figure 1**)] and the jet as the aeroacoustical source will be adequately described in this chapter.

Fluid flow brings about vortices and then generates the sound as well. However, one of the essential characteristics of wind instruments is the resonance, which is an acoustic mechanism amplifying very small perturbations to periodic disturbances

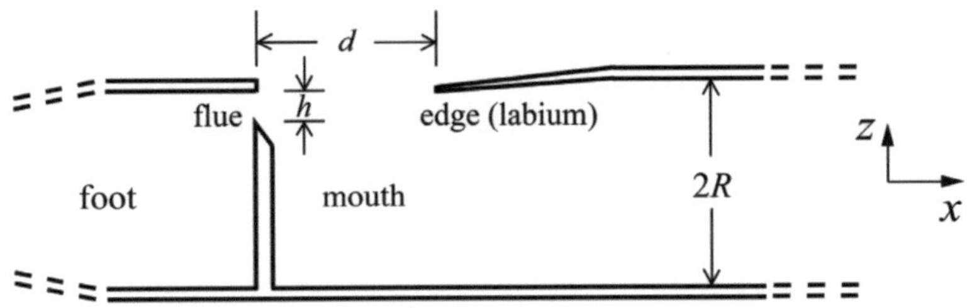

Figure 1.
An organ flue pipe as a typical example of the musical flue instruments and its important parameters. d, the flue-to-edge distance (or cutup, jet length); h, the jet thickness (or height of the flue exit); R, the pipe inner radius. The origin of the coordinate system is located at the center of the flue exit surface.

with large amplitudes. Any synchronization is then required, and it is realized by the suitable *phase relation* between the flow (or the jet) driving wind (or flue) instruments and the acoustic wave propagating in the instruments. For example, in the case of flue instruments, if the air flow enters into the pipe at the instant when the acoustic pressure near the edge takes a relatively large positive value, acoustic power (given by the product of the alternating volume flow and the acoustic pressure) becomes positive, and the sound is sustained.

However, a big dissatisfaction to the above viewpoint is the assumption of the existence of the sound at the starting point. Therefore, exactly saying, acoustical theory above is not sound generation theory but sound *regeneration* theory. The viewpoint of positive feedback between the jet and the pipe and the time-domain formulation based on the pipe reflection function [1] are both sound regeneration theory [2]. Musical instrument acoustics has treated such regeneration theories and phenomena as chief objects of research. This is because the resonance is acoustically essential, and we may consider that the resonance controls fluid movement as the energy source. It will be open to the charge of being imperfect combustion that sound existence is presupposed at the starting point when we try to answer how flue instruments produce their sounds.

Then, if we introduce a thesis, "the vortex itself is the true sound source," of the vortex sound theory [3] to flue instruments, is the problem solved? Flue instruments do not seem to be such an obedient subject. Certainly, the vortex sound theory is satisfactorily valid to the edge tone, where the jet-edge system has no pipe that gives the resonance or the acoustic feedback; instead the fluid-dynamical feedback between the edge and the flue (flow issuing slit) is a main mechanism of sound generation.

Moreover, there are a few non-negligible differences other than the acoustic resonance between the edge tone and the pipe tone (or flue tone). First is the amplitude magnitude when the jet oscillates against the edge. The oscillation amplitude of the edge-tone jet is as small as two to three times the jet thickness. On the other hand, the pipe-tone jet in an organ pipe often exceeds 10 times the jet thickness. The edge in an organ pipe (or flue instruments) is just a partition wall which separates the inside from the outside of the pipe. It may be said that the direct jet-edge interaction time is quite short compared with a tonal period in flue instruments. Large vortices *visible* behind the pipe edge are, so to speak, odds and ends of the jet driving the pipe. We should pay more attention toward *invisible* (for our naked eyes) vortices carried along the jet to the edge.

Second is the difference in the jet-edge configuration. The configuration is usually symmetrical in the edge tone. In other words, the jet center surface corresponds to the edge tip. Alternate small vortices continuously appear above and beneath the edge. On the other hand, the edge is usually displaced upward in organ pipes (see **Figure 1**). The flute may have such asymmetry depending on the player. This jet-edge divergence is called the *offset*, which is one of important parameters to adjust the tone color of flue instruments.

Also, we should relevantly notice largely different flow-acoustic interactions involved in various vortex-related sound generations. A thin jet and a sharp edge are used for the edge tone [4–6]. A thick (or semi-infinite) jet usually drives a wall-mounted cavity to produce its resonance called the cavity tone [7–9]. A thin jet drives a sharp edge (called the labium) of the resonant pipe to produce an organ pipe tone [1, 2, 10, 11]. A thin jet drives a thick edge with an angle of about 60° in the flute [1]. A thin jet issuing from a flue with the chamfer drives a sharp edge in the recorder [12, 13]. In addition, jet velocity widely extends from a few meters per second to about 50 m/s for these tone productions. Flow condition is laminar or turbulent. The Strouhal number $St = fd/U_0$ (f the sounding frequency; d the flue-to-edge distance; and U_0 the jet velocity at the flue exit) extends from about 0.05 to 5. Generally, a thin edge tends to enhance higher harmonics. As St has higher values, the jet flow, which drives the resonant pipe, tends to break down into *discrete vortices* [10, 14].

Although the vortex is essential in flow-excited sound generation, it may operate as an important source of acoustic energy dissipation in various flow-acoustic interactions [3, 15–17]. In the context of musical instruments, the *acoustically induced vortex* shedding at the edge is a key damping mechanism to determine the final amplitude of the steady-state flue instrument tones [16, 17]. Hence, sound dissipation and generation in flow-acoustic interactions are widely dominated by the vortex shedding at the edge [3].

Howe [18] assumes that a compact vortex core with relatively large size appearing alternately just above and below the pipe edge is created by the interaction between the jet velocity vector $\boldsymbol{U}$ and the cross-flow velocity (acoustic reciprocating velocity) vector $\boldsymbol{u}$ at the mouth opening formed between the flue and the edge. This vortex core with the vorticity $\boldsymbol{\omega}$ $(= \nabla \times \boldsymbol{U})$ is then considered to drive the air column in the pipe. The sound excitation by this periodic vortex shedding at the edge is controlled by the product of the aeroacoustic source term div $(\boldsymbol{\omega} \times \boldsymbol{U})$ and the potential function representing the irrotational cross-flow $\boldsymbol{u}$ at the mouth.

This discrete-vortex model of Howe is successfully applied to analyze and evaluate both cavity-tone generation [9] and tone generation in flue instruments [10, 14] when the jet is thick and the condition $d/h < 2$ (d the width of the mouth opening or the flue-to-edge distance and h the jet thickness) is satisfied. On the other hand, when the condition $d/h > 2$ is satisfied for thin jets, a jet-drive model on the basis of the intrinsic jet instability [19, 20] is applied instead of the discrete-vortex model [21]. This jet-drive model has been developed in the field of acoustics [1, 2, 11, 20, 22–27].

Although the jet-drive model has been proven to be effective for an explanation of sound generation by the thin jet, there remain rooms for improvement in applying the vortex sound theory for another explanation of sound generation by the thin jet in flue instruments because small vortices may be produced along the boundaries by the mixing process between the jet flow and the surrounding still air. The boundary layer consisting of small vortices is called the *vortex layer*, which can act as the source of an accelerating force to oscillate the jet. Based on such a viewpoint, the vortex-layer model was proposed recently [28].

In Section 2, the jet-drive, discrete-vortex, and vortex-layer models are described. Acoustically induced vortices (simply, acoustic vortices) on sound dissipation are discussed with the aid of flow visualization in Section 3. The jet vortex layer on sound generation in an organ pipe is visualized by the particle image velocimetry (PIV), and the microstructure of the vortex layer is demonstrated in

Section 4, and some examples of jet vortices are also introduced from experiments and simulations. Conclusions are given in Section 5.

2. Models on sound generation in flue instruments

2.1 Jet-drive model

2.1.1 Volume-flow drive vs. pressure (momentum) drive

Jet motion in an organ pipe model when the jet drive is operating at the steady state is depicted in **Figure 2**. The air jet smoked with incense sticks is observed by a stroboscope and recorded on a VTR (8-mm video cassette) as analog data [2]. The pipe length L is 500 mm, the flue-to-edge distance (cutup length) d 10.2 mm, and the jet thickness h at the flue exit 2.2 mm [2, 17]. The blowing pressure is 200 Pa (the jet velocity at the flue exit is estimated from Bernoulli's law to be 18.3 m/s). The sounding frequency is 285 Hz. One period T of the jet motion is divided by 9 in **Figure 2**.

As shown in **Figure 2**, the jet oscillates up and down. It does not break into vortices but keeps a diaphragm-like shape in the jet-drive operation. Large vortex-like air observed above the edge will not take a part in the sound generation. It is a kind of odds and ends of the jet driving the pipe. Also, it should be noticed that the jet behaves like an amplifying wave as inferred from the first six frames [29].

When the jet enters the pipe passing through the mouth area between the flue exit and the edge, the jet provides the pipe with the acoustic volume flow $q(t)$ that is roughly approximated by the product of jet velocity U_e, jet breadth b, and jet lateral displacement $\xi_e(t)$ (these quantities are given at the edge):

$$q(t) \approx -U_e b \xi_e(t), \tag{1}$$

where t is the time. The minus sign is needed from the definition that $\xi_e(t)$ is positive outward and $q(t)$ is positive inward. Eq. (1) defines the volume-flow model that was first proposed by Helmholtz [30] and utilized by many researchers afterwards for small amplitudes of jet oscillation [1, 2, 10–13, 20, 21, 24–27]. At the same time, the jet provides the pipe with the acoustic pressure produced by the momentum exchange with still air in the pipe:

$$p(t) = \rho U_e^2 (S_j/S_p), S_j \approx b\xi_e(t), \tag{2}$$

where ρ is the air density, S_j the temporally varying cross section of the jet entering the pipe from the edge, and S_p the pipe cross section. Eq. (2) defines the jet pressure of the jet momentum model, which was first proposed by Rayleigh [19] and utilized by many researchers afterwards [1, 2, 10–13, 21–27]. Opposing Helmholtz, Rayleigh insisted that the momentum drive should be effective. This is based on that the pipe is open and the acoustic power is produced by the product of the acoustic particle velocity near the pipe edge and the driving pressure given by the jet. However, the acoustic pressure considerably remains near the pipe edge due to the end correction. As a result, the volume-flow drive of Helmholtz is usually predominant except for the jet drive with very high blowing pressures [1, 2, 10, 21, 24, 25, 27].

The jet-drive model based on the volume-flow drive and the pressure drive was first formulated by Elder [24] by deriving the so-called jet momentum equation and then simplified by Fletcher [25]. They assumed a small control volume with length

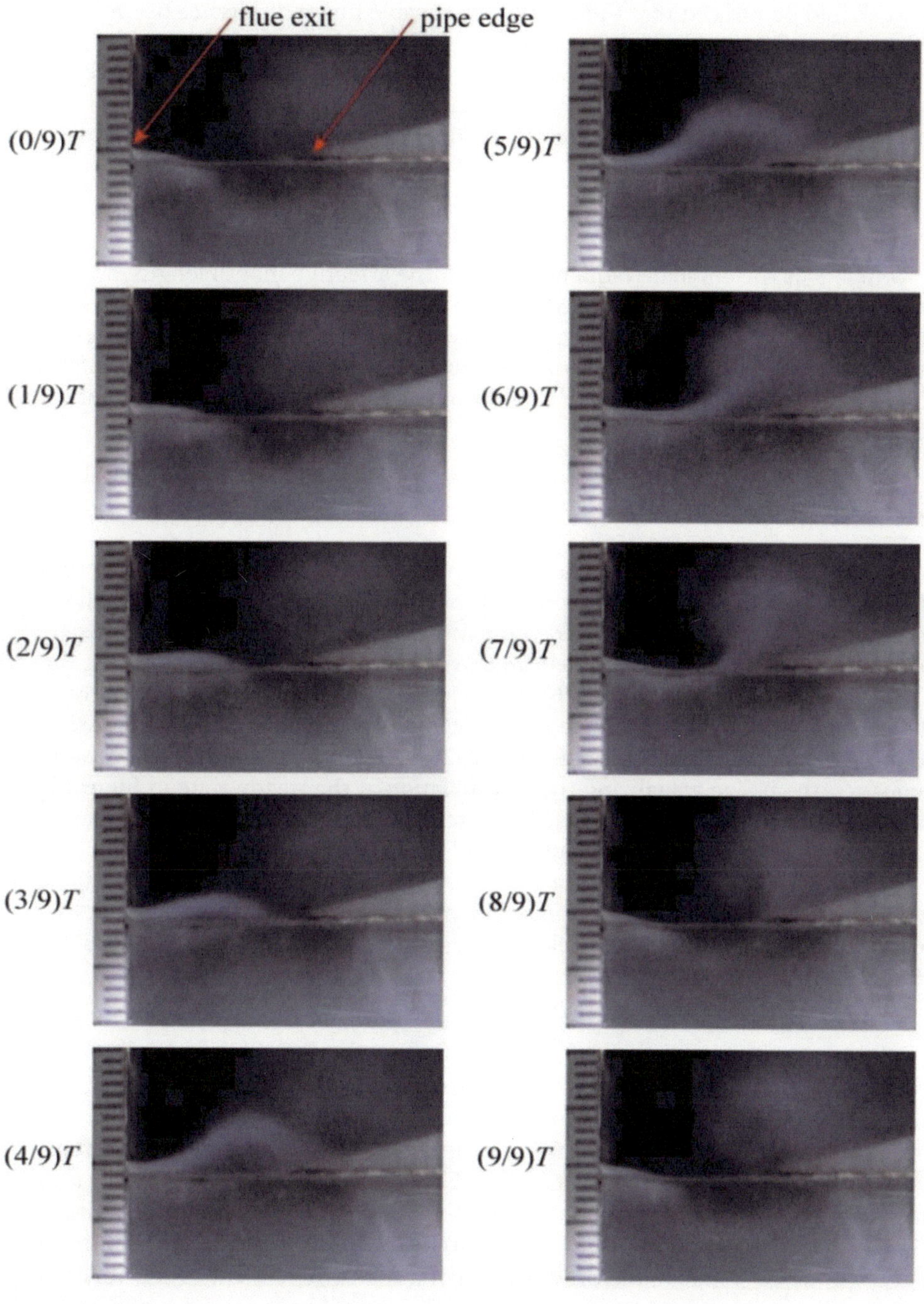

Figure 2.
Stroboscopically visualized jet oscillation at the steady state caused by the jet drive of an organ pipe model made with the acrylic resin.

Δx below the pipe edge. The turbulent mixing takes place over this control volume. The loss of jet momentum there will result in the simple pressure rise at the inner plane of the control volume. The net force on the control volume due to this pressure rise can then be equated to the rate at which jet momentum changes in the control volume [24]. In other words, just as "the momentum difference equals to the force impulse," the momentum-flow-rate difference gives the force that accelerates the mass of the control volume.

It should be noticed that there is an appreciable phase difference between the volume-flow drive and the pressure drive. This phase difference is not well understood from Eqs. (1) and (2). The acoustic impedance or admittance should be introduced to connect these equations. According to Fletcher [1], this phase difference ϕ, which gives the phase lag of the pressure drive, is given by:

$$\phi = -\tan^{-1}(U_e/\omega\Delta L), \tag{3}$$

where ω is the angular frequency and ΔL the effective mouth length including the open end correction. Since $\omega\Delta L > U_e$ in usual cases, ϕ is quite small. However, as mentioned above, ϕ becomes appreciably large when the high blowing pressure is applied. See [1, 2, 10, 24–27] for more discussion on the complex jet/pipe interaction in flue instruments and on the conditions of the phase and amplitude for sound regeneration.

2.1.2 Jet wave and its amplification

Each frame in **Figure 2** does not show the path of air particle, but corresponds to the snapshot of the position of air particle at a given instant. Therefore, each frame indicates the *streak line* in the fluid-dynamical sense. On the other hand, the jet particle pass (*pass line*) may be determined as soon as the jet issues from the flue to the acoustic field in the mouth [26]. If the jet pass line can be determined, the jet deflection shape as the streak line may be estimated by considering the transit time of the particle issued from the flue exit [2, 26]. However, in the field of musical acoustics, the image visualizing the jet oscillation at the steady state in **Figure 2** has been called the *jet wave*, which seems to give the transverse displacement (in z direction) of the jet (cf. **Figure 1**).

It is assumed that the jet displacement may be expressed as a superposition of a progressive wave due to the jet instability [19, 20] and a spatially uniform oscillation induced by acoustic velocity u through the mouth as follows [1, 31]:

$$\xi(x,t) = (u/\omega)\{-\cosh(\mu x)\sin[\omega(t - x/U_{ph})] + \sin(\omega t)\}, \tag{4}$$

where $\xi(x,t)$ denotes the transverse displacement of the jet at distance x from the flue exit. Also, μ and U_{ph} denote the amplification factor and phase speed of the instability wave, respectively, while u is the transverse acoustic velocity of the mouth field. The jet displacement $\xi_e(t)$ in Eqs. (1) and (2) is given by $\xi(x = d, t)$. It should be noted that $\xi(x)$ and u are positive in the external (positive z) direction. Both μ and U_{ph} are functions of x in a rigorous sense. Also, it is important that ω in Eq. (4) corresponds to the fundamental of a sound generated. When very high blowing pressures are applied, the effect of the second harmonic is significant and the jet strikes the edge downward twice a period. Such an effect is excluded in Eq. (4). Also, an important parameter, jet thickness h is not included in Eq. (4), and a thin jet ($d/h > 2$) is assumed. Although Eq. (4) lacks its physical basis and experimental confirmation, it is a simple and practical representation of the jet deflection that describes our current knowledge. So, we postulate in this chapter that Eq. (4) is valid in the acoustical sense.

The envelope of positive and negative peak displacements is yielded from Eq. (4) as follows:

$$\xi_{env}(x) = \pm(u/\omega)\left[\cosh^2(\mu x) - 2\cosh(\mu x)\cos(\omega x/U_{ph}) + 1\right]^{1/2}. \tag{5}$$

Since the $\cos(\omega x/U_{ph})$ dependence diminishes as μx becomes larger than one, it may be reasonable to approximate the second term in square bracket of Eq. (5) as $2\cosh(\mu x)$. We then obtain the following simple expression [29]:

$$\xi_{env}(x) \approx \pm(u/\omega)[\cosh(\mu x) - 1]. \quad (6)$$

Figure 3 shows the transverse displacements of the jet oscillation and their approximated envelopes (indicated by the broken line) at the steady state which are calculated by Eqs. (4) and (6), respectively. The following parameter values are supposed: the mouth-field strength $u/\omega = 1$ (mm), $\mu = 0.20$ (mm^{-1}), ω=2π • 133 (rad/s), $k = \omega/U_{ph} = 0.073 + 0.01x$ (mm^{-1}), $U_{ph}(x = 0) = U_0 = 11.5$ (m/s). The slowdown of U_{ph} is assumed so that k may be proportional to x in order to make calculation easier. For more detailed information, refer to [29].

The jet displacements from t = (3/10)T to t = (8/10)T in **Figure 3** almost correspond to those from t = (2/9)T to t = (7/9)T in **Figure 2**, though the flue-to-edge distance d is not the same (d = 10.2 mm in **Figure 2** and d = 15.8 mm in **Figure 3**). This good correspondence proves the effectiveness of the displacement model based on Eq. (4).

It is possible to directly estimate μ from Eq. (4) by applying it to the experimental data. However, such an approach needs exact information about U_{ph} and a time reference. Another much simpler method to estimate μ is to apply Eq. (6) to the experimental data. The result is shown in **Figure 4** for the first mode of an organ pipe model with L = 50.0 cm and d = 10.2 mm (cf. **Figure 2**) [29]. If a high-speed digital video camera is used instead of a stroboscope, the frames showing the jet waves such as given in **Figure 2** are digitally memorized, and then the digital superposition of these data yields a direct superposition of jet waves.

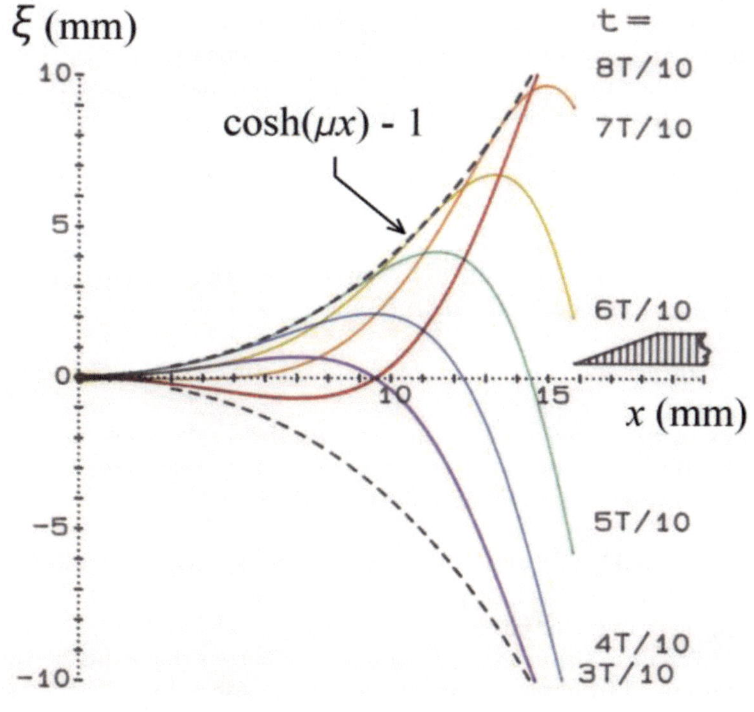

Figure 3.
The jet displacements calculated at the instants from 3 T/10 to 8 T/10 and their approximated envelope (the negative envelope is also indicated).

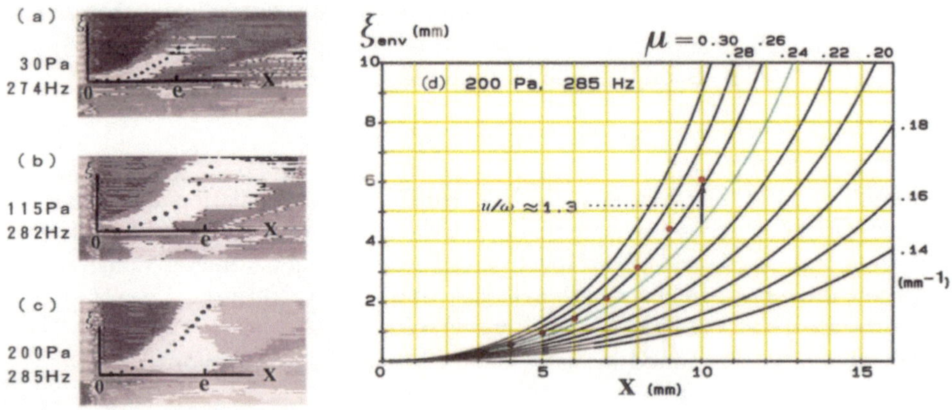

Figure 4.
Digitally superposed jet waves for different blowing pressures (a)–(c) and an illustration of how to derive the amplification factor μ and the mouth-field strength u/ω (d). The envelope of jet center-planes is estimated by the red dotted line that best fits to the green template curve with μ = 0.24 mm⁻¹ and u/ω = 1. The amplitude ratio of the dotted line to the green line determines u/ω = 1.3 mm.

Digitally superposed jet waves are shown in frames (a), (b), and (c) for different blowing pressures in **Figure 4**, where the x axis is drawn straightforwardly from the flue center labeled as "0," and label "e" indicates the edge position. Frame (d) on the right of **Figure 4** illustrates how the jet envelope function is derived. The envelope of jet center-planes (indicated by dotted lines in **Figure 4**) is almost parallel to the outer fringe of the superposed jet waves as long as significant spreading of the jet can be neglected by using thin smokes for visualization. More details on the estimation of μ from flow visualization are given in [29].

In **Figure 4(d)** the estimated envelope is shown by dots on a template, that is, curves of jet envelope function $\cosh(\mu x) - 1$ for various μ values assuming that $u/\omega = 1$. **Figure 4(d)** suggests a close fit of the dotted line to an envelope curve with $\mu = 0.24\ \text{mm}^{-1}$ when the magnitude ratio of the dotted line to that curve is about 1.3. This ratio determines $u/\omega = 1.3$ mm from Eq. (6).

Flow visualization suggests the following general trends from the result summarized in **Figure 5** on a particular experimental model of the organ pipe:

1. The amplification factor μ tends to decrease and saturate to a given value as the oscillation of each mode shifts toward higher blowing velocities, although the data on the second mode are not sufficient.

2. The averaged amplification factor is roughly estimated as $0.24\ \text{mm}^{-1}$.

3. The mouth-field strength u/ω, which means the displacement amplitude of the acoustic field at the mouth, tends to increase and saturate to a given value as the oscillation of each mode shifts toward higher blowing velocities.

It should be noted here that the estimate of μ based on Eq. (6) tends to be a little larger (about 10%) than that based on Eq. (4). However, this estimation error is roughly equivalent to the resolution of the experimental data [29].

In order to confirm the validity of our digital superposition explained above, u/ω was determined from measurements of the acoustic particle velocity u with a hot-wire anemometer (its sensing part is 1-mm long and 5 μm in diameter) and of the sounding frequency $\omega/2\pi$ (about 280 Hz in the first mode) with a microphone located inside the pipe. It is important to avoid exposing the hot-wire probe to the

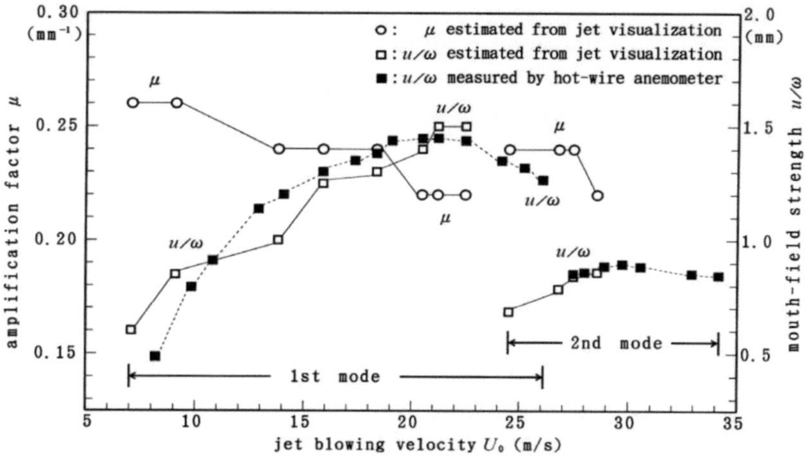

Figure 5.
The amplification factor μ and the mouth-field strength u/ω of the organ pipe jet as functions of the jet velocity U_0 at the flue exit [29]. The symbols ○ and □ indicate μ and u/ω estimated from the digital superposition based on the jet visualization, respectively. The symbol ■ indicates u/ω measured by a hot-wire anemometer and a microphone.

jet flow for the measurement of u because $u \ll U_0$. Therefore, after carrying out a preliminary experiment to select proper positions of the probe, the rms acoustic particle velocity $u_{\rm rms}$ was measured as the rms output voltage $V_{\rm rms}$ of the hot-wire anemometer ($u = \sqrt{2}u_{\rm rms}$) [29]. The result of u/ω is indicated by the closed square in **Figure 5**.

Comparing the values of u/ω measured using the hot-wire anemometer with those estimated from flow visualization, we may see a good agreement between them. This agreement implies that the method of deriving the envelope function of the jet wave is valid and sufficiently accurate. However, we had particular difficulty in obtaining a smooth jet-wave envelope near saturation, and the estimated data of μ and u/ω were lacking in **Figure 5**. This was due to other jet waves which were generated by the second harmonic and superposed upon the jet waves generated by the fundamental.

The origin of the jet-wave amplification is the jet instability. The applicability of the spatial and temporal theories on the jet instability [1, 32–34] to organ pipe jets can be discussed. If we assume a Poiseuille flow at the flue exit and a subsequent Bickley jet, the spatial theory [32, 33] seems to be relevant to organ pipe jets [29].

2.1.3 Jet-drive model for large jet displacements

The jet-drive model described above has supposed small displacements of the jet at the pipe edge. However, as demonstrated in **Figures 2–4**, the jet displacement is too large to apply Eqs. (1) and (2) to the sound generation in flue instruments in rigorous sense. According to Dequand et al. [21] and Verge et al. [27], a jet-drive model reasonable for large jet displacements is explained and roughly formulated below.

As understood from **Figures 2** and **3**, the passage time of the jet from one side to the other side of the edge seems to be very short compared to the oscillation period. In other words, the jet seems to be instantaneously switching from the inside to the outside of the pipe and vice versa. Then, the jet volume flow may be assumed to be split into two complementary antiphase monopole sources $q_{\rm in}$ [$= |q| \exp(i\omega t)$)] and $q_{\rm out}$ ($= -q_{\rm in}$) whose temporal waveforms are rectangular pulses with the same amplitude $|q|$ [27]. These sources are supposed to be placed at a distance ϵ from the edge tip at the lower and upper sides of the edge.

The acoustic pressure $p_j(t)$ is derived from the potential difference across the mouth induced by two monopole sources [21, 27]:

$$p_j(t) = -\rho(\delta_j/bd)(dq/dt), \tag{7}$$

where ρ is the air density, b the jet (or mouth) breadth, d the flue-to-edge distance (or the mouth length), and δ_j the effective distance between the two monopole sources. If $\epsilon \ll d$, we have from [27].

$$\delta_j/d \approx (4/\pi)\sqrt{2\epsilon/d}. \tag{8}$$

In the limit of thin jets $(d/h \gg 1)$, where h denotes the jet thickness at the flue exit, $\epsilon = h_e$ (the jet thickness at the edge) is assumed [27].

The power generated by the source is calculated by assuming that the source is in phase with the acoustic volume flow $q_m(t) = (d\xi_m/dt)db = udb$ through the mouth opening, where ξ_m denotes the particle displacement over the mouth opening [21]. This q_m is supposed to be a local two-dimensional incompressible flow. The above in-phase relation between the pressure source p_j and the acoustic volume flow q_m gives the condition for which the oscillation amplitude has a maximum as a function of the blowing pressure.

The time average over an oscillation period T of the power generated by the jet drive above is given as follows [21]:

$$\langle \Pi_{jet} \rangle = \langle p_j q_m \rangle \approx (8/\pi T)\rho\sqrt{2h_e/d}U_0 h_e bd|d\xi_m/dt|. \tag{9}$$

In addition to the thin jet assumption $(d/h \gg 1)$, we have to suppose that the jet does not break down into discrete vortices. This is only reasonable for the first hydrodynamic mode $(S_t = fd/U_0 < 0.3)$. The validity of Eq. (9) will be discussed in Section 3 after deriving the acoustic energy loss due to vortex shedding at the edge.

2.2 Discrete-vortex model

2.2.1 Discrete-vortex model based on the vortex shedding at the edge

On the basis of the two-dimensional theory, Howe [18] proposed a discrete-vortex model on sound generation in flute-like instruments. He assumed that a compact vortex core appearing alternately just above and beneath the edge was created by the interaction with the acoustic cross-flow velocity $\boldsymbol{u}$ [$d\xi_m/dt$ in Eq. (9) corresponds to one-dimensional (z direction) component] at the mouth opening (see **Figure 6**). That is, instead of the jet oscillation over the mouth explained in the previous section, a point vortex is produced at the edge. Then, this vortex core is assumed to drive the air column in the pipe. A discrete-vortex model for thick jets assumes that a discrete vortex is generated from the flow separation at the flue exit corner [9, 10, 21], while Howe [18] attached greater importance to the flow separation (vortex shedding) at an opposing sharp edge due to the acoustic cross-flow.

The sound excitation by the periodic vortex shedding at the edge is controlled by the product of the aeroacoustic source term $\mathrm{div}(\boldsymbol{\omega} \times \boldsymbol{v})$ and the potential function $\phi(\boldsymbol{y})$ representing the irrotational cross-flow into and out of the mouth as expressed by the following integral [18]:

$$I(t) = \int \mathrm{div}(\boldsymbol{\omega} \times \boldsymbol{v})\phi(\boldsymbol{y})\mathrm{d}^3\boldsymbol{y} = -\int (\boldsymbol{\omega} \times \boldsymbol{v}) \bullet \nabla\phi(\boldsymbol{y})\mathrm{d}^3\boldsymbol{y}, \tag{10}$$

where the vorticity $\boldsymbol{\omega}$ is defined as rot $\boldsymbol{v}$ and the velocity $\boldsymbol{v}$ is the superposition of the jet mean flow velocity $\boldsymbol{U}$ directing against the edge and the time-dependent cross-flow velocity $\boldsymbol{u}$ that is specified by reciprocating potential flow through the mouth opening (see **Figure 6**). That is, $\boldsymbol{v}$ is given by:

$$\boldsymbol{v} = \boldsymbol{U} + \boldsymbol{u}. \tag{11}$$

Also, $\boldsymbol{y}$ denotes the source region (vortical field) of the pipe mouth over which the integral of Eq. (10) is performed. The specific form of $\phi(\boldsymbol{y})$ is given as $\phi_2^*(\boldsymbol{y})$ of Eq. (10.42) in [18].

According to [3], the power density supplied from the acoustic field to the vortical field around the edge is given by:

$$(1/2)(\partial/\partial t)(\rho \boldsymbol{v}^2) = \rho(\boldsymbol{\omega} \times \boldsymbol{v}) \bullet \boldsymbol{u}. \tag{12}$$

Therefore, if $(\boldsymbol{\omega} \times \boldsymbol{v}) \bullet \boldsymbol{u} > 0$, it may be said that the energy of the acoustic field is absorbed by the vortical field. As a result, an *acoustically induced vortex* or *acoustic vortex* is generated [16, 17]. The generation of it satisfies the phase relation in which a clockwise-rotating vortex appears above the edge when $\boldsymbol{u}$ directs into the pipe [see **Figure 12(b)** in Section 3.2.1]. This phase relation is contrary to that illustrated in **Figure 6**. A discrete vortex of Howe's type has not been observed in sound generation of flue instruments where jets are used in normal conditions [16, 17, 20–27, 29].

Howe [3, 35] then proposed the *acoustic dissipation formula*:

$$\Pi_{\mathrm{D}}(t) \approx \iiint \rho(\boldsymbol{\omega} \times \boldsymbol{v}) \bullet \boldsymbol{u}\mathrm{d}V, \boldsymbol{v} \approx \boldsymbol{U}. \tag{13}$$

This equation determines the rate of dissipation of acoustic energy, where $\nabla\phi$ in Eq. (10) is now simply denoted by $\boldsymbol{u}$. Also, V denotes a volume enclosing the vorticity formed in the flow field. This Π_{D} of Eq. (13) can be negative in oscillation systems: If the phase of vorticity production enables a steady transfer of energy to the oscillation from a mean flow, the self-sustained oscillation can be maintained.

Hence, the *acoustic generation formula* may be proposed:

$$\Pi_{\mathrm{G}}(t) \approx -\iiint \rho(\boldsymbol{\omega} \times \boldsymbol{U}) \bullet \boldsymbol{u}\mathrm{d}V, \boldsymbol{\omega} = \mathrm{rot}\boldsymbol{U}, \tag{14}$$

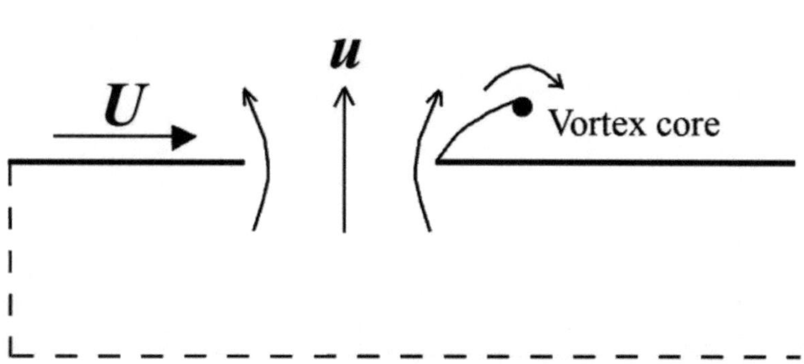

Figure 6.
Conceptual sketch of a discrete-vortex model (the vortex shedding at the edge) for sound generation in flue instruments proposed by Howe [18].

where the vorticity ω is simply given as rot$\boldsymbol{U}$. If the time average $\langle \Pi_G(t) \rangle$ is positive, the vorticity production from the jet flow supplies the acoustic power to the resonant pipe.

2.2.2 Discrete-vortex model based on the vortex shedding at the flue exit

When the jet is thick, the jet flow is not fully deflected into the resonant pipe. It is then difficult to apply the jet-drive model to thick jets. As the jet becomes thicker and thicker, the two shear layers at both sides of the jet tend to behave independently of each other. Meissner [9] described both shear layers in terms of discrete vortices (see **Figure** 7). He used the jet with $h = 2.7$ mm (not so very thick). This jet excited a cavity resonator (its cross section: 40 mm × 28 mm; its depth: 12, 14, 16, 18, and 20 cm). The distance from the nozzle exit to the opposing orifice edge corresponding to the flue-to-edge distance d was set to be 8 mm ($d/h = 2.96$). The orifice width was 28 mm, and the edge thickness 2 mm.

Meissner [9] found experimental results as follows: In stage I (5.9 m/s < U_0 < 8.3 m/s), the frequency increased fast with the jet speed, and a frequency increment was proportional to the jet speed just as in the edge-tone generation [36, 37]. Similar phenomenon often appears at the very first stage in sound generation of flue instruments [1, 23]. In stage II (8.3m/s < U_0 < 16.1 m/s), an increase in the frequency was still observed, but a frequency growth was much smaller. The experimental results obtained for different cavity depths correlated reasonably well because data points corresponding to this stage approximately lay in one curve [9].

The cavity-tone generator shown in **Figure** 7 can be considered as a simplified model of the ocarina. It is assumed that vorticity generation begins immediately after the jet issues from the nozzle exit due to flow separation. The vorticity of both shear layers is concentrated into line vortices traveling along straight lines with the convection velocity U_c. In the case of asymmetric vortex formation as shown in **Figure** 7, a configuration of vortices will be similar to that in the conventional Kármán vortex street. Thus, U_c may be approximated as that of an infinite street [9].

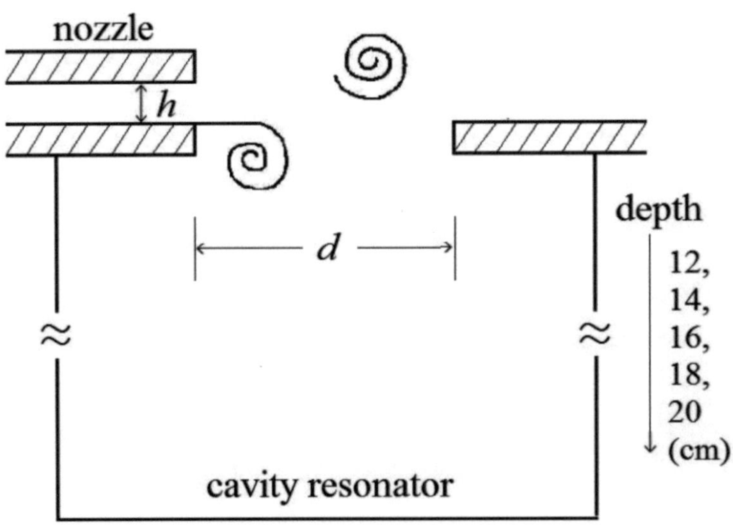

Figure 7.
Vortex shedding in a cavity-tone generator [9], which can be considered as a simplified mechanical model of the ocarina. h = 2.7 mm and d = 8 mm.

$$U_c = (U_0/2)\tanh(\pi h/\lambda_v), \tag{15}$$

where λ_v is the distance between successive vortices in the lower and the upper line vortices and h is the distance between both lines which is equal to the jet thickness. It is also noted that the circulation Γ of the vortex increases linearly with the time according to $\Gamma(t) = (1/2)U_0^2 t$.

In the case of flue instruments with thick jets ($d/h < 2$) [21], a new vortex is formed at the inner shear layer (on the resonant pipe side) each time the acoustic velocity $d\xi_m/dt$ changes sign from directed toward the outside to directed toward the inside of the resonator (acoustic pressure in the resonator takes the minimum). A new vortex is formed at the outer shear layer half an oscillation period ($T/2$) later (acoustic pressure in the resonator takes the maximum). In the steady state of oscillation, the circulation of the jth vortex (j = 1, 2, 3, ….) at the inner shear layer $\Gamma_{\text{in}}^{(j)}$ and at the outer shear layer $\Gamma_{\text{out}}^{(j)}$ can be written as [21].

$$\Gamma_{\text{in}}^{(j)}(t) = (1/2)U_0^2[t-(j-1)T] \text{ for } (j-1)T \le t \le jT, \tag{16}$$

$$\Gamma_{\text{in}}^{(j)}(t) = (1/2)U_0^2 T \text{ for } t > jT, \tag{17}$$

$$\begin{aligned}\Gamma_{\text{out}}^{(j)}(t) &= -(1/2)U_0^2[t-(2j-1)(T/2)],\\ &\text{for } (2j-1)(T/2) \le t \le (2j+1)(T/2)\end{aligned} \tag{18}$$

$$\Gamma_{\text{out}}^{(j)}(t) = -(1/2)U_0^2 T \text{ for } t > (2j+1)(T/2). \tag{19}$$

The acoustic power generated by the vortices is calculated by Eq. (14). The average of it over the oscillation period T is as follows:

$$\langle \Pi_{\text{vortex}} \rangle = -(1/T)\int_0^T \int_{V_S} \rho(\boldsymbol{\omega} \times \boldsymbol{U}) \bullet \boldsymbol{u}\, dV dt, \tag{20}$$

where the volume integration is taken over the source region of volume V_S. The vorticity field $\boldsymbol{\omega} = \text{rot}\boldsymbol{U}$ takes into account the contribution of each vortex at the shear layers [21]:

$$\boldsymbol{\omega}(x,y,t) = \sum_j \left[\Gamma_{\text{in}}^{(j)}(t)\delta\left(\boldsymbol{x} - \boldsymbol{x}_{\text{in}}^{(j)}(t)\right) + \Gamma_{\text{out}}^{(j)}(t)\delta\left(\boldsymbol{x} - \boldsymbol{x}_{\text{out}}^{(j)}(t)\right)\right], \tag{21}$$

where $\boldsymbol{x}$ defines a two-dimensional (2-D) point (x, y), $\boldsymbol{x}_{\text{in}}^{(j)}(t)$ the position of the jth vortex at the inner shear layer, and $\boldsymbol{x}_{\text{out}}^{(j)}(t)$ the position of the jth vortex at the outer shear layer.

It is first necessary to know the position and the circulation of vortices in order to calculate $\langle \Pi_{\text{vortex}} \rangle$ from Eqs. (20) and (21). It was done by time-domain simulations in [38]. For the sake of paper space, see the details described in [38].

2.3 Flow visualization and some discussion

2.3.1 Jet-wave drive vs. discrete-vortex drive

Dequand et al. [21] visualized the steady-state periodic flow in the mouth of the resonator by applying a standard Schlieren technique. They used three types of flute-like mouth configuration with a common edge of 60°, a common sharp edged flue exit, a common flue-to-edge distance d = 24 mm, and different flue channel height (i.e., jet thickness) h = 4, 14, and 30 mm (d/h = 6, 1.7, and 0.8). The pipe length L was 552 mm. The initial jet velocity U_0 had, respectively, 16.3, 14, and 14.5 m/s (S_t = 0.19, 0.22, and 0.22). **Figure 8** summarizes their result by rough illustration, though the flow inside the pipe is not so clear in [21].

Frames in the left column [(a), (c), and (e)] show flow conditions at the phase of $u = 0$ (the instant from the positive to the negative), while frames in the right column [(b), (d), and (f)] show flow conditions at the next phase of $u = 0$ (the instant from the negative to the positive). Note that the positive u indicates the upward (outward) acoustic velocity here, but it indicates the downward (inward) acoustic velocity in [21]. Also, the flow visualization of each case is shown by eight frames consisting of one oscillation period [21].

The jet-wave drive illustrated in **Figure 8(a)** reveals that the jet enters into the pipe at the instant when the acoustic pressure p (t) is maximum because the phase

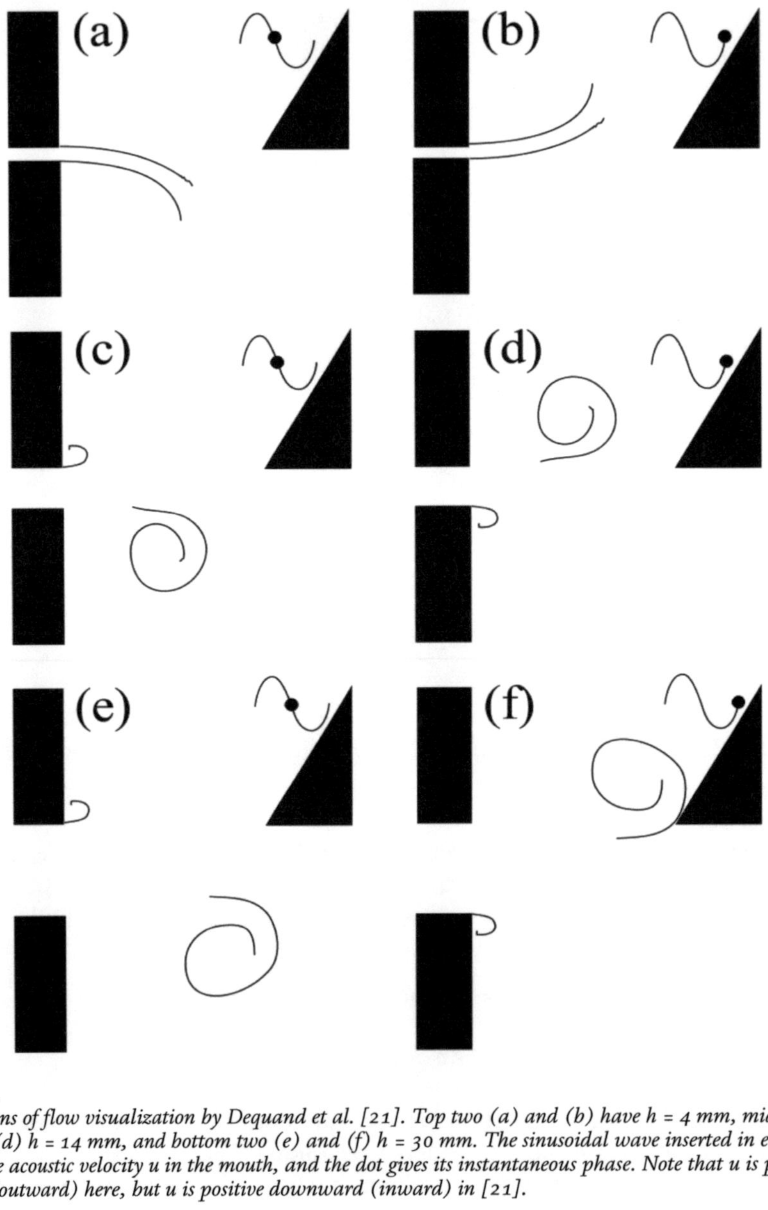

Figure 8.
Illustrations of flow visualization by Dequand et al. [21]. Top two (a) and (b) have h = 4 mm, middle two (c), and (d) h = 14 mm, and bottom two (e) and (f) h = 30 mm. The sinusoidal wave inserted in each frame depicts the acoustic velocity u in the mouth, and the dot gives its instantaneous phase. Note that u is positive upward (outward) here, but u is positive downward (inward) in [21].

of p (t) is delayed from u (t) by 90° at the resonance. Therefore, the positive acoustic power $\langle p(t) \bullet q(t) \rangle$ is generated, where $q(t)$ denotes the acoustic volume flow into the pipe [see Eq. (1)]. The positive acoustic power is also generated at the instant given in **Figure 8(b)**. As the result, the volume-flow drive in the jet-drive model is satisfied very well.

The discrete-vortex drive illustrated in **Figure 8(e)** indicates that the upper vortex is just created at the upper flue exit corner and the lower vortex reaches to the pipe edge in a fully developed shape. Also, **Figure 8(f)** indicates that the lower vortex is just created at the lower flue exit corner and the upper vortex is reached to the pipe edge in a fully developed shape. As a result, the positive acoustic power given by Eq. (20) or Eq. (14) is generated [see **Figure 9(b)** below].

Two illustrations of **Figure 8(c)** and **(d)** correspond to a boundary condition between the jet-wave drive and the discrete-vortex drive. Then, the lower and upper large vortices are located halfway between the flue exit and the edge. This is probably due to an opposing effect between both drives.

Figure 9 summarizes the phase relation between the physical quantities involved in the jet-wave drive (a) and the discrete-vortex drive (b). The red dot on the sinusoidal curve of u corresponds to the phase of u in **Figure 8**. The time scale is converted to the integer by $12(t/T)$. The magnitude of physical quantities is arbitrary. The jet-wave drive indicates the antiphase relation between the jet displacement $\xi_e(t)$ at the edge and the acoustic pressure $p(t)$ in the resonant pipe. This result endorses the acoustic power generation in good manner (note that u and ξ_e are defined positive upward).

In the discrete-vortex drive, the horizontal arrow connects the vortex creation at the flue exit and the vortex arrival at the edge as shown in **Figure 9(b)**. Since the upper vortex rotates anticlockwise, the vector direction of $\boldsymbol{\omega} \times \boldsymbol{U}$ in Eq. (20) is upward (positive z direction). During the passage of the upper vortex from the flue exit to the pipe edge, the acoustic cross-flow $\boldsymbol{u}$ in z direction has negative values. Hence, $\langle \Pi_{\text{vortex}} \rangle$ of Eq. (20) takes a positive value during the latter half of an oscillation period. Similarly, the lower vortex produces positive $\langle \Pi_{\text{vortex}} \rangle$ during the former half of an oscillation period. As a result, discrete vortices will create positive $\langle \Pi_{\text{vortex}} \rangle$ in an oscillation period. In other words, the vortex configurations illustrated in **Figure 8(e)** and **(f)** may create the acoustic power for sound generation in flue pipes.

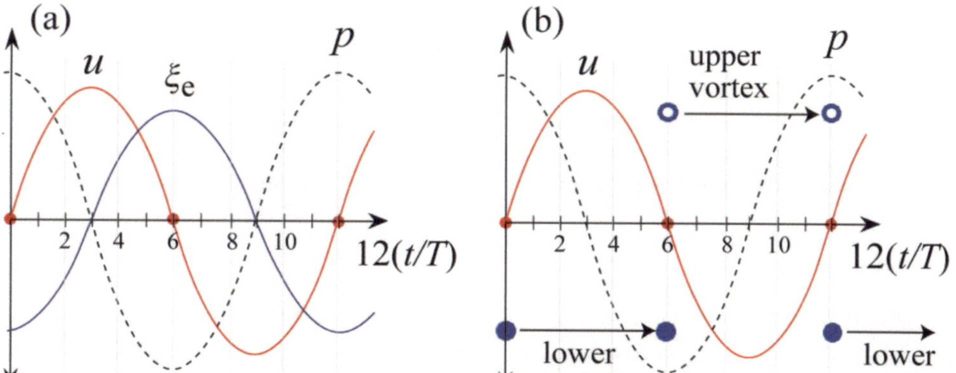

Figure 9.
*Phase relation between the physical quantities involved in the jet-wave drive (a) and the discrete-vortex drive (b). Their amplitudes are arbitrary. The red dot on the curve of u corresponds to the phase of u in **Figure 8**. The horizontal arrow in (b) connects the vortex creation at the flue exit and the vortex arrival at the pipe edge. Note that the positive direction of u and ξ_e is upward (outward) and the phase of p is delayed from u by 90° at the resonance.*

2.3.2 Edge tone vs. pipe tone

The edge tone is a dipole source, whose acoustic pressure directly correlates with the vortex generation. That is, when the jet impinges the edge by moving from the downward to the upward, a vortex rotating clockwise is produced just below the edge, and another vortex rotating anticlockwise exists downstream above the edge. This configuration of the vortex pair generates the maximum acoustic pressure above the edge and the minimum acoustic pressure below the edge. When the jet impinges the edge by moving from the upward to the downward, a vortex rotating anticlockwise is produced just above the edge, and another vortex rotating clockwise exists downstream below the edge. This configuration of the vortex pair generates the maximum acoustic pressure below the edge and the minimum acoustic pressure above the edge [2, 39]. Although Eq. (14) cannot be applied to the edge tone since there is no acoustic feedback (i.e., $\boldsymbol{u} = 0$), the acceleration $\boldsymbol{\omega} \times \boldsymbol{U}$ given by vortex rotation and jet velocity should be involved in the edge-tone generation. As a result, the edge tone satisfies:

$$\text{phase}\left[p_{\mathrm{e}}(t)\right] - \text{phase}[\xi_{\mathrm{e}}(t)] \approx \pm\pi/2, \tag{22}$$

where $p_{\mathrm{e}}(t)$ denotes the acoustic pressure below or above the edge. This phase relation is clearly different from that of the pipe tone shown in **Figure 9(a)**:

$$\text{phase}[p(t)] - \text{phase}[\xi_{\mathrm{e}}(t)] \approx \pi. \tag{23}$$

The difference between the edge tone and the pipe tone is reflected in Eqs. (22) and (23) in good manner.

2.3.3 Feedback loop gain and time delay of the jet wave in the jet-drive model

Let us consider the feedback loop to find out the time delay of the jet wave which fulfills the phase condition for sound generation. As mentioned in Section 2.1.2, the jet particle pass may be determined as soon as the jet issues from the flue to the acoustic field in the mouth [26]. At that instant, the initial transverse displacement $\xi_{\mathrm{f}}(t)$ of the jet at the flue exit is supposed to be non-zero and related with the acoustic velocity $u_{\mathrm{f}}(t)$ at the flue exit as follows [38, 40]:

$$\xi_{\mathrm{f}}(t)/h = u_{\mathrm{f}}(t)/U_0, \tag{24}$$

where h and U_0 are the jet thickness and jet velocity at the flue exit, respectively. The starting point of the feedback loop is $u_{\mathrm{f}}(t)$, which creates $\xi_{\mathrm{f}}(t)$. The jet displacement travels to the pipe edge as the jet wave, and we have at the edge [21].

$$\xi_{\mathrm{e}}(t) = \mathrm{e}^{\mu d}\xi_{\mathrm{f}}(t - \tau_{\mathrm{jw}}) = (h/U_0)\mathrm{e}^{\mu d}u_{\mathrm{f}}(t - \tau_{\mathrm{jw}}), \tag{25}$$

which yields the acoustic pressure $p_{\mathrm{j}}(t)$ at the pipe entrance according to Eq. (7) in which $q(t) \approx -U_{\mathrm{e}}b\xi_{\mathrm{e}}(t)$ from Eq. (1). Note that Eq. (25) largely simplifies Eq. (4) by considering the essential elements (spatial amplification and phase velocity) of the jet wave. The quantity τ_{jw} denotes the time delay of the jet wave when it travels from the flue exit to the edge ($\tau_{\mathrm{jw}} = d/U_{\mathrm{ph}}$).

The acoustic pressure $p_{\mathrm{j}}(t)$ drives the pipe and yields its resonance. As a result, $u_{\mathrm{f}}(t)$ at the starting point is fed back through the input admittance $Y(\omega)$ of the pipe. The Fourier transform of $p_{\mathrm{j}}(t)$ is thus given by:

$$P_j(\omega) = (\rho h \delta_j / d)(\mathrm{i}\omega) e^{\mu d} e^{-\mathrm{i}\omega\tau_{jw}} U_f(\omega), \quad (26)$$

where $U_f(\omega)$ is the Fourier transform of $u_f(t)$ and $U_e = U_0$ is assumed by neglecting the jet spreading for simplicity [41]. The feedback loop gain $G(\omega)$ is thus defined as

$$G(\omega) = Y(\omega)[P_j(\omega)/U_f(\omega)]. \quad (27)$$

Hence, the phase condition for the self-sustained (feedback) oscillation is:

$$\text{phase}[G(\omega)] = \text{phase}[Y(\omega)] + \pi/2 - \omega\tau_{jw} = \pm 2m\pi \quad (28)$$

That is, the time delay τ_{jw} of the jet wave must satisfy:

$$\begin{aligned} \omega\tau_{jw} &= \text{phase}[Y(\omega)] + \pi/2 + 2m\pi \\ &= \text{phase}[Y(\omega)] + \left(m + \frac{1}{4}\right)2\pi, \end{aligned} \quad (29)$$

where $-2m\pi$ is abandoned because $\omega\tau_{jw}$ is always positive and m (= 0, 1, 2, ….) denotes the hydrodynamic mode number. Usually, sound generation in flue pipes occurs for $m = 0$. Since phase$[Y(\omega)] = \pi/2$ at the pipe resonance (see **Figure 9**), we finally have:

$$\tau_{jw} = \pi/\omega = T/2 \quad (30)$$

for the first mode $m = 0$. Therefore, it may be said that flue instruments are well excited when the time delay of the jet wave is around half an oscillation period. More detailed discussion is given in [38, 41]. Although the amplitude condition for sound generation can be calculated from Eq. (27), we do not have the space enough to do that.

2.3.4 Time delay of vortex convection in the discrete-vortex model

According to **Figure 9(b)**, the upper and lower vortices created at the flue exit arrive at the pipe edge with a time delay of $T/2$, respectively. As explained in Section 2.3.1 the convection of these two vortices may create the acoustic power $\langle \Pi_{\text{vortex}} \rangle$ defined by Eq. (20) for sound generation in flue pipes. Thus, the time delay of vortex convection τ_{dv} in the discrete-vortex model is:

$$\tau_{dv} = d/U_c = T/2. \quad (31)$$

This Eq. (31) just corresponds to Eq. (30) in the jet-drive model. Therefore, both models provide the same dependence of the oscillation frequency on the jet velocity.

Although τ_{dv} is easily derived from **Figure 9(b)**, it will be desirable to consider τ_{dv} based on the phase balance such as in Eq. (29). The sound generation by the periodic pulse-like force (produced by each vortex arrival) at the edge will be maximum when the pulse is in phase with the maximum of acoustic velocity u at the edge. Since the vortex arrival occurs at the zero-crossing of u [see **Figure 9(b)**], the instant of the maximum u is given by $\tau_{dv} - T/4$, which should be balanced with the delay due to the input admittance $Y(\omega)$ of the resonant pipe. Then, we have the following phase balance by allowing a time delay of m periods:

$$\tau_{\mathrm{dv}} - \left(m + \frac{1}{4}\right)T = \mathrm{phase}[Y(\omega)]/\omega. \tag{32}$$

This equation is the same as Eq. (29) when it is divided by ω. A similar derivation of Eq. (32) is given in [38]. When m = 0, Eq. (31) is given from Eq. (32).

2.3.5 *Aspect ratio d/h of the jet*

The aspect ratio d/h (jet length/jet thickness) is an essential parameter that discriminates the jet-drive model from the discrete-vortex model as indicated in **Figure 8** based on [21]. The value of d/h = 1.7 set up in **Figure 8(c)** and **(d)** seems to be more favored by the discrete-vortex model. The critical aspect ratio that discriminates both models is d/h = 2.3 in more rigorous sense [21, 38].

Dequand et al. [21] calculated $|\mathrm{d}\xi_{\mathrm{m}}/\mathrm{d}t|_{\max}/U_0$ as a function of d/h for both models, and experimental data on the flue exit with chamfered edges and the pipe edge with 15° were plotted on the calculated diagram (see Figure 12 in [21]). In very rough sense, the solution curve of the jet-drive model is proportional to $(d/h)^{-1}$, and that of the discrete-vortex model is proportional to $(d/h)^{1/2}$. The crossing of the two curves occurs near d/h = 2.3, and the experimental data approximately fit the discrete-vortex solution below this crossing and better fit the jet-drive solution above the crossing. Also, the experimental curve of data points indicates the maximum value at the crossing or the critical aspect ratio. It was experimentally confirmed on four kinds of the flue-edge geometry (see Figure 10 in [21]). Furthermore, Auvray et al. [38] extended similar calculation for different oscillation regimes (m = 0, 1; $f = f_1, f_2$) (see Figure 8 in [38]). According to [38], the critical aspect ratio depends on the hydrodynamic jet mode m. For an eolian regime (m = 1, $f = f_2$), the critical aspect ratio is much larger ($d/h \cong 13$).

2.4 Vortex-layer model

Howe [18] and Dequand et al. [21] proposed the discrete-vortex model driven by thick jets (d/h < 2) as explained in Section 2.2. However, there seems to be a room for the vortex even in the sound generation of flue instruments driven by thin jets that satisfy d/h > 2. Since an actual jet has a velocity profile as indicated by the broken line in **Figure 10(a)**, the vorticity can be formed along the boundary between the jet and the surrounding fluid. As the result, a layer (or sheet) of vorticity is organized along an immediate vicinity of the jet. The upper layer consists of the positive vorticity (the counterclockwise-rotating tiny vortices), and the lower layer consists of the negative vorticity (the clockwise-rotating tiny vortices). This physical picture depicted in **Figure 10(a)** may be called the vortex-layer model on the sound generation in musical flue instruments [2, 28].

It should be carefully noted that actual sound generation is three-dimensional (3-D) as inferred from the volume integral of Eq. (14), but our vortex-layer model illustrated in **Figure 10(a)** assumes the two-dimensionality (2-D). This 2-D assumption corresponds to the 2-D assumption of $\boldsymbol{U}$ and $\boldsymbol{u}$ that has been conventionally made in acoustical models. See Figures 10 and 11 in [28] on the 3-D nature of $\boldsymbol{U}$ and $\boldsymbol{u}$. The PIV observation on a plane sheet by the laser (see Section 4.1 and **Figure 14**) is based on the 2-D assumption above.

Since $\boldsymbol{u}$ is periodic, the time average $\langle \Pi_{\mathrm{G}}(t)\rangle$ should be null if $\boldsymbol{\omega} \times \boldsymbol{U}$ is stationary in time as shown in **Figure 10(b)**. However, since the actual jet has any fluctuation, if $\boldsymbol{\omega} \times \boldsymbol{U}$ has a component that changes temporarily in accordance with temporal change of $\boldsymbol{u}$, non-zero value of $\langle \Pi_{\mathrm{G}}(t)\rangle$ may be expected from the unbalance between

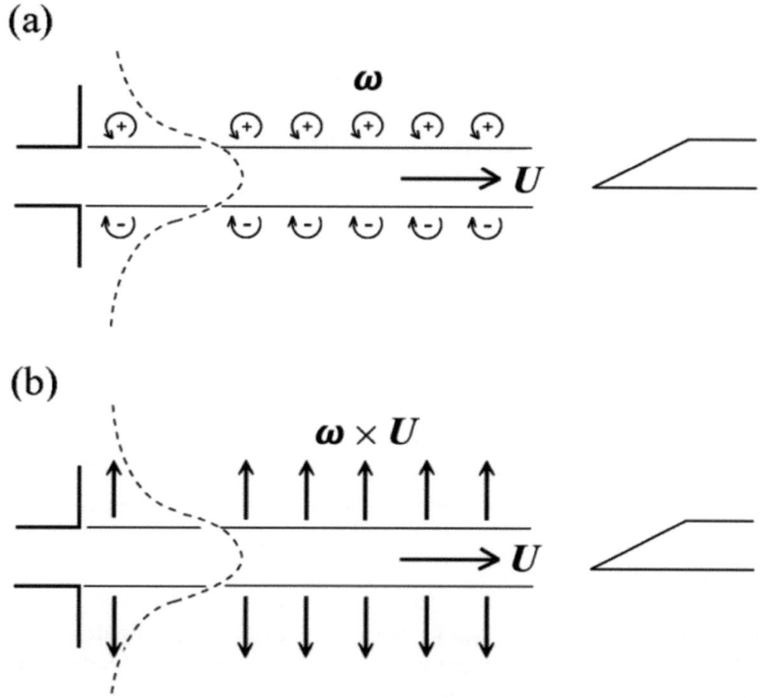

Figure 10.
Conceptual sketch of the vortex-layer model on sound generation in flue instruments [2, 28]: (a) the vortex layers (consisting of tiny vortices) along both sides of the jet flow and (b) the generation and cancelation of the aeroacoustical source term $\boldsymbol{\omega} \times \boldsymbol{U}$*. The dashed line depicts a lateral profile of the jet velocity* $\boldsymbol{U}$*.*

the upper and lower vortex layers. Hence, the sound generation in flue pipes may be yielded by the interaction between the jet vortex layer and the acoustic mouth flow. In this sense, $\boldsymbol{\omega} \times \boldsymbol{U}$ and $(\boldsymbol{\omega} \times \boldsymbol{U}) \bullet \boldsymbol{u}$ may be called the *aeroacoustical source term* (with the same dimension as the acceleration) and the *acoustic generation term*, respectively.

Helmholtz [30] already suggested the importance of the jet vortex layer. His vortical surface (or stratum) that has a very unstable equilibrium acts as "an accelerating force with a periodically alternating direction" to reinforce the inward and outward velocity at the pipe entrance. Interestingly enough, this physical picture of Helmholtz is very similar to the jet vortex-layer model shown in **Figure 10** [2, 28]. It should be recognized that the volume-flow drive first proposed by Helmholtz [30] is based on his physical concept of the vortex layers formed along the jet flow.

3. Vortices on sound dissipation

3.1 Sound dissipations in linear acoustics

Let us briefly discuss the mechanisms of sound dissipation (or absorption) in flue instruments because the self-sustained musical instruments must overcome the acoustic dissipations involved in them. At first let us consider within the field of linear acoustics and start from sound dissipation in free space.

3.1.1 Classical absorption and molecular absorption in free space

In free space, the classical sources of dissipation are internal friction and heat conduction. Both phenomena tend to equalize the local variations of the particle velocity and temperature accompanying the acoustic wave [42]. As a result, the acoustic energy is removed from the acoustic wave.

The equations on dissipation due to internal friction were derived by G. Stokes in 1845 and those on dissipation due to heat conduction by G. Kirchhoff in 1868. A plane sound wave is exponentially damped in the direction of propagation (x direction): $e^{-\alpha x}$. The coefficient α_F of the dissipation due to internal friction is [42]

$$\alpha_F = \left(2\omega^2/3c^3\right)(\eta/\rho), \tag{33}$$

where c is the propagating sound velocity, η the dynamic viscosity, and ρ the air (or generally, gas) density. The coefficient α_H due to heat conduction in gases is of the same order of magnitude as α_F and is proportional to ω^2 [42]. The distance within which the sound level falls by 1 dB (an amplitude decrease of about 11%) due to classical absorption is very large in air (1-kHz wave gives 5 km; 10-kHz wave 50 m) [42]. Therefore, classical dissipation is almost negligibly weak.

The major source of strong dissipation in free space is molecular sound absorption. The translational and rotational energies of gas molecule are very quickly increased by a sudden impact, while the oscillatory energy builds up gradually at the expense of the translational and rotational energies [42]. The delay in reaching thermal equilibrium is called relaxation, and its time constant is called the relaxation time τ_R . The source of molecular absorption in air is oscillatory relaxation of oxygen. The relaxation frequency (defined as $1/2\pi\tau_R$) of pure oxygen is very low (about 10 Hz). However, the water vapor content of air greatly shortens the relaxation time and shifts the absorbing range into the audio frequencies (see Figure 3.7 in [42]). The acoustic dissipation in moist air is significantly greater than the classical absorption given by $\alpha_F + \alpha_H$.

3.1.2 Sound dissipation at the internal wall of a long pipe

Next let us consider the dissipation in the confined air instead of in free air. If a sound wave propagates in a long pipe where sound reflection can be neglected, it suffers additional losses because of internal friction and heat conduction in the boundary layer next to the wall. The acoustic particle velocity parallel to the pipe axis is zero at the internal wall surface because of friction (called no-slip condition). Its maximum value is not reached until the distance from the wall amounts to a quarter of viscosity wavelength λ_{vw} (see Figure 3.10 in [42]). This λ_{vw} characterizes the thickness of the wall boundary layer and is given by

$$\lambda_{vw} = 2\pi\delta = 2\pi\sqrt{2\eta/\omega\rho} \approx \left(1.4\ \text{cms}^{-1/2}\right)/\sqrt{f}, \tag{34}$$

where δ is the skin depth and the equation of the right-hand side is for the air with the kinematic viscosity $\eta/\rho = 1.5 \times 10^{-5}\ \text{m}^2\ \text{sec}^{-1}$ [42].

The losses occurring in the wall boundary layer due to viscous friction and heat conduction (the wall is considered as a surface with a constant temperature and the thermal change followed by the acoustic wave should be null at the wall surface) attenuate sound waves in pipes. A parameter to appropriately express the sound attenuation in a pipe is the ratio of the pipe radius R to the boundary layer thickness (or the skin depth) δ. This ratio defines the quality factor Q or the inverse of the loss factor ζ (generally, $Q = \zeta^{-1} = R/\delta$). The values of Q for viscous friction and heat conduction are, respectively, given as [1, 43].

$$Q_v = (\omega\rho/2\eta)^{1/2}R, Q_t = (\omega\rho C_p/2\kappa)^{1/2}(\gamma - 1)^{-1}R, \tag{35}$$

where η denotes the viscosity of the air, C_p the specific heat at constant pressure, κ the thermal conductivity, and γ the ratio of specific heats. The ratio of thermal loss to viscous loss is given by ${Q_t}^{-1}/{Q_v}^{-1} = 0.46$ for the air.

The attenuation constant α_w of a round pipe is a function of the frequency f (in hertz) and the pipe radius R (in meters) [1, 42, 43]:

The attenuation constant α in total is given by adding ${Q_{rad}}^{-1}$ in Eq. (36):

$$\begin{aligned} \alpha_w &= (\omega/2c)\left({Q_v}^{-1} + {Q_t}^{-1}\right) \approx 3.1 \times 10^{-5}\left(f^{1/2}/R\right)\text{in m}^{-1} \\ &= 2.7 \times 10^{-4}\left(f^{1/2}/R\right)\text{in dBm}^{-1}, \end{aligned} \tag{36}$$

where α_w is approximately evaluated in m^{-1} and dBm^{-1}. The conversion is done by the relation $\alpha_w(\text{in dBm}^{-1}) = 20\log e \times \alpha_w(\text{m}^{-1}) = 8.68\ \alpha_w(\text{m}^{-1})$ based on the exponential decay. For example, the modern flute with $R \approx 10$ mm indicates $\alpha_w \approx 0.85\ \text{dBm}^{-1}$ for $f = 1000$ Hz. This attenuation is much larger than that occurs in free air, but still small enough. For example, tubes many meters long formerly were used on ships to transmit commands from the bridge to the engine room [42]. This large but still small enough magnitude of α_w is the right reason why musical flue instruments and other wind instruments work out well. In order to suppress sound propagation in tubes (or in air conditioning systems) from the viewpoint of noise control, the tube walls should be covered with sound-absorbing material.

3.1.3 Finite cylindrical pipe: acoustic resonance and sound radiation

Since most musical flue instruments are of finite length, the sound wave that propagates in the instrument bore is reflected at both open ends. As a result, the acoustic resonance occurs if the energy enough to overcome all dissipations is supplied to the bore. The acoustical condition of the bore is characterized by the input impedance or admittance in which wall boundary losses defined by α_w is involved. However, the reflection is not complete, and a little of the acoustic energy confined in the bore escapes to free space. This is sound radiation, which is another source of sound dissipations in the bore.

If the resonance condition is given by $kL = n\pi$ ($k = \omega/c$ denotes the wave number and $n = 1, 2, 3, \ldots$.) and the source strength of radiation at each open end is the same, we have the value of Q for sound radiation as follows [43]:

$$Q_{rad} = \left(n/{\omega_n}^2\right)\left(\pi c_0 c/R^2\right), \tag{37}$$

where ω_n is the angular frequency at the nth mode resonance, c_0 the sound speed in free space, and c the sound speed in the bore (c is a little smaller than c_0 due to wall boundary losses [1, 43]).

Generally, the loss factor or the inverse of Q is defined as [43, 44].

$$Q^{-1} = \frac{1}{2\pi}\frac{\text{time average of the power lost from the bore}}{\text{time average of the power stored in the bore}}. \tag{38}$$

Therefore,

$$\langle \Pi_{rad}\rangle / \langle \Pi_B\rangle = 2\pi {Q_{rad}}^{-1}, \tag{39}$$

where $\langle \Pi_{rad} \rangle$ denotes the time average of the power lost from the bore by sound radiation and $\langle \Pi_B \rangle$ the time average of the power stored in the bore. For f = 300 Hz and R = 25 and 15 mm, we have $Q_{rad} = 164$ and 454, respectively. We can thus understand that the radiated power is very little from the evaluated values $\langle \Pi_{rad} \rangle / \langle \Pi_B \rangle = 0.038$ and 0.014. This implies that the first priority is clearly offered to the resonance in wind instruments. Sound radiation is only the faint leakage of the power stored in the bore. In spite of it, we can easily hear instrument tones.

$$\alpha = \alpha_w + \alpha_{rad} = (\omega/2c)\left(Q_v^{-1} + Q_t^{-1} + Q_{rad}^{-1}\right), \tag{40}$$

where α_{rad} indicates the attenuation constant due to sound radiation. Also, the total Q^{-1} defined by Eq. (38) is equal to $\left(Q_v^{-1} + Q_t^{-1} + Q_{rad}^{-1}\right)$. Since Q (called the quality factor) indicates the sharpness or height of the resonance, it is adequate to show Q instead of Q^{-1} for wind instruments. **Figure 11** depicts Q as a function of f for the bores of the clarinet, flute, and bass flute. We may well understand that wind instruments with cylindrical bores have appreciably high Q values over their playing ranges.

3.2 Acoustically induced vortices as the final dissipation agent

The above description in 3.1 on sound dissipations is correct within the scope of linear acoustics. Then, as the input energy from the player continues to increase, the output energy (viz., the sound level) from the instrument keeps increasing. However, in actual wind instruments, the saturation of the output energy necessarily occurs. In other words, sound generation is nonlinear.

An important source of the saturation in flue instruments is acoustically induced vortices (simply, acoustic vortices) at the pipe edge. These acoustic vortices work as the final dissipation agent that determines the final amplitude of the saturated sound.

3.2.1 Visualization of acoustic vortices and their modeling

Jet and vortex behaviors during attack transients in organ pipe models were studied intensively using a high-speed video camera and a smoked jet in [17]. Experimental procedures are described in [17, 29]. **Figure 12(a)** and **(b)** is the visualization result which shows the exterior vortex (a) is rotating clockwise and the interior vortex (b) is rotating anticlockwise (the blowing pressure is about

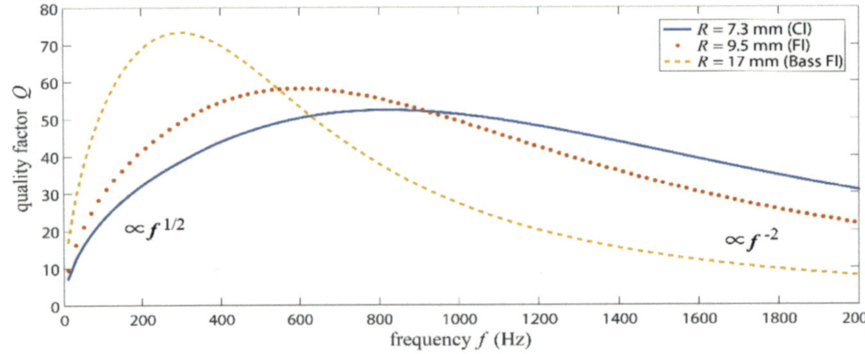

Figure 11.
The quality factor Q as a function of f for cylindrical bore instruments. The bore radius R = 7.3, 9.5, and 17 mm for the clarinet, flute, and bass flute, respectively.

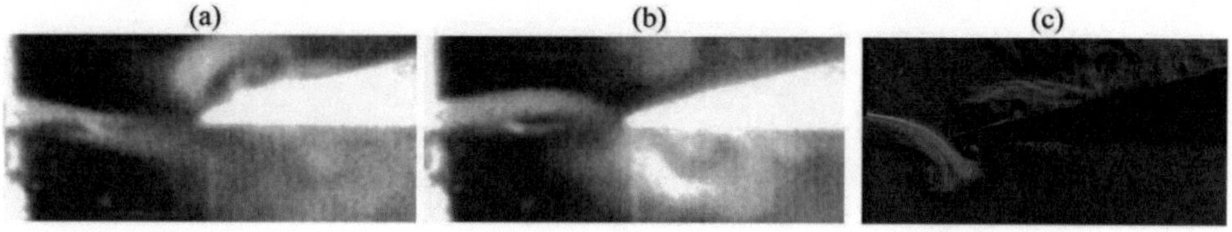

Figure 12.
Visualized examples of acoustic vortices. (a) and (b): Visualized using a high-speed video camera and a smoked jet when the jet blowing pressure is 150 Pa and the sounding frequency 283 Hz [17]; (c): Visualized by means of a standard Schlieren technique when jet blowing pressure is 270 Pa and the sounding frequency 477 Hz [16]. Note that the jet-edge configuration and the buildup of the blowing pressure in frames (a) and (b) are quite different from those in frame (c). Also, frames (a) and (b) appear just before the steady state (at the "pre-steady state" in [17]), while frame (c) appears at the steady state.

150 Pa *just before* the steady state and the sounding frequency 283 Hz). These vortices are making up an acoustic dipole. We can see the same rotation of the interior vortex in **Figure 12(c)**. That vortex was recognized as acoustic vortex in [16] (the steady-state blowing pressure was 270 Pa and the sounding frequency 477 Hz). Therefore, the vortices in **Figure 12(a)** and **(b)** may be regarded as acoustic vortices, too. An interior vortex is produced in **Figure 12(a)** when the jet just crosses the edge from the outside to the inside, while that in **Figure 12(c)** is produced when the jet is deflected to the inside. This difference may be due to various causes. A long flue-to-edge distance d (=10.2 mm), almost null offset of the edge, a very slow buildup of the blowing pressure, and a low final blowing pressure were used in [17], while a much shorter d, a large offset, a quick buildup of the blowing pressure, and a much higher final pressure were used in [16].

These acoustic vortices shedding from the edge are considered to serve as a significant source of the sound energy dissipation in large-amplitude nonlinear oscillation [16]. According to **Figure 5** [the same organ pipe model was used in **Figures 5, 12(a)** and **(b)**], the acoustic particle velocity at the mouth is estimated as 2.3 m/s for the jet blowing velocity of 15.8 m/s (corresponding to the jet blowing pressure of 150 Pa). The acoustic velocity is thus about 15% of the flow velocity and seems to be large enough to cause nonlinear oscillations.

A physical modeling of the acoustic vortex generation in organ flue pipes is shown in **Figure 13** in comparison with the hydrodynamic vortex generation. A typical hydrodynamic vortex formed above the edge at the starting transient rotates anticlockwise as shown in **Figure 13(a)**. At this time the vorticity vector $\boldsymbol{\omega}$ is in the negative y direction and $\boldsymbol{\omega} \times \boldsymbol{U}$ is in the positive z direction (upwards) when the jet velocity $\boldsymbol{U}$ is in the positive x direction. On the other hand, an acoustic vortex formed above the edge [see **Figure 12(a)**] rotates clockwise as shown in **Figure 13(b)**. Then, $\boldsymbol{\omega} \times \boldsymbol{U}$ is in the negative z direction. Since the jet oscillates from the upward to the downward in **Figure 12(a)**, the acoustic particle velocity $\boldsymbol{u}$ takes negative maximum amplitude as known from **Figure 9(a)**. This condition is indicated by the dashed line around the mouth area in **Figure 13(b)**. As a result, the inner product $(\boldsymbol{\omega} \times \boldsymbol{U}) \bullet \boldsymbol{u}$ becomes positive and the absorption of sound energy by the vortex is caused according to Eq. (13). Half a period later, an acoustic vortex rotating anticlockwise is formed below the edge as shown in **Figures 12(b)** and **13(c)**, and $\boldsymbol{\omega} \times \boldsymbol{U}$ as well as $\boldsymbol{u}$ is in the positive z direction. Hence, $(\boldsymbol{\omega} \times \boldsymbol{U}) \bullet \boldsymbol{u}$ is positive again as shown in **Figure 13(c)**, and sound energy absorption takes place at the pre-steady state or the steady state.

Although the jet deflection shown in **Figure 12(c)** is negative, the jet might be moving upward [the phase of ξ_e may be around $12(t/T) = 1$ in **Figure 9(a)**] and then $\boldsymbol{u}$ as well as $\boldsymbol{\omega} \times \boldsymbol{U}$ is possibly upward as shown in **Figure 13(c)**. Also, the effects of the pressure drive [cf. Eqs. (2) and (3)] should be considered because of

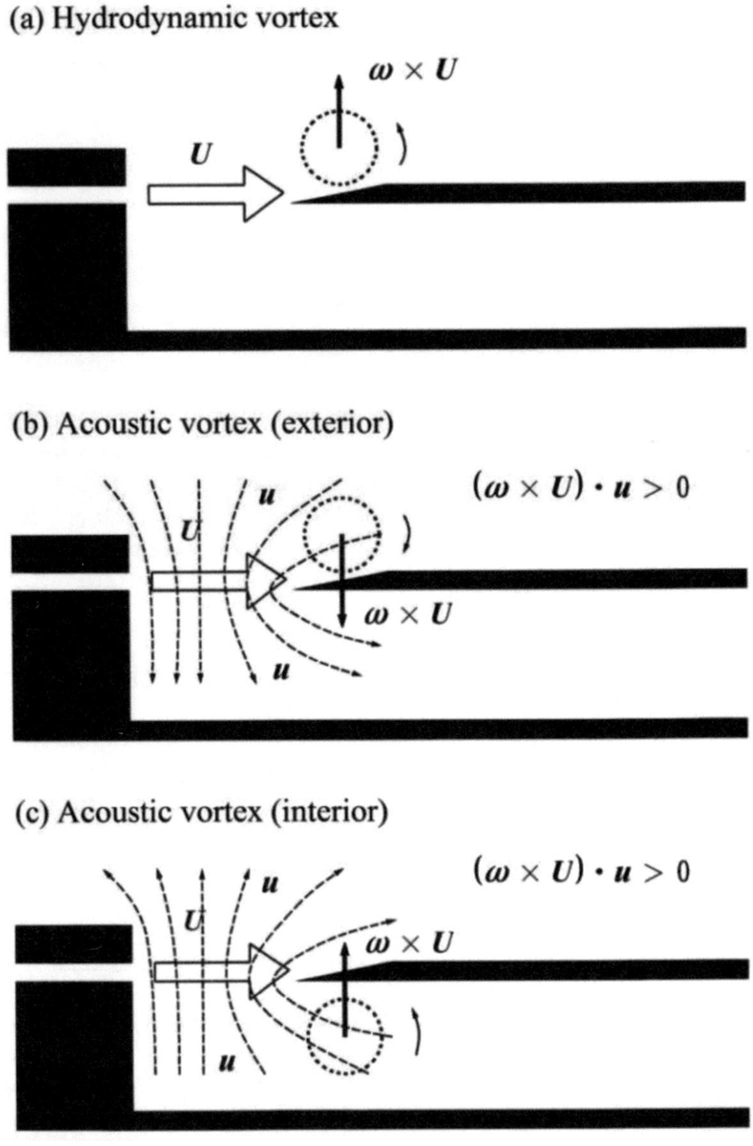

Figure 13.
Schematic of vortex formation in organ flue pipes. (a): A hydrodynamic vortex formed at the initial phase of the starting transient; (b) and (c): An acoustic vortex formed at the pre-steady state [17].

quite high blowing pressure. The phase lag due to the pressure drive can make the acoustic velocity in the case of **Figure 12(c)** more positive as inferred from **Figure 8(a)**. Then, $(\boldsymbol{\omega} \times \boldsymbol{U}) \bullet \boldsymbol{u} > 0$ will be realized in better fashion.

Interestingly enough, the acoustic vortices shown in **Figure 12(a)** and **(b)** were not observed at the steady state in [17]. Instead of that, we observed a steadily deflecting jet, particularly its penetration into the pipe as captured in Figure 13 in [17]. According to this result, we may consider that the acoustic vortex is formed to lead the finally saturated amplification of the jet stability wave by absorbing the final excess in the acoustic energy generation occurring at the pre-steady state. The acoustic vortex may be then conveyed by the jet flow into the region where the vorticity can no longer continue to interact with the acoustic field [35]. Since the completely steady state has already reached the energy balance, any more acoustic vortices seem to be not needed. Instead, the acoustic vortices will be strongly needed just before the completely steady state or at the pre-steady state. Also, the acoustic vortex should be discussed from the common viewpoint of acoustic power dissipation and radiation of high-amplitude jet noise at duct termination [3, 15, 35, 44].

3.2.2 Acoustic power balance between vortex layers and acoustic vortices

Acoustic power generation by the unbalance between the upper and lower vortex layers (cf. Section 2.4) will be balanced with acoustic power dissipations by the wall boundary effects (cf. Section 3.1.2), sound radiation (cf. Section 3.1.3), and acoustic vortices in the sense of time average:

$$\langle \Pi_{\mathrm{G}}(t)\rangle = \langle \Pi_{\alpha}(t)\rangle + \langle \Pi_{\mathrm{D}}(t)\rangle, \tag{41}$$

where $\Pi_{\mathrm{G}}(t)$ is given by Eq. (14) with $\boldsymbol{\omega} = \boldsymbol{\omega}_{\mathrm{vl}}$ concerning the vortex layer, $\Pi_{\alpha}(t)$ is the power lost from the bore that is given by the total attenuation constant of Eq. (40), and $\Pi_{\mathrm{D}}(t)$ is given by Eq. (13) with $\boldsymbol{\omega} = \boldsymbol{\omega}_{\mathrm{av}}$ concerning the acoustic vortex. A more exact description of $\Pi_{\mathrm{G}}(t)$ derived from the unbalance between the upper and lower vortex layers will be given in Section 4.1.

3.2.3 Acoustic losses due to vortex shedding at the edge

In the framework of the jet-drive model, Dequand et al. [21] assumed that the separation of the acoustic flow $q_{\mathrm{m}}(t) = (\mathrm{d}\xi_{\mathrm{m}}/\mathrm{d}t)db$ occurs at the edge by following Verge et al. [12]. This acoustic flow separation causes a free jet [45].

Although they neglect the effects of the separation of the jet flow and their viewpoint is different from the modeling illustrated in **Figure 13**, it seems to be worth taking into consideration. The effects of vortices can be represented by a fluctuating pressure p_{v} across the mouth [12, 21]:

$$p_{\mathrm{v}} = -(1/2)\rho\left(q_{\mathrm{m}}/c_{\mathrm{v}}db\right)^2 \mathrm{sign}\left(q_{\mathrm{m}}\right), \tag{42}$$

where c_{v} (= 0.6) is the vena contracta factor of the free jet. The time-averaged power losses due to the acoustic vortex shedding at the edge is then given as [21].

$$\langle \Pi_{\mathrm{lost}}\rangle = \langle p_{\mathrm{v}} q_{\mathrm{m}}\rangle = -(1/2T)\left(\rho db/c_{\mathrm{v}}^2\right)\int_0^T (\mathrm{d}\xi_{\mathrm{m}}/\mathrm{d}t)^2|\mathrm{d}\xi_{\mathrm{m}}/\mathrm{d}t|\mathrm{d}t, \tag{43}$$

where it is assumed that the dissipation occurs during the entire period T.

Therefore, the power dissipation given by Eq. (43) may be considered as an upper limit approximation, and by neglecting Π_{α} in Eq. (41), it can be roughly balanced with the power generation by the jet drive given by Eq. (9) [12, 21]:

$$\langle \Pi_{\mathrm{jet}}\rangle + \langle \Pi_{\mathrm{lost}}\rangle \approx 0. \tag{44}$$

If the integral in the right-hand side of Eq. (43) can be replaced with a product of $(\mathrm{d}\xi_{\mathrm{m}}/\mathrm{d}t)_{\max}{}^2|\mathrm{d}\xi_{\mathrm{m}}/\mathrm{d}t|$ and an appropriate division of T by supposing a rectangular-like waveform of $\xi_{\mathrm{m}}(t)$, we have the following relation between the maximum acoustic velocity $(\mathrm{d}\xi_{\mathrm{m}}/\mathrm{d}t)_{\max}$ over the mouth, the Strouhal number $S_{\mathrm{t}} = fd/U_0$, and the aspect ratio d/h of the jet [12, 21]:

$$\left[(\mathrm{d}\xi_{\mathrm{m}}/\mathrm{d}t)_{\max}/U_0\right]^2 \sim S_{\mathrm{t}}(h/d)^{3/2}. \tag{45}$$

This interesting non-dimensional relation was almost confirmed by the experiment on thin jets ($d/h > 2$) for the four different flue-edge geometries (see Figure 11 in [21]). The maximum of non-dimensional amplitude $(d\xi_m/dt)_{max}/U_0$ reached for the edge with an angle of 60° is 20% higher than that obtained for the edge with an angle of 15°. This difference in amplitude can reflect the difference between the flute and the recorder. The recorder with a sharper edge probably brings about stronger losses due to vortex shedding at the edge.

4. Vortices on sound generation

In this section let us consider what the cause of the jet oscillation is for thin jets ($d/h > 2$). Fletcher's displacement model of Eq. (4) [1, 31] has no definite physical basis, and Coltman's velocity model [22, 23] lacks in quantitative analysis. The present author proposed an acceleration model based on the pressure difference between the upper and lower surfaces of the jet [26]. Although this model could not involve the effects of the jet instability [20], it could successfully predict the possibility of underwater organ pipes [46]. Therefore, another acceleration model based on the vorticity generation is greatly expected [28].

4.1 Vortex layer along the jet visualized by PIV

A great advantage of PIV is to yield global and quantitative information on the flow-acoustic interaction. The PIV was already successfully applied to the experimental research of the edge-tone generation [6], where the complicated jet-edge interaction was investigated to accurately localize the vortex cores (dipole sources) just before the edge. Also, it was applied to measure the flow velocity and acoustic particle velocity [47]. Measurements of both quantities are required to consider sound generation based on the vortex sound theory.

4.1.1 Measurement requirements

Since the vortex sound theory hypothesizes an irrotational potential flow for $\boldsymbol{u}$ [18], the measurement should meet the requirement for this potential flow. However, actual acoustic cross-flow $\boldsymbol{u}$ in flue instruments tends to yield a non-potential flow through any nonlinear process at large-amplitude conditions. Hence, the measurement should be carried out at low amplitudes to assure the potential flow. Moreover, the PIV cannot evaluate the contribution from the harmonics but only estimates the instantaneous flow magnitude on a plane sheet illuminated by the laser. Hence, the acoustic cross-flow should be measured based on the acoustical field with a waveform as sinusoidal as possible.

Both requirements of (1) a potential flow and (2) a sinusoidal flow for $\boldsymbol{u}$ are not easily fulfilled at the same time in rigorous manner. A practical way seems to be a measurement at appropriate low amplitudes (the drive at *piano* level was better than that at *pianissimo* level [28]). Bamberger [48–50], who first introduced the PIV into the field of musical acoustics, carried out his measurement at *mezzo-forte* level by driving a flute at a very high pitch of about 1150 Hz. This sounding condition seems to satisfy the sinusoidal flow condition because a very high tone of the flute almost consists of only the fundamental. This is due to the cutoff frequency around 2 kHz of the column resonance of the modern flute [51]. However, it is uncertain whether the requirement of the irrotational potential flow on $\boldsymbol{u}$ is satisfied or not. On the other hand, an organ pipe with a low-frequency resonance (at about 195 Hz) is driven at low amplitudes to satisfy the requirement of the potential and sinusoidal flow on $\boldsymbol{u}$ simultaneously [28].

Since $|\boldsymbol{u}| \ll |\boldsymbol{U}|$, it is very difficult to measure the distribution of $\boldsymbol{u}$ over the mouth area using the PIV when the pipe is driven by the air jet. Therefore, both measurements of $\boldsymbol{u}$ and $\boldsymbol{U}$ should be separately carried out. Of course, $\boldsymbol{U}$ cannot be measured without using the jet. On the other hand, $\boldsymbol{u}$ can be measured by resonating the pipe externally, for example, by using an inverse exponential horn [28, 49]. A larger cross section of this horn is firmly fitted to the loudspeaker diaphragm, and a smaller cross section is coupled to the pipe end with a distance larger than the end correction to maintain the resonance pattern of the pressure distribution along the air column. The loudspeaker is driven by an oscillator to generate a sinusoidal wave in the pipe with the same frequency and amplitude as those when the jet drives the pipe. The organ pipe is thus driven by this loudspeaker horn system when $\boldsymbol{u}$ is measured.

Also, in order to experimentally examine the generation of the vortex sound based on Eq. (14), both measurements of $\boldsymbol{u}$ and $\boldsymbol{U}$ must be carried out at the same condition as exactly as possible. That is, these vectors must be measured at the same phase of the generating sound and at the same measurement area by using the same organ pipe. However, since $\boldsymbol{u}$ and $\boldsymbol{U}$ cannot be measured simultaneously, the jet drive and the loudspeaker horn drive must be switched as quickly as possible while maintaining the same sounding condition and the same measurement condition. The phase-locked PIV measurement on $\boldsymbol{u}$ and $\boldsymbol{U}$ (see **Figure 14**) is thus essentially important to evaluate Eq. (14). Since $|\boldsymbol{u}| \ll |\boldsymbol{U}|$, $\boldsymbol{u}$ should be first measured using the horn drive at a given phase of the sound, and the $\boldsymbol{U}$ is measured at the same phase by quickly switching the horn drive to the jet drive.

4.1.2 Measurement procedures

PIV measurement of $\boldsymbol{u}$ and $\boldsymbol{U}$ was carried out twice (Trials 1 and 2) in [28]. The fundamental frequencies of the pipe tone were 192.0 Hz and 192.1 Hz, respectively (the cutoff frequency of the horn was designed to be about 150 Hz). The averaged sound levels were 59.0 dB and 59.3 dB, respectively. When the averaged sound level was 57.8 dB (the *pianissimo* level), the third and fifth harmonics were not negligible as compared with these trials. On the other hand, when the averaged sound level was 60.2 dB, the second harmonic was only 10 dB lower than the fundamental.

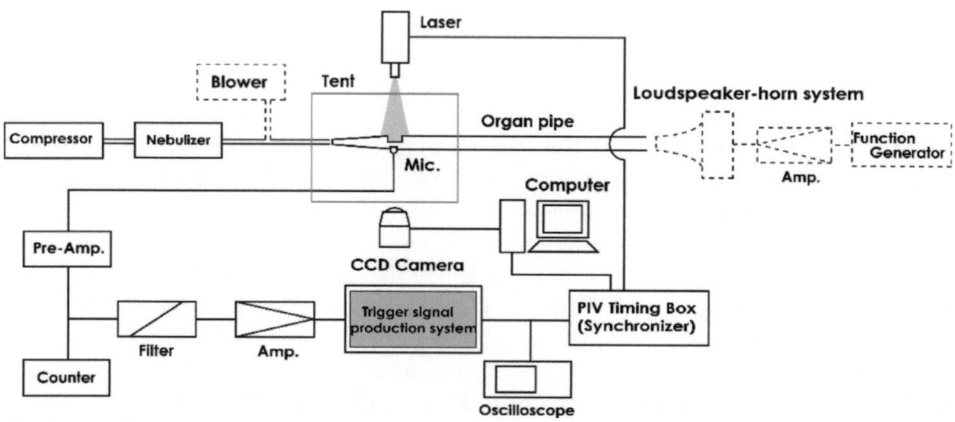

Figure 14.
Experimental setup based on the phase-locked PIV system (the PIV itself is manufactured by Dantec dynamics). The blower and the loudspeaker horn system are alternately used for the jet drive to measure the jet velocity and for the horn drive to measure the acoustic cross-flow velocity, respectively [28].

When the level was 69.4 dB, the second harmonic showed almost the same magnitude as the fundamental. As a result, two trials above with the *piano* level excitation seem to yield good conditions satisfying the measurement requirement for $\boldsymbol{u}$. It should be noted that all these tones with their levels from 57.8 dB to 69.4 dB are produced by the same first mode resonance.

The phase lock of the PIV system is easily implemented if the external trigger signal is produced to activate the laser and the CCD camera. This is because the PIV system can set the trigger delay almost arbitrarily through the software embedded in the trigger signal production system shown in **Figure 14**. The trigger delay was set to be $(1/12)T$ times n ($n = 0, 1, 2, \ldots, 11$), where T denotes the period of the pipe tone. For more details on the production of the external trigger, refer to [28]. As a result, the phase-locked measurement of $\boldsymbol{u}$ and $\boldsymbol{U}$ is carried out at the specific phases (Phase 0, Phase 1, Phase 2, ..., Phase 11). Note that Phase 0 is defined by the buildup of the positive trigger pulse when the trigger delay is not applied.

In the experiment a metallic organ pipe, which was made by a German organ builder, was measured [28]. Its cross-sectional structure (in x-z plane) around the mouth is already shown in **Figure 1**, and its important geometry is as follows: the physical pipe length $L = 793$ mm, the pipe inner diameter $2R = 43.6$ mm, the flue-to-edge distance $d = 8.8$ mm, the jet thickness $h = 0.75$ mm, and the mouth breadth $b = 31.9$ mm. The value of d/h is 11.7, much larger than 2. It should be noted that the edge is not very sharp like a wedge but plate-like as an extension of pipe wall, although the edge tip is 0.4 mm thick compared to the pipe wall that is 1.0 mm thick. Such an edge and flue are common in metal flue pipes as illustrated in Figure 17.6 of [1].

4.1.3 Calculation of the acoustic generation formula

The PIV can derive the vorticity map from the jet velocity distribution. The vorticity $\boldsymbol{\omega}$ ($=$ rot $\boldsymbol{U}$) at a field point is calculated from the 2-D velocities at four discrete points surrounding the point of interest. Therefore, the aeroacoustical source term $\boldsymbol{\omega} \times \boldsymbol{U}$ and the acoustic power generation term $(\boldsymbol{\omega} \times \boldsymbol{U}) \bullet \boldsymbol{u}$ can be calculated from the measurement of velocity fields $\boldsymbol{u}$ and $\boldsymbol{U}$ (see [28] on their measurement results, which are spared in this chapter).

The vorticity map given at Trial 1 is illustrated in **Figure 15(a)**. Since the 2-D velocity $\boldsymbol{U}$ was measured in x-z plane, the vorticity vector has only y direction component $[\boldsymbol{\omega} = (0, \omega_y, 0)]$:

$$\omega_y(x,z) = \frac{\partial U_x}{\partial z}(x,z) - \frac{\partial U_z}{\partial x}(x,z). \tag{46}$$

The vorticity is formed along the upper and lower boundaries of the jet. The upper layer possesses the positive vorticity (the counterclockwise rotation of small vortices) and the lower layer the negative vorticity [cf. **Figure 10(a)**]. These layers may be called *vorticity layers* or simply *vortex layers*. At Phases 7 and 10, large-scaled positive vortices are indicated before and above the edge, but these do not seem to be important because the magnitudes of $\boldsymbol{U}$ and $\boldsymbol{u}$ there are very small. As a result, the effects of these vortices almost completely disappear as shown in **Figure 15(c)**. It should be correctly recognized that vortex shedding from the edge tip (cf. **Figure 6**) is never observed in **Figure 15(a)**. This implies that Howe's vortex-shedding model may not be applicable to the sound generation in organ flue pipes that are usually driven by thin jets.

The resulting aeroacoustical source term $\boldsymbol{\omega} \times \boldsymbol{U}$ is displayed in **Figure 15(b)**, where the upper and lower layers of the vorticity yield the positive z and negative z

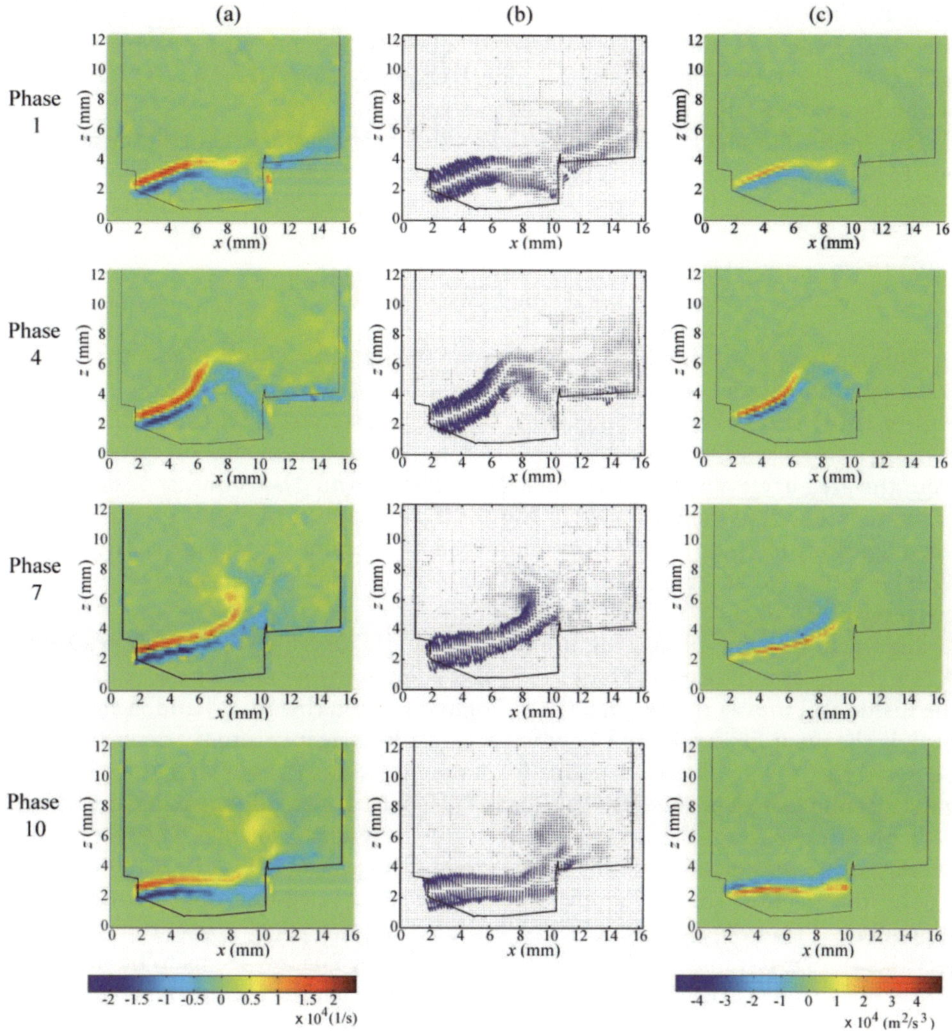

Figure 15.
Aerodynamical quantities derived from the phase-locked PIV measurement at the organ pipe mouth: (a) vorticity $\boldsymbol{\omega}$; (b) aeroacoustical source term (acceleration) $\boldsymbol{\omega} \times \boldsymbol{U}$; (c) acoustic generation term $(\boldsymbol{\omega} \times \boldsymbol{U}) \bullet \boldsymbol{u}$. The positions of the flue exit and the edge tip correspond to (x, z) = (2.0 mm, 2.2 mm) and (x, z) = (10.5 mm, 3.8 mm), respectively [28].

components of $\boldsymbol{\omega} \times \boldsymbol{U}$, respectively. The maximum magnitude of $\boldsymbol{\omega} \times \boldsymbol{U}$, which is about 2×10^5 (m/s^2), can be observed near the flue. The opposing vectors of $\boldsymbol{\omega} \times \boldsymbol{U}$ along both layers seem to almost cancel each other as depicted in **Figure 10(b)**. However, since $\boldsymbol{\omega} \times \boldsymbol{U}$ has a very large acceleration, an imperfect cancelation (or a slight unbalance) between $\boldsymbol{\omega} \times \boldsymbol{U}$ vectors along both layers yields a significant effect in acoustical events.

4.1.4 Generation of the acoustic power from the vortical field

Acoustic generation term $(\boldsymbol{\omega} \times \boldsymbol{U}) \cdot \boldsymbol{u}$ defined by Eq. (14) is shown in **Figure 15(c)**. Since $\boldsymbol{u}$ indicates the outflow at Phases 1 and 4, and inflow at Phases 7 and 10, $(\boldsymbol{\omega} \times \boldsymbol{U}) \cdot \boldsymbol{u}$ takes the opposite sign along the vortex layer between these phases. This sign inversion occurs near Phases 0 and 6 [28]. The maximum magnitude of $(\boldsymbol{\omega} \times \boldsymbol{U}) \cdot \boldsymbol{u}$ appears near 1–2 millimeters downstream from the flue at Phases 3 and 8 as about

4.2×10^4 (m^2/s^3). The jet crosses the edge from the inside at Phase 3 and from the outside at Phase 8 as inferred from **Figure 15(c)**. Although the magnitude of $(\boldsymbol{\omega} \times \boldsymbol{U}) \cdot \boldsymbol{u}$ is relatively small near the edge, it should be noted that $(\boldsymbol{\omega} \times \boldsymbol{U}) \cdot \boldsymbol{u}$ originally has very large values in acoustical sense.

Since the volume integral defined by Eq. (14) is not easily executed, the acoustic power generation from the vortex layer is estimated from the following surface integral by assuming the 2-D property (see Figure 11 in [28]) of $\boldsymbol{u}$ and $\boldsymbol{U}$:

$$\partial \Pi_G(t)/\partial y \approx -\iint \rho(\boldsymbol{\omega} \times \boldsymbol{U}) \cdot \boldsymbol{u} \mathrm{d}x\mathrm{d}z. \tag{47}$$

This surface integral, which may be called the *instantaneous 2-D vortex sound power*, is carried out at each phase, and the result is represented in **Figure 16(a)** and **(b)** concerning Trials 1 and 2, respectively. The integral area is restricted to $1 \leq x \leq 13$ mm and $1 \leq z \leq 9$ mm to reduce the calculation error caused from the area irrelevant to the acoustic generation term. Also, another scale of the ordinate is added to the right side of **Figure 16** in order to give a rough estimate of the magnitude of $\Pi_G(t)$. Since the mouth breadth b of our organ pipe is 31.9 mm, $\Pi_G(t)$ is estimated as $[\partial \Pi_G(t)/\partial y] \times 0.030$ (m) by assuming almost perfect 2-D property of $\boldsymbol{u}$ and $\boldsymbol{U}$.

Significant double-peak structure of Π_G is clearly demonstrated in **Figure 16(a)**. A larger peak is indicated at Phases 2, 3, and 4 when the jet crosses the edge from the inside and moves to the outside (cf. **Figure 15**). On the other hand, a smaller peak is shown at Phases 10 and 11 when the jet enters deeply into the pipe. It should be noted that the jet crosses the edge from the outside at Phase 8 and almost null vortex sound power is generated at Phase 8. Hence, this smaller peak occurs in a little phase delay from the impingement of the jet against the edge. The temporal average of Π_G estimated from the 2-D vortex sound power in **Figure 16(a)** will take a definitely positive value. Therefore, it may be recognized that the acoustic power is generated from the jet vortex layers. Although **Figure 16(b)** shows the characteristics similar to those of **Figure 16(a)**, the value at Phase 3 seems to be too large and erroneous because of the instant when the jet impinges against the edge [28].

4.1.5 Dominant area for the acoustic power generation and receptivity problem

The maps of the vorticity and aeroacoustical source term definitely indicate much larger magnitudes at the flue side as shown in **Figure 15(a)** and **(b)**, respectively. On the other hand, the acoustic flow velocity takes much larger magnitudes

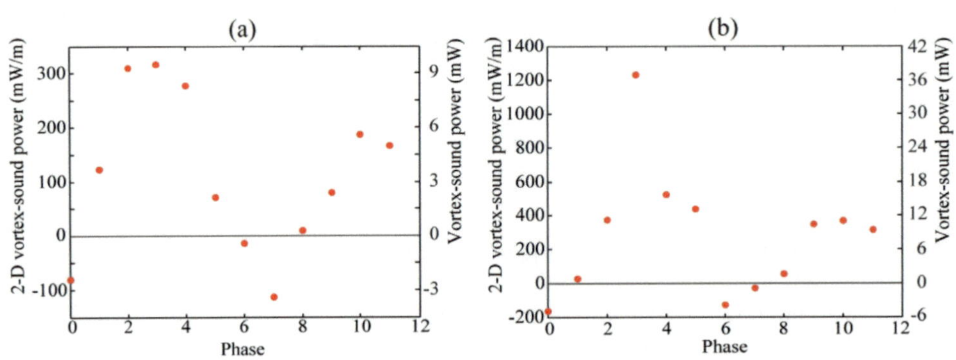

Figure 16.
The 2-D vortex sound power (left ordinate) and vortex sound power (right ordinate) as a function of the phase: (a) trial 1 and (b) trial 2 [28].

at the edge side (see Figures 5(a) and 7 in [28]). It should be then discussed which side is more dominant for the acoustic power generation.

The area for the surface integral of Eq. (47) is now set to be $2 \leq x \leq 11$ mm and $1 \leq z \leq 6$ mm. Then, this area is divided into two at $x = 7$, 8, and 9 mm. Hence, we have six sub-areas with the same z extent. The calculation result is demonstrated in Figure 12 of [28]. A very sharp contrast is displayed between area 5 ($2 \leq x \leq 9$ mm) and area 6 ($9 \leq x \leq 11$ mm): Area 5 yields larger negative values of $\partial \Pi_G(t)/\partial y$ at Phases 3, 4, and 5; area 6 yields much larger positive values of $\partial \Pi_G(t)/\partial y$ at the same phases. Hence, it may be concluded that such a small area as area 6 (very close to the edge) is most responsible for the acoustic power generation whose instantaneous contributions are given from Phases 2 to 5.

Also, the phase relation between the jet displacement, the acoustic velocity, and the acoustic pressure at the edge can be considered based on the PIV measurement results. The result is the same as **Figure 9(a)** (see Figure 13 in [28]). The dominant sound generation in our PIV experiment occurs with a phase lag of about $60 - 90°$ from the jet impingement against the edge [28]. This seems to verify that our experiment satisfies the requirements for the volume-flow model.

Coltman [52] discussed the activating force for the jet wavy motion. This is the most difficult problem in the flue instrument acoustics and is defined as the problem of the *receptivity* (the generation of jet oscillation by acoustic flow perturbations at the flue exit) [10]. Our present study based on the PIV measurement demonstrated that the aeroacoustic source term $\boldsymbol{\omega} \times \boldsymbol{U}$ (having the dimension of the acceleration) associated with the vortex-layer formation along the jet could activate the jet oscillation in an organ pipe. More precisely, an incomplete cancelation (or a net unbalance) of $\boldsymbol{\omega} \times \boldsymbol{U}$ between both sides of the jet can activate (oscillate) the jet.

Since this $\boldsymbol{\omega} \times \boldsymbol{U}$ can also activate the jet motion in the edge-tone generation [3–6, 18], the vortex-layer formation may be regarded as the fluid-dynamical mechanism common to the edge-tone generation and the pipe-tone generation. This fluid-dynamical model, which is a leading candidate to solve the receptivity problem, may be referred to as the *acceleration unbalance model* [28]. In the acoustical framework, this model can lead the volume-flow drive in an organ pipe driven by a thin jet with relatively low blowing pressures. Helmholtz might have envisaged such a physical picture as mentioned in Section 2.4.

4.2 Vortices from the jet visualized by direct numerical simulations

Sound generation in flue instruments is the revelation of the fluid *compressibility* in *low* Mach number state. This is a contradicting phenomenon in fluid-dynamical sense. Because of this, direct numerical simulations based on the Navier–Stokes equation could not achieve satisfactory outcomes [2]. However, in the 2010s we had many outstanding results from various viewpoints [53–55]. They are mentioning the roles of vortices, but do not have resolution enough to discuss the almost invisible (to our naked eyes) microstructure in the jet and its boundaries, particularly vortices in jet vortex layers, which seem to be a key point to solve the receptivity problem in the near future.

For the sake of page limitation, the description here is restricted to an essential point given by Eq. (24), which manifests the importance of the acoustic velocity $u_f(t)$ at the flue exit. By reformulating Fletcher's displacement model given by Eq. (4), Onogi et al. [56] proposed another formula that decomposed the jet oscillation into hydrodynamic and acoustic displacements, which were simulated on the basis of the 3-D compressible Navier–Stokes equations. They supposed the non-zero initial amplitude at the flue exit and the variable oscillation center with the flow direction for the jet displacement, although Coltman [52] strongly denied Fletcher's displacement model. Their simulation results (see Figures 7, 9, and 10 and Table IV in [56]) seem to confirm the non-zero amplitude at the flue exit, and the acoustic feedback effects on the jet wave may be given at its starting point.

5. Conclusions

Vortices on sound generation are clearly revealed in edge tones (with thin jets, without any resonators) and cavity tones (with thick, almost semi-infinite, jets, with cavity resonators). Although visible, relatively large vortices are seen in flue instruments driven by thick jets, these are in rare cases. Usually flue instruments are driven by thin jets [(jet length d/jet thickness h) > 2]. Any visible, discrete vortices do not appear at the flue exit and at the pipe edge in those cases. Instead, vortex layers are formed along the jet upper and lower boundaries, and the acceleration unbalance between them drives the jet as a whole in flue instruments.

The jet-wave drive (or the volume-flow drive) and the vortex-layer drive by thin jets assure sound generation in good manner when the jet enters into the pipe at the instant when the acoustic pressure is maximum. In the discrete-vortex drive by thick jets, the acoustic cross-flow (particle velocity) takes positive and negative values during the passage of the lower and upper vortices from the flue exit to the pipe edge, respectively. These vortex configurations can create sound power during the former and latter halves of an oscillation period.

On the other hand, acoustically induced vortices universally appear as the final dissipation agent. Their role in acoustic energy balance near the saturated state in flue instruments should be reconfirmed in more detail to exactly judge whether the acoustic vortex is generated just at the saturated state or just before the saturated state (at the pre-saturated state).

The receptivity problem is a key point to elucidate the sounding mechanism in flue instruments from the fluid-dynamical viewpoint. The initial amplitudes of acoustic quantities at the flue exit are regarded as the starting point for the acoustic feedback effects upon the jet wave. The vortex-layer model above will then be expected to solve this problem with the aid of direct aeroacoustical simulations.

Acknowledgements

The present author expresses his appreciation to three European scientists: Dr. Judit Angster of Fraunhofer-Institute fur Bauphysik, in Stuttgart, for her long-term support to carry out the PIV measurement; Prof. Avraham Hirschberg of Technishe Universiteit Eindhoven for his kind offer of the picture used as **Figure 12(c)** and helpful comments to the author's journal papers from the aeroacoustical viewpoint; and Prof. Andreas Bamberger of Freiburg University for his effective comments and suggestions on the PIV. Also, the author thanks Keita Arimoto and Takayasu Ebihara of Yamaha Corporation, in Hamamatsu, Japan, for their sincere support and appropriate comments to the manuscript.

Author details

Shigeru Yoshikawa
Kyushu University, Dazaifu, Japan

*Address all correspondence to: shig@lib.bbiq.jp

References

[1] Fletcher NH, Rossing TD. The Physics of Musical Instruments. 2nd ed. New York: Springer-Verlag; 1998. p. 690

[2] Yoshikawa S. Fluid-dynamical models of musical wind instruments. In: Yoshikawa S, Wada H, editors. Aeroacoustics of Sound Sources. Tokyo: Corona Publishing Co. Ltd; 2007. pp. 207-252. (in Japanese)

[3] Howe MS. Acoustics of Fluid-Structure Interactions. New York: Cambridge University Press; 1998. p. 570

[4] Powel A. On the edge-tone. Journal of the Acoustical Society of America. 1961;**33**:395-409

[5] Holger KD, Wilson TA, Beavers GS. Fluid mechanics of the edge-tone. Journal of the Acoustical Society of America. 1977;**62**:1116-1128

[6] Lin J-C, Rockwell D. Oscillations of a turbulent jet incident upon an edge. Journal of Fluids and Structures. 2001; **15**:791-826

[7] Rockwell D, Naudascher E. Review–self-sustaining oscillations of flow past cavities. Journal of Fluids and Engineerings, Transactions of ASME. 1978;**100**:152-165

[8] Elder SA. Self-excited depth-mode resonance for a wall-mounted cavity in turbulent flows. Journal of the Acoustical Society of America. 1978;**63**:877-890

[9] Meissner M. Aerodynamically excited acoustic oscillations in cavity resonator exposed to an air jet. Acta Acustica united with Acustica. 2002;**88**: 170-180

[10] Fabre B, Hirschberg A. Physical modeling of flue instruments: A review of lumped models. Acta Acustica united with Acustica. 2000;**86**:599-610

[11] Elder SA. The mechanism of sound production in organ pipes and cavity resonators. Journal of the Acoustical Society of Japan (E). 1992;**13**:11-23

[12] Verge MP, Fabre B, Hirschberg A, Wijnands APJ. Sound production in a recorder-like instrument. Part 1: Dimensionless amplitude of the internal acoustic field. Journal of the Acoustical Society of America. 1997;**101**:2914-2924

[13] Segoufin C, Fabre B, Verge MP, Hirschberg A, Wijnands APJ. Experimental study of the influence of the mouth geometry on sound production in a recorder-like instrument: Windway length and chamfers. Acta Acustica united with Acustica. 2000;**86**:649-661

[14] Hirschberg A. Aeroacoustics of wind instruments. In: Hirschberg A, Kergomard J, Weinrich A, editors. Mechanics of Musical Instruments. Wien: Springer-Verlag; 1995. pp. 291-369

[15] Bechert DW. Sound absorption caused by vorticity shedding, demonstrated with a jet flow. Journal of Sound and Vibration. 1980;**70**:389-405

[16] Fabre B, Hirschberg A, Wijnands APJ. Vortex shedding in steady oscillation of a flue organ pipe. Acta Acustica united with Acustica. 1996;**82**:863-877

[17] Yoshikawa S. A pictorial analysis of jet and vortex behaviours during attack transients in organ pipe models. Acta Acustica united with Acustica. 2000;**86**: 623-633

[18] How MS. Contributions to the theory of aerodynamic sound, with application to excess jet noise and the theory of the flute. Journal of Fluid Mechanics. 1975;**71**:625-673

[19] Rayleigh L. The Theory of Sound. New York: Macmillan; 1894. Reprinted by Dover; 1945. pp. 376-414

[20] Cremer L, Ising H. Die selbsterregten Schwingungen von Orgelpfeifen. Acoustica. 1967/1968;**19**: 143-153

[21] Dequand S, Willems JFF, Leroux M, Vullings R, van Weert M, Thieulot C, et al. Simplified models of flue instruments: Influence of mouth geometry on the sound source. Journal of the Acoustical Society of America. 2003;**113**:1724-1735

[22] Coltman JW. Sounding mechanism of the flute and organ pipe. Journal of the Acoustical Society of America. 1968; **44**:983-992

[23] Coltman JW. Jet drive mechanism in edge tones and organ pipes. Journal of the Acoustical Society of America. 1976; **60**:725-733

[24] Elder SA. On the mechanism of sound production in organ pipes. Journal of the Acoustical Society of America. 1973;**54**:1554-1564

[25] Fletcher NH. Jet drive mechanism in organ pipes. Journal of the Acoustical Society of America. 1976;**60**:481-483

[26] Yoshikawa S, Saneyoshi J. Feedback excitation mechanism in organ pipes. Journal of the Acoustical Society of Japan (E). 1980;**1**:175-191

[27] Verge MP, Causse R, Fabre B, Hirschberg A, Wijnands APJ, van Steenbergen A. Jet oscillations and jet drive in recorder-like instruments. Journal of the Acoustical Society of America. 1994;**80**:403-419

[28] Yoshikawa S, Tashiro H, Sakamoto Y. Experimental examination of vortex-sound generation in an organ pipe: A proposal of jet vortex-layer formation model. Journal of Sound and Vibration. 2012;**331**:2558-2577

[29] Yoshikawa S. Jet-wave amplification in organ pipes. Journal of the Acoustical Society of America. 1998;**103**:2706-2717

[30] Helmholtz HLF. On the Sensations of Tone. Berlin: Springer; 1863. Reprinted by Dover; 1954. pp. 88-94 (translated by A. J. Ellis in English)

[31] Fletcher NH. Sound production by organ flue pipes. Journal of the Acoustical Society of America. 1976;**60**: 926-936

[32] Mattingly GE, Criminale WO Jr. Disturbance characteristics in a plane jet. Physics of Fluids. 1971;**14**:2258-2264

[33] Michalke A. On spatially growing disturbances in an inviscid shear layer. Journal of Fluid Mechanics. 1965;**23**: 521-544

[34] Drazin PG, Howard LH. Hydrodynamic stability of parallel flow of in-viscid fluid. Advances in Applied Mechanics. 1966;**9**:1-89

[35] Howe MS. The dissipation of sound at an edge. Journal of Sound and Vibration. 1980;**70**:407-411

[36] Brown GB. The vortex motion causing edge tones. Proceedings of the Physical Society of London. 1937;**49**: 493-507

[37] Brown GB. The mechanism of edge–tone production. Proceedings of the Physical Society of London. 1937;**49**: 508-521

[38] Auvray R, Ernoult A, Fabre B, Lagrée P-Y. Time-domain simulation of flute-like instruments. Comparison of jet-drive and discrete-vortex models. Journal of the Acoustical Society of America. 2014;**136**:389-400

[39] Tsuchida J, Fujisawa T, Yagawa G. Direct numerical simulation of aerodynamic sounds by a compressible CFD scheme with node-by-node finite elements. Computational Methods on Applied Mechanical Engineering. 2006; **195**:1896-1910

[40] de la Cuadra P. The sound of oscillating air jet. Physics, modeling and simulation in flute-like instruments. Ph. D. thesis. University of Stanford; 2005

[41] Auvray R, Lagrée P-Y, Fabre B. Regime change and oscillation thresholds in recorder-like instruments. Journal of the Acoustical Society of America. 2012;**131**:1574-1585

[42] Meyer E, Neumann E-G. Attenuation of sound. In: Physical and Applied Acoustics. New York: Academic Press; 1972. pp. 95-115

[43] Yoshikawa S. Energy dissipations in underwater and aerial organ pipes. Journal of the Acoustical Society of Japan (E). 1984;**5**:181-192

[44] Salikuddin M, Ahuja KK. Acoustic power dissipation on radiation through duct termination: Experiments. Journal of Sound and Vibation. 1983;**91**:479-502

[45] Ingard KU, Ising H. Acoustic nonlinearity of an orifice. Journal of the Acoustical Society of America. 1967;**42**: 6-17

[46] Yoshikawa S. Underwater organ pipes. Journal of the Acoustical Society of Japan (E). 1984;**5**:211-221

[47] Hann DB, Greated CA. The measurement of flow velocity and acoustical particle velocity using particle-image velocimetry. Measurement Science and Technology. 1997;**8**:1517-1522

[48] Bamberger A. Investigations and recent results on flutes with PIV. In: Proceedings of the Forum Acusticum. Sevilla, Spain; 2002. MUS -04-003-IP

[49] Bamberger A. Vortex sound of flutes observed with particle image velocimetry. In: Proceedings of 18th International Congress on Acoustics. Kyoto, Japan; 2004. pp. 1417-1420

[50] Bamberger A. Vortex sound in flutes using flow determination with Endo- PIV. In: Proceedings of the Forum Acusticum. Budapest, Hungary; 2005. pp. 665-670

[51] Wolfe J, Smith J. Cutoff frequencies and cross fingering in baroque, classical, and modern flutes. Journal of the Acoustical Society of America. 2003; **114**:2263-2272

[52] Coltman JW. Jet behavior in the flute. Journal of the Acoustical Society of America. 1992;**92**:74-83

[53] Miyamoto M, Ito Y, Iwasaki T, Akamura T, Takahashi K, Tamaki T, et al. Numerical study on acoustic oscillation of 2D and 3D flue organ pipe like instruments with compressible LES. Acta Acustica united with Acustica. 2013;**99**:154-171

[54] Giordano N. Simulation studies of the recorder in three dimensions. Journal of the Acoustical Society of America. 2014;**135**:906-916

[55] Yokoyama H, Miki A, Onitsuka H, Iida A. Direct numerical simulation of fluid-acoustic interactions in a recorder with tone holes. Journal of the Acoustical Society of America. 2015;**138**: 858-873

[56] Onogi K, Yokoyama H, Iida A. Analysis of jet oscillations with acoustic radiation in the recorder by direct aeroacoustic simulations. Journal of the Acoustical Society of America. 2019; **146**:1427-1437

Section 2

Mathematical Modelling and Simulations

Chapter 4

A Fully Discrete SIPG Method for Solving Two Classes of Vortex Dominated Flows

Lunji Song

Abstract

To simulate incompressible Navier–Stokes equation, a temporal splitting scheme in time and high-order symmetric interior penalty Galerkin (SIPG) method in space discretization are employed, while the local Lax-Friedrichs flux is applied in the discretization of the nonlinear term. Under a constraint of the Courant–Friedrichs–Lewy (CFL) condition, two benchmark problems in 2D are simulated by the fully discrete SIPG method. One is a lid-driven cavity flow and the other is a circular cylinder flow. For the former, we compute velocity field, pressure contour and vorticity contour. In the latter, while the von Kármán vortex street appears with Reynolds number $50 \leq Re \leq 400$, we simulate different dynamical behavior of circular cylinder flows, and numerically estimate the Strouhal numbers comparable to the existing experimental results. The calculations on vortex dominated flows are carried out to investigate the potential application of the SIPG method.

Keywords: Navier–stokes equations, von Kármán vortex street, discontinuous Galerkin method, interior penalty

1. Introduction

The Navier–Stokes equations are a concise physics model of low Knudsen number (i.e. non-rarefied) fluid dynamics. Phenomena described with the Navier–Stokes equations include boundary layers, shocks, flow separation, turbulence, and vortices, as well as integrated effects such as lift and drag. Analytical solutions of real flow problems including complex geometries are not available, therefore numerical solutions are necessary. The Navier–Stokes equation has been investigated by many scientists conducting research on numerical schemes for approximation solutions (see [1–8]).

There exist two ways to provide reference data for such problems: One consists in the measurement of quantities of interest in physical experiments and the other is to perform careful numerical studies with highly accurate discretizations. With the prevalence and development of high-performance computers, advanced numerical algorithms are able to be tested for the validation of approaches and codes and for high-order convergence behavior of delicate discretizations.

Among discontinuous Galerkin (DG) methods, primal schemes and mixed methods are distinguished. The former depend on appropriate penalty terms of the discontinuous shape functions, while the latter rely on the mixed methods as the original second-order or higher-order partial differential equations are written as a system of first order partial differential equations with designed suitable numerical fluxes. Interior penalty discontinuous Galerkin methods (IPDG) are known as the representative of primal schemes whereas local discontinuous Galerkin methods (LDG) belong to the class of mixed methods [9]. About IPDG, symmetric interior penalty Galerkin (SIPG) and non-symmetric interior penalty Galerkin (NIPG) methods were first introduced originally for elliptic problems by Wheeler [10] and Rivière et al. [11]. Then, we presented some numerical analysis and simulations on nonlinear parabolic problems [12]. Recently, some work based on the SIPG and NIPG methods has been successfully applied to the steady-state and transient Navier–Stokes Equations [2, 13, 14], with careful analysis being conducted on optimal error estimates for the velocity.

The physics of Navier–Stokes flows are non-dimensionalized by Mach number M and Reynolds number Re,

$$M = \frac{\mathbf{u}_\infty}{\mathbf{a}}, \tag{1}$$

$$Re = \frac{\rho \mathbf{u}_\infty D}{\mu}, \tag{2}$$

where ρ is the density of the fluid, and μ is the dynamic viscosity. The kinematic viscosity ν is the ratio of μ to ρ. At low Knudsen numbers, Navier–Stokes surface boundary conditions are effectively no-slip (i.e. zero velocity). Diffusion of momentum from freestream to surface no-slip velocities forms boundary layers decreasing in thickness as Reynolds number increases. Thus, the range of characteristic solution scale increases as the Reynolds number increases. Nonlinear convective terms coupled with the strong velocity gradients in the Navier Stokes equations drive fluid flow at even moderate Reynolds numbers to inherently unsteady behavior. Rotational flow is measured in terms of the vorticity ω, defined as the curl of a velocity vector $\mathbf{v}$,

$$\omega = \nabla \times \mathbf{v} \tag{3}$$

The related concept of circulation Γ is defined as a contour integral of vorticity by

$$\Gamma = \oint_{\partial S} \mathbf{v} \cdot \mathbf{ds} = -\iint_{\mathbf{S}} \omega \cdot \hat{\mathbf{n}} \mathbf{dS} \tag{4}$$

The concept of a vortex is that of vorticity concentrated along a path [15].

Lid driven cavity flows are geometrically simple boundary conditions testing the convective and viscous portions of the Navier Stokes equation in an enclosed unsteady environment. The cavity flow is characterized by a quiescent flow with the driven upper lid providing energy transfer into the cavity through viscous stresses. Boundary layers along the side and lower surfaces develop as the Reynolds number increases, which tends to shift the vorticity center of rotation towards the center. A presence of the sharp corner at the downstream upper corner increasingly generates small scale flow features as the Reynolds number increases. Full cavity flows remain a strong research topic for acoustics and sensor deployment technologies.

For non-streamlined blunt bodies with a cross-flow, an adverse pressure gradient in the aft body tends to promote flow separation and an unsteady flow field. The velocity field develops into an oscillating separation line on the upper and lower surfaces. This manifests as a series of shed vortices forming and then convecting downstream with the mean flow. The von Kármán vortex street is named after the engineer and fluid dynamicist Theodore Kármán (1963; 1994). Vortex streets are ubiquitous in nature and are visibly seen in river currents downstream of obstacles, atmospheric phenomena, and the clouds of Jupiter (e.g. The Great Red Spot). Shed vortices are also the primary driver for the zig-zag motion of bubbles in carbonated drinks. The bubble rising through the drink creates a wake of shed vorticity which impacts the integrated pressures causing side forces and thus side accelerations. The physics of sound generation with an Aeoleans harp operates by alternating vortices creating harmonic surfaces pressure variations leading to radiated acoustic tones. Tones generated by vortex shedding are the so-called Strouhal friction tones. If the diameter of the string, or cylinder immersed in the flow is D and the free stream velocity of the flow is $\mathbf{u}_\infty$ then the shedding frequency f of the sound is given by the Strouhal formula

$$St = \frac{fD}{\mathbf{u}_\infty}, \tag{5}$$

where $f = T^{-1}$, and St is the Strouhal number named after Vincent Strouhal, a Czech physicist who experimented in 1878 with wires experiencing vortex shedding and singing in the wind (Strouhal, 1878; White, 1999). The Strouhal formula provides an experimentally derived shedding frequency for fluid flow. Therefore, we are interested in an investigation of Stouhal numbers of incompressible flow at different Reynolds number.

Let Ω be a bounded polygonal domain in $\mathbb{R}^2$. The dynamics of an incompressible fluid flow in 2D is described by the Navier–Stokes equations, which include the equations of continuity and momentum, written in dimensionless form [8] as follows:

$$\frac{\partial \mathbf{u}}{\partial t} - \nu\Delta\mathbf{u} + (\mathbf{u}\cdot\nabla)\mathbf{u} + \nabla p = \mathbf{f}, \qquad \text{in} \quad \Omega\times(0,T) \tag{6}$$

$$\nabla\cdot\mathbf{u} = 0, \qquad \text{in} \quad \Omega\times(0,T) \tag{7}$$

$$\mathbf{u}|_{t=0} = \mathbf{u}_0, \tag{8}$$

subject to the boundary conditions on $\partial\Omega$:

$$\alpha\mathbf{u} + (1-\alpha)\frac{\partial\mathbf{u}}{\partial\mathbf{n}} = \mathbf{u}_\infty.$$

Here the parameter α has the limit values of 0 for the free-slip (no stress) condition (Neumann) and 1 for the no-slip condition (Dirichlet); $\mathbf{u} = (u, v)$ is the velocity; t is the time; and p is the pressure. In general, the external force $\mathbf{f}$ is not taken into account in Eq. (6).

Using the divergence free constraint, problem (6)–(8) can be rewritten in the following conservative flux form [16]:

$$\frac{\partial \mathbf{u}}{\partial t} - \nu\Delta\mathbf{u} + \nabla\cdot F + \nabla p = \mathbf{f}, \qquad \text{in} \quad \Omega\times(0,T) \tag{9}$$

$$\nabla\cdot\mathbf{u} = 0, \qquad \text{in} \quad \Omega\times(0,T) \tag{10}$$

$$\mathbf{u}|_{t=0} = \mathbf{u}_0, \tag{11}$$

with the flux F being defined as

$$F(\mathbf{u}) = \mathbf{u}\otimes\mathbf{u} = \begin{bmatrix} u^2 & uv \\ uv & v^2 \end{bmatrix}. \tag{12}$$

and $\mathbf{u} \otimes \mathbf{v} = u_i v_j, i, j = 1, 2$. Indeed, it holds

$$\mathcal{N}(\mathbf{u}) = \begin{pmatrix} \dfrac{\partial(u^2)}{\partial x} + \dfrac{\partial(uv)}{\partial y} \\ \dfrac{\partial(uv)}{\partial x} + \dfrac{\partial(v^2)}{\partial y} \end{pmatrix} = \nabla \cdot F(\mathbf{u}).$$

A locally conservative DG discretization will be employed for the Navier–Stokes Eq. (9)–(11). We denote by $\mathcal{E}_h$ a shape-regular triangulation of the domain $\overline{\Omega}$ into triangles, where h is the maximum diameter of elements. Let Γ_h^I be the set of all interior edges of $\mathcal{E}_h$ and Γ_h^B be the set of all boundary edges. Set $\Gamma_h = \Gamma_h^I \cup \Gamma_h^B$. For any nonnegative integer r and $s \geq 1$, the classical Sobolev space on a domain $E \subset \mathbb{R}^2$ is

$$W^{r,s}(E) = \{v \in L^s(E) : \forall |m| \leq r, \partial^m v \in L^s(E)\}.$$

We define the spaces of discontinuous functions

$$V = \{\mathbf{v} \in L^2(\Omega)^2 : \quad \forall E \in \mathcal{E}_h, \ \mathbf{v}|_E \in \left(W^{2,4/3}(E)\right)^2\},$$
$$M = \{q \in L^2(\Omega) : \quad \forall E \in \mathcal{E}_h, \ q|_E \in W^{1,4/3}(E)\}.$$

The jump and average of a function ϕ on an edge e are defined by:

$$[\phi] = \left(\phi|_{E_k}\right)|_e - \left(\phi|_{E_l}\right)|_e,$$
$$\{\phi\} = \frac{1}{2}\left(\left(\phi_{E_k}\right)|_e + \left(\phi|_{E_l}\right)|_e\right).$$

Further, let $\mathbf{v}$ be a piecewise smooth vector-, or matrix-valued function at $\mathbf{x} \in e$ and denote its jump by

$$[\mathbf{v}] := \mathbf{v}^+ \cdot \mathbf{n}_{E^+} + \mathbf{v}^- \cdot \mathbf{n}_{E^-},$$

where e is shared by two elements E^+ and E^-, and an outward unit normal vector $\mathbf{n}_{E^+}$ (or $\mathbf{n}_{E^-}$) is associated with the edge e of an element E^+ (or E^-). The tensor product of two tensors $\mathbf{T}$ and $\mathbf{S}$ is defined as $\mathbf{T} : \mathbf{S} = \sum_{i,j} T_{ij} S_{ij}$.

Let $\mathbb{P}_N(E)$ be the set of polynomials on an element E with degree no more than N. Based on the triangulation, we introduce two approximate subspaces $\mathbf{V}_h(\subset V)$ and $M_h(\subset M)$ for integer $N \geq 1$:

$$\mathbf{V}_h = \left\{\mathbf{v} \in L^2(\Omega)^2 : \quad \forall E \in \mathcal{E}_h, \ \mathbf{v}_h \in (\mathbb{P}_N(E))^2\right\},$$
$$M_h = \{q \in L^2(\Omega) : \quad \forall E \in \mathcal{E}_h, \ q \in \mathbb{P}_{N-1}(E)\}.$$

We mainly cite the content of [17], in which was motivated by the work of Girault, Rivière and Wheeler in a series of papers [2, 14]. Some projection methods [6, 18] have been developed to overcome the incompressibility constraints $\nabla \cdot \mathbf{u} = 0$. An implementation of the operator-splitting idea for discontinuous Galerkin elements was developed in [2]. We appreciate the advantages of the discontinuous Galerkin methods, such as local mass conservation, high order of approximation, robustness and stability. In this work, we will make use of the underlying physical nature of incompressible flows in the literature and extend the interior penalty discontinuous

Galerkin methods to investigate dynamical behavior of vortex dominated lid-driven and cylinder flows.

The chapter is organized following [17]. In Section 2, a temporal discretization for the Navier–Stokes equation is listed with operator-splitting techniques, and subsequently, the nonlinear term is linearized. Both pressure and velocity field can be solved successively from linear elliptic and Helmholtz-type problems, respectively. In Section 3, a local numerical flux will be given for the nonlinear convection term and an SIPG scheme will be used in spacial discretization for those linear elliptic and Helmholtz-type problems with appropriate boundary conditions, and in Section 4, simulation results are presented for a lid-driven cavity flow up to $Re = 7500$ and a transient flow past a circular cylinder, while numerical investigation on the Strouhal-Reynolds-number has been done, comparable to the experimental values from physics. Finally, Section 5 concludes with a brief summary.

2. Temporal splitting scheme

We consider here a third-order time-accurate discretization method at each time step by using the previous known velocity vectors. Let Δt be the time step, $M = \frac{T}{\Delta t}$, and $t_n = n\Delta t$. The semi-discrete forms of problem (6)–(8) at time t_{n+1} is

$$\frac{\gamma_0 \mathbf{u}^{n+1} - \alpha_0 \mathbf{u}^n - \alpha_1 \mathbf{u}^{n-1} - \alpha_2 \mathbf{u}^{n-2}}{\Delta t} - \nu \Delta \mathbf{u}^{n+1} + \nabla p^{n+1} \\ = -\beta_0 \mathcal{N}(\mathbf{u}^n) - \beta_1 \mathcal{N}(\mathbf{u}^{n-1}) - \beta_2 \mathcal{N}(\mathbf{u}^{n-2}) + \mathbf{f}(t_{n+1}), \tag{13}$$

$$\nabla \cdot \mathbf{u}^{n+1} = 0, \tag{14}$$

which has a timestep constraint based on the CFL condition (see [19]):

$$\Delta t \approx O\left(\frac{\mathcal{L}}{\mathcal{U} N^2}\right),$$

where $\mathcal{L}$ is an integral length scale (e.g. the mesh size) and $\mathcal{U}$ is a characteristic velocity. Because the semi-discrete system (13)–(14) is linearized, thus, a time-splitting scheme can be applied naturally, i.e., the semi-discretization in time (13)–(14) can be decomposed into three stages as follows.

- The first stage

When $\mathbf{u}^n$ and $\mathbf{u}^{n-1}$ $(n \geq 1)$ are known, the following linearized third-order formula can be used

$$\frac{\gamma_0 \tilde{\mathbf{u}} - \alpha_0 \mathbf{u}^n - \alpha_1 \mathbf{u}^{n-1} - \alpha_2 \mathbf{u}^{n-2}}{\Delta t} = -\beta_0 \mathcal{N}(\mathbf{u}^n) - \beta_1 \mathcal{N}(\mathbf{u}^{n-1}) + \mathbf{f}(t_{n+1}) \tag{15}$$

with the following coefficients for the subsequent time levels $(n \geq 2)$

$$\gamma_0 = \frac{11}{6}, \quad \alpha_0 = 3, \alpha_1 = -\frac{3}{2}, \quad \alpha_2 = \frac{1}{3}, \quad \beta_0 = 2, \quad \beta_1 = -1.$$

Especially, by using the Euler forward discretization at the first time step $(n = 0)$, we can get a medium velocity field $\mathbf{u}^1$ by

$$\frac{\mathbf{u}^1 - \mathbf{u}^0}{\Delta t} = -\mathcal{N}\left(\mathbf{u}^0\right) + \mathbf{f}(t_1),$$

and $\mathbf{u}^2$ by

$$\frac{\gamma_0 \mathbf{u}^2 - \alpha_0 \mathbf{u}^1 - \alpha_1 \mathbf{u}^0}{\Delta t} = -\beta_0 \mathcal{N}\left(\mathbf{u}^1\right) - \beta_1 \mathcal{N}\left(\mathbf{u}^0\right) + \mathbf{f}(t_2),$$

which adopts the following coefficients to construct a second-order difference scheme for the time level ($n = 2$) in (15)

$$\gamma_0 = \frac{3}{2}, \quad \alpha_0 = 2, \alpha_1 = -\frac{1}{2}, \quad \alpha_2 = 0, \quad \beta_0 = 2, \quad \beta_1 = -1.$$

Note that the coefficients in (15) are adjustable, but high-order time discrete schemes need to be verified with stability analysis.

- The second stage

The pressure projection is as follows

$$\gamma_0 \frac{\tilde{\tilde{\mathbf{u}}} - \tilde{\mathbf{u}}}{\Delta t} = -\nabla p^{n+1}. \tag{16}$$

To seek p^{n+1} such that $\nabla \cdot \tilde{\tilde{\mathbf{u}}} = 0$, we solve the system

$$-\Delta p^{n+1} = -\frac{\gamma_0}{\Delta t} \nabla \cdot \tilde{\mathbf{u}}, \tag{17}$$

with a Neumann boundary condition being implemented on the boundaries as

$$\begin{aligned} \frac{\partial p^{n+1}}{\partial \mathbf{n}} = \mathbf{f}^{n+1} - \beta_0 \mathbf{n} \cdot \left(\frac{\partial \mathbf{u}^n}{\partial t} + \nabla \cdot F(\mathbf{u}^{\mathbf{n}}) - \nu \Delta \mathbf{u}^n \right) \\ -\beta_1 \mathbf{n} \cdot \left(\frac{\partial \mathbf{u}^{n-1}}{\partial t} + \nabla \cdot F\left(\mathbf{u}^{\mathbf{n}-1}\right) - \nu \Delta \mathbf{u}^{n-1} \right) \\ -\beta_2 \mathbf{n} \cdot \left(\frac{\partial \mathbf{u}^{n-2}}{\partial t} + \nabla \cdot F\left(\mathbf{u}^{\mathbf{n}-2}\right) - \nu \Delta \mathbf{u}^{n-2} \right) := G_n. \end{aligned}$$

One can compute the vorticity $\omega^n = \nabla \times \mathbf{u}^n$ at time $t^n = n \cdot \Delta t$. Then we use p^{n+1} to update the intermediate velocity $\tilde{\tilde{\mathbf{u}}}$ by (16).

- The third stage is completed by solving

$$\gamma_0 \frac{\mathbf{u}^{n+1} - \tilde{\tilde{\mathbf{u}}}}{\Delta t} = \nu \Delta \mathbf{u}^{n+1},$$

which can be written as a Helmholtz equation for the velocity

$$-\Delta \mathbf{u}^{n+1} + \frac{\gamma_0}{\nu \Delta t} \mathbf{u}^{n+1} = \frac{\gamma_0}{\nu \Delta t} \tilde{\tilde{\mathbf{u}}}. \tag{18}$$

From the three stages given above, we notice that (15) in the semi-discrete systems is presented in a linearized and explicit process, moreover, (17) and (18) are obviously a type of elliptic and Helmholtz problems at each time step as $n \geq 2$. We decouple the incompressibility condition and the nonlinearity, then the pressure and velocity semi-discretizations (17)–(18) will be formulated by the interior penalty discontinuous Galerkin methods in spacial discretizations in the next section.

3. The spatial discretizations

For spacial approximations, assume that piecewise polynomials of order N are employed, then the approximation space can be rewritten as $\mathbf{V}_h = \oplus_{k=1}^{K} \mathbb{P}_N(E_k)^2$. In the approximating polynomial space for the velocity or pressure restricted to each element, a high-order nodal basis can be chosen, consisting of Lagrange interpolating polynomials defined on a reference simplex introduced in [20, 21]. We let $\mathbf{u}$ be approximated by $\mathbf{u}_h \in \mathbf{V}_h$ and adopt a suitable approximation for the term F, i.e., $F(\mathbf{u}) \approx F(\mathbf{u}_h)$, where $F(\mathbf{u}_h)$ also can be represented as the L^2-projection of $F(\mathbf{u}_h)$ on each element of $\mathcal{T}_h$. Multiplying the nonlinear term by a test function $\mathbf{v}_h \in \mathbf{V}_h$, integrating over the computational domain, and applying integration by parts, we have

$$\int_{\Omega} (\nabla \cdot F) \cdot \mathbf{v}_h d\mathbf{x} = -\sum_{E_k} \int_{E_k} (F \cdot \nabla) \cdot \mathbf{v}_h dx + \sum_{e \in \Gamma_h} \int_e \mathbf{n}_e \cdot [F \cdot \mathbf{v}_h] ds, \tag{19}$$

where the term $(F \cdot \nabla) \cdot \mathbf{v}_h$ equals to $F_{ij} \frac{\partial v_{hi}}{\partial x_j}$, for $i,j = 1,2$ and the indexes i,j correspond to the components of the related vectors. On each edge $e \in \partial E_1 \cap \partial E_2$ shared by two elements, to ensure the flux Jacobian of purely real eigenvalues, we may define $\lambda^+_{E_1,e}$, $\lambda^-_{E_2,e}$ the largest eigenvalue of the Jacobians $\frac{\partial}{\partial \mathbf{u}}(F \cdot \mathbf{n}_e)\big|_{\overline{\mathbf{u}}_{E_1}}$ and $\frac{\partial}{\partial \mathbf{u}}(F \cdot \mathbf{n}_e)\big|_{\overline{\mathbf{u}}_{E_2}}$, respectively, where $\overline{\mathbf{u}}_{E_1}$ and $\overline{\mathbf{u}}_{E_2}$ are the mean values of u_h on the elements E_1 and E_2, respectively. The global Lax-Friedrichs flux is generally more dissipative than the local Lax-Friedrichs flux, therefore, we primarily consider the local flux on each edge. Although the Lax-Friedrichs flux is perhaps the simplest numerical flux and often the most efficient flux, it is not the most accurate scheme. A remedy of the problem is to employ high-order finite elements. By replacing the integrand in the surface integral as

$$\mathbf{n}_e \cdot [F \cdot \mathbf{v}_h] = \mathbf{n}_e \cdot \{F\} \cdot [\mathbf{v}_h] + \frac{\lambda_e}{2} [\mathbf{u}_h] \cdot [\mathbf{v}_h],$$

with $\lambda_e = \max\left(\lambda^+_{E_1,e}, \lambda^-_{E_2,e}\right)$, one can get a DG discretization for the nonlinear term in (19) by the local Lax-Friedrichs flux.

For the pressure correction step (17) and the viscous correction step (18), we use the SIPG method to approximate the correction steps. Choosing the orthonormal Legendre basis and the Legendre-Gauss-Lobatto quadrature points gives a well-conditioned Vandermonde matrix and the resulting interpolation well behaved, which greatly simplifies the formulas. The C^0 continuity condition of the basis in the discontinuous Galerkin formulation is not required. Enforcing a weak continuity on the interior edges by a penalty term, we have for (17)

$$a\left(p_h^{n+1}, \phi_h\right) = L_p(\phi_h), \quad \phi \in M_h,$$

where

$$a\left(p_h^{n+1}, \phi_h\right) = \sum_{E_k \in \mathcal{E}_h} \int_{E_k} \nabla p_h^{n+1} \cdot \nabla \phi_h dx - \sum_{e_k \in \Gamma_h} \int_{e_k} \left\{ \frac{\partial p_h^{n+1}}{\partial \mathbf{n}} \right\} [\phi_h] ds$$

$$- \sum_{e_k \in \Gamma_h} \int_{e_k} \left\{ \frac{\partial \phi_h}{\partial \mathbf{n}} \right\} [p_h^{n+1}] ds + \sum_{e_k \in \Gamma_h} \frac{\sigma_e}{|e_k|^\beta} \int_{e_k} [p_h^{n+1}][\phi_h] ds,$$

$$\text{and} \quad L_p(\phi_h) = \sum_{E_k \in \mathcal{E}_h} \int_{E_k} \frac{\gamma_0}{\Delta t} \nabla \cdot \tilde{\mathbf{u}} \phi_h dx + \sum_{e_k \in \partial\Omega} \int_{e_k} \phi_h G_n.$$

In general, σ_e shall be chosen sufficiently large to guarantee coercivity, more accurately, the threshold values of σ_e in [22] are given for $\beta = 1$ in the above formula, which is referred to an SIPG scheme. Especially, as $\beta > 1$, the scheme is referred to an over-penalized scheme and the threshold values of σ_e are presented in [23, 24]. Analogously, the SIPG discretization for (18) is given by

$$a\left(\mathbf{u}_h^{n+1}, \mathbf{v}_h\right) + \frac{\gamma_0}{\nu \Delta t}\left(\mathbf{u}_h^{n+1}, \mathbf{v}_h\right)_\Omega = L_{\mathbf{u}}(\mathbf{v}_h), \quad \forall \mathbf{v}_h \in \mathbf{V}_h,$$

where

$$a\left(\mathbf{u}_h^{n+1}, \mathbf{v}_h\right) = \sum_{E_k \in \mathcal{E}_h} \int_{E_k} \nabla \mathbf{u}_h^{n+1} : \nabla \mathbf{v}_h d\mathbf{x} - \sum_{e_k \in \Gamma_h} \int_{e_k} \{\nabla \mathbf{u}_h^{n+1}\} \mathbf{n} \cdot [\mathbf{v}_h] ds \tag{20}$$

$$-\varepsilon \sum_{e_k \in \Gamma_h} \int_{e_k} \{\nabla \mathbf{v}_h\} \mathbf{n} \cdot \left[\mathbf{u}_h^{n+1}\right] ds + \sum_{e_k \in \Gamma_h} \frac{\sigma_e}{|e_k|^\beta} \int_{e_k} \left[\mathbf{u}_h^{n+1}\right] \cdot [\mathbf{v}_h] ds, \tag{21}$$

and

$$L_{\mathbf{u}}(\mathbf{v}_h) = \left(\frac{\gamma_0}{\nu \Delta t} \tilde{\tilde{\mathbf{u}}}_h, \mathbf{v}_h\right)_\Omega - \sum_{e_k \in \partial\Omega} \int_{e_k} \left(\varepsilon \nabla \mathbf{v}_h \cdot \mathbf{n}_e - \frac{\sigma_e}{|e_k|^\beta} \mathbf{v}_h \right) \mathbf{u}_0.$$

where $\beta = 1, \varepsilon = 1$. Note that the parameter ε can be 1, −1, and 0, the scheme (20) becomes SIPG, NIPG and IIPG, respectively. The SIPG scheme exhibits a stiffness matrix with symmetric structure. As a DG method, the SIPG scheme has some attractive advantages of DG methods including high order $hp-$ approximation, local mass conservation, robustness and accuracy of DG methods for models with discontinuous coefficients and easy implementation on unstructured grids, while the flexibility of p-adaptivity (different orders of polynomials might be used for different elements) in DG methods has become competitive for modeling a wide range of engineering problems.

4. Numerical results

We present a lid-driven flow problem to verify the efficiency and robustness of the interior penalty discontinuous Galerkin method, and then investigate a flow past a cylinder with walls or without a wall, as well as the relationship between the Strouhal number and the Reynold number. Throughout the section, time steps $\Delta t \leq 1E - 03$ are taken.

Example 1. The lid-driven boundary conditions are given by:

$$\begin{cases} u(x,1) = 1; & u(1,y) = 0; & u(0,y) = 0; & u(x,0) = 0; \\ v(x,1) = 1; & v(1,y) = 0; & v(0,y) = 0; & v(x,0) = 0. \end{cases}$$

Here the mesh size of the initial coarse grid is 0.2 and then it is uniformly refined three times with piecewise discontinuous elements being applied into the fully discrete SIPG approach.

The boundary condition at the vertex is a jump from zero velocity on the edge to a unit velocity on the upper edge. Nature prevents this singularity with a boundary layer forming along all walls, making the vertex velocity zero. It is reasonable to adopt adaptive meshes for solving those singularity problems. Here, we apply the semi-implicit SIPG method with approximation polynomials of order $N = 3$ in a locally refined mesh in **Figure 1** to solve the incompressible flow. In **Figure 2**, the velocity profiles of (u, v) through the geometric center of the cavity are plotted with $Re = 1000, 5000, 7500$ taken. From **Figures 3–5**, with different Reynolds numbers taken up to 7500, the vorticity field exhibits the expected characteristics of a driven cavity flow consisting of a region of vortical flow centrally located. Energy enters the cavity through the viscous boundary later formed by velocity gradients on the upper driven edge. Convection distributes flow properties throughout the domain.

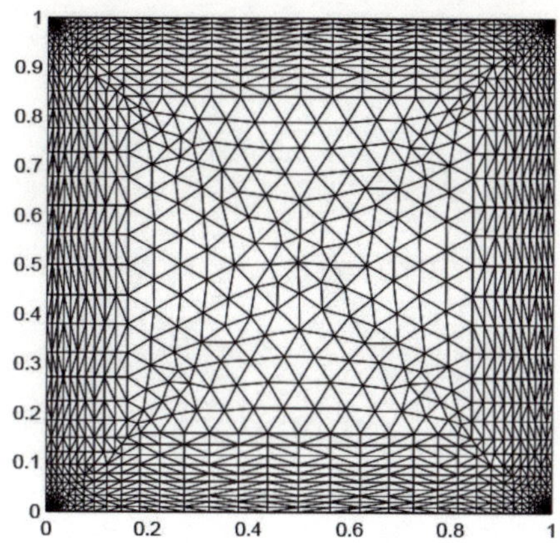

Figure 1.
An initial locally refined mesh.

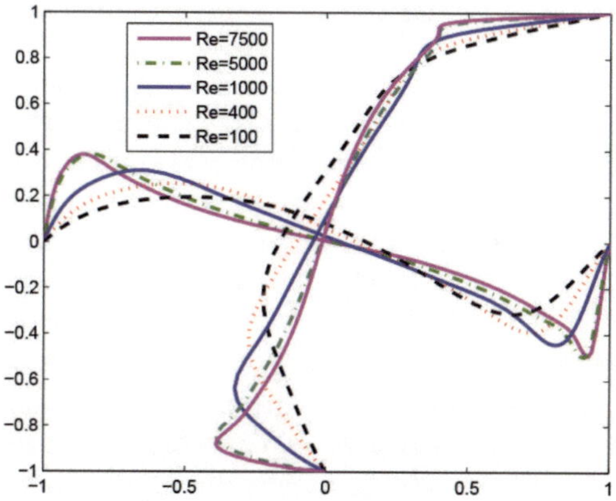

Figure 2.
Velocity profiles (u, v) through geometric center of the cavity for $Re = 100, 400, 1000, 5000, 7500$.

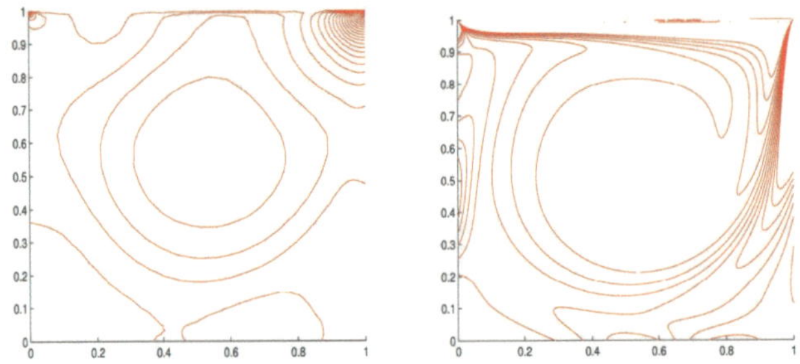

Figure 3.
Re =1000, N = 4, mesh #2. Left: pressure contour; right: vorticity contour.

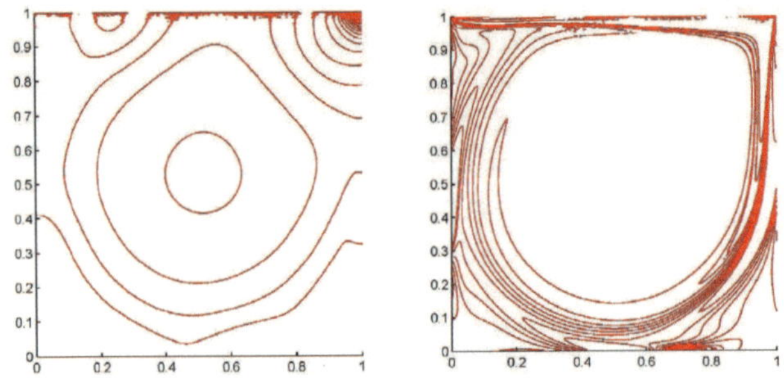

Figure 4.
Re =5000, N = 3, the initial mesh refined. Left: pressure contour; right: vorticity contour.

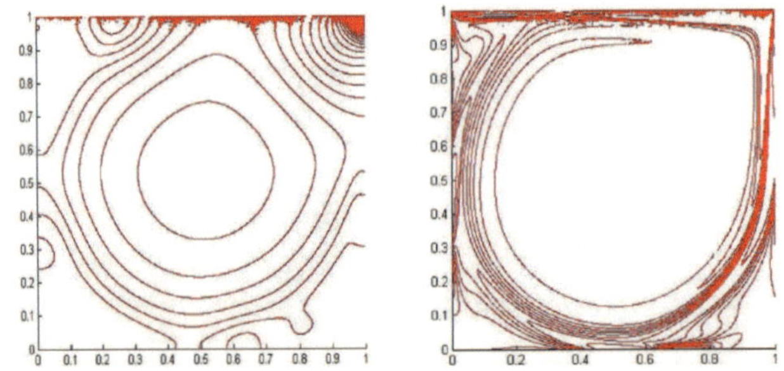

Figure 5.
Re =7500, N = 3, the initial mesh refined. Left: pressure contour; right: vorticity contour.

Moreover, a video on the dynamical evolution of vorticity isolines ($Re = 1000, N = 4$) can be browsed through a website (Available from: https://youtu.be/UfGWvnoiW58). These numerical simulations are performed for the Navier–Stokes equations which illustrate the effectiveness of the DG method.

Example 2. We simulate a channel flow past a circular cylinder with a radius 0.05 at the origin $(0, 0)$ for $Re = 100$ by the discontinuous Galerkin method in the domain $(-1, 3) \times (-0.5, 0.5)$. The free stream velocity on the inflow boundary is $\mathbf{u}_\infty = (1, 0)$, while the outflow boundary is $\frac{\partial \mathbf{u}}{\partial \mathbf{n}} = 0$. To the boundary conditions on the upper and lower sides, we present two different conditions for comparison (see **Figure 6**), which are wall ($u = 0, v = 0$) and homogeneous Neumann boundary conditions ($u = 1, \frac{\partial v}{\partial n} = 0$), respectively. The homogeneous Neumann boundary condition is a special non-reflecting case, where the boundary flux is zero. For reference, the density of the fluid is given by $\rho = 1\ kg \cdot m^{-3}$ and a locally refined mesh ($\max h = 0.088$) will be used for the simulations.

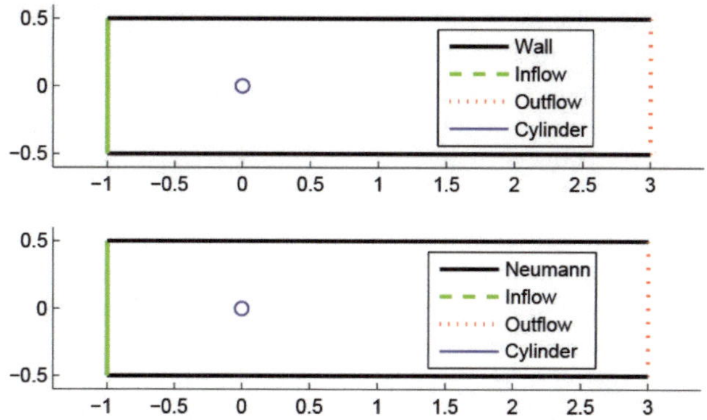

Figure 6.
Cylinder flow in a channel N = 3. Top: with walls; bottom: without a wall.

Cylinder flow contains the fundamentals of unsteady fluid dynamics in a simplified geometry. The flow properties and unsteadiness are well defined through years of experimental measurements across a wide range of Reynolds numbers [25], making the cylinder an ideal validation testcase for unsteady numerical fluid dynamics simulations.

Verification in a numerical domain requires insights from physics for a proper comparison to experimental and theoretical data. In **Figure** 7, we observe that the boundary layers forms along the upper and lower walls. From continuity of mass, the presence of a boundary layer decreasing the flow velocity near the wall requires an increase in the centerline average flow velocity. The cylinders wake provides a similar increase in centerline velocities. This implies a non-intuitive reality that drag can increase velocities within constrained domains. This effect is compensated for in wind tunnel test [26] environments topologically similar to **Figure** 7 with a constant mass flow rate and no-slip walls. Drela [15] develops an analysis for 2D wind tunnels resulting in an effective coefficient of drag of

$$C_d = \left(1 - \frac{1}{2}\frac{c}{H}C_d - \frac{\pi}{2}\frac{A}{H^2}\right)C_{d_{un}},$$

and an effective Reynolds number of

$$Re = \left(1 + \frac{1}{4}\frac{c}{H}C_d + \frac{\pi}{6}\frac{A}{H^2}\right)Re_{un},$$

where the *un* subscript represents the uncorrected value, H represents the domain height, A represents the cylinder area and c represents the cylinder radius. Drelas analysis does not specifically include the boundary layer forming on the

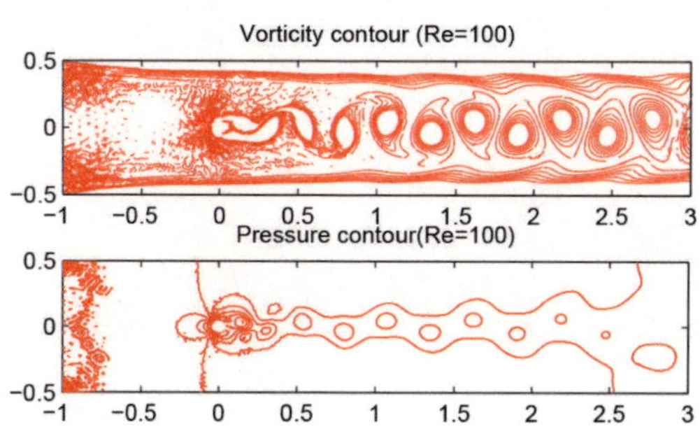

Figure 7.
Cylinder flow with walls, Re = 100. Top: vorticity contour; bottom: pressure contour.

upper and lower walls. The flow physics associated with wall boundary layer drag differs from cylinder drag in that the wall drag is a distributed effect of monotonically increasing drag with downstream distance rather than a conceptual point source of drag. The wall boundary layer tends to provide a steady acceleration of flow within the interior flow domain (i.e. non-boundary layer portion) leading to an effective buoyancy drag. A secondary feature of the wall boundary layer is that downstream flow features such as vortices are convected at a higher perturbation velocity compared to the initial upstream velocity. For numerical validation of raw experimental data, either the wind tunnel geometry should exactly match the numerical geometry, or the numerical geometry should be corrected using the concepts introduced above to match the actual wind tunnel geometry. Alternatively, the open-air corrected values should be used for validation. The above analysis provides insight into the domain height necessary to reduce volume blockage (c/H) and wake buoyancy (A/H^2) effects.

Alternatively, the flow without walls in **Figure 8** has no interference of the boundary layers along the channel on the up and bottom boundaries, thus the pressure contours expend after flow passing through the cylinder. We also compared the components of the velocity profiles along the $x-$ direction in **Figure 9**, and observed that the boundary layers are produced in the top picture rather than in the bottom one. If the effect of the boundary layers disappeared, the velocity in the $x-$ direction would reduce dispersively, in other words, the vortex lifespan is less than those produced in the channel with walls. The velocity profiles in the $y-$ direction have been given in **Figure 10**.

We localize the domain around the cylinder and refine the mesh, then show the vorticity startup behavior in **Figure 11** as well as the pressure **Figure 12**. Upon startup, two vortices of opposite direction are formed on the upper and lower

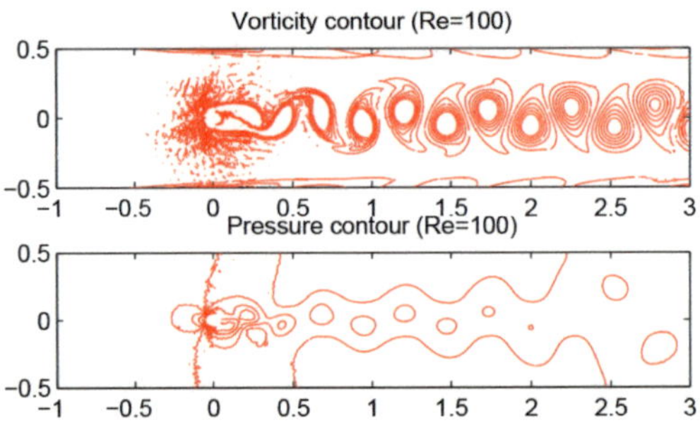

Figure 8.
Flow past a cylinder in a channel without a wall, Re = 100. Above: vorticity contour; below: pressure contour.

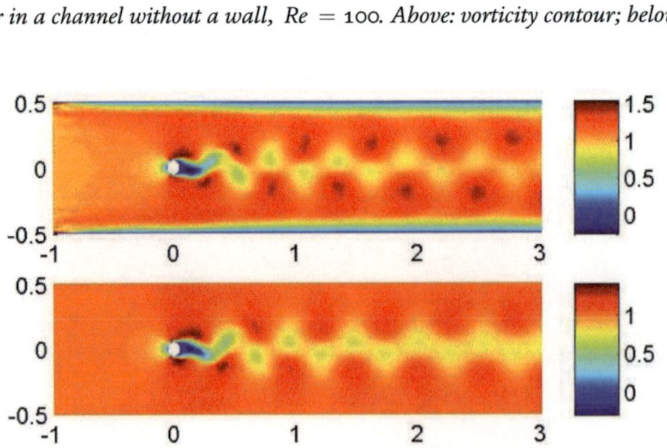

Figure 9.
Re = 100. Above: velocity in x-direction with walls; below: velocity in x-direction without a wall.

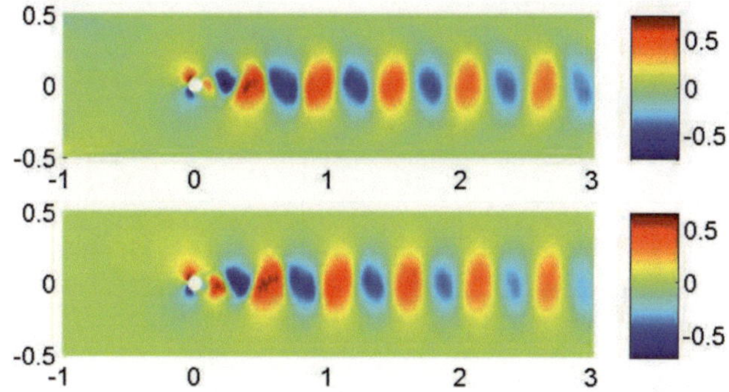

Figure 10.
Re $= 100$. *Above: velocity in y-direction with walls; below: velocity in y-direction without a wall.*

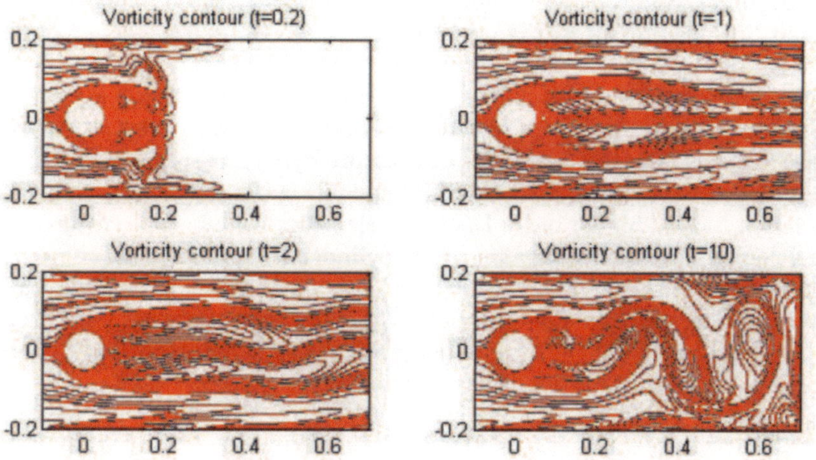

Figure 11.
Vorticity contours of flow past a cylinder without a wall at different time.

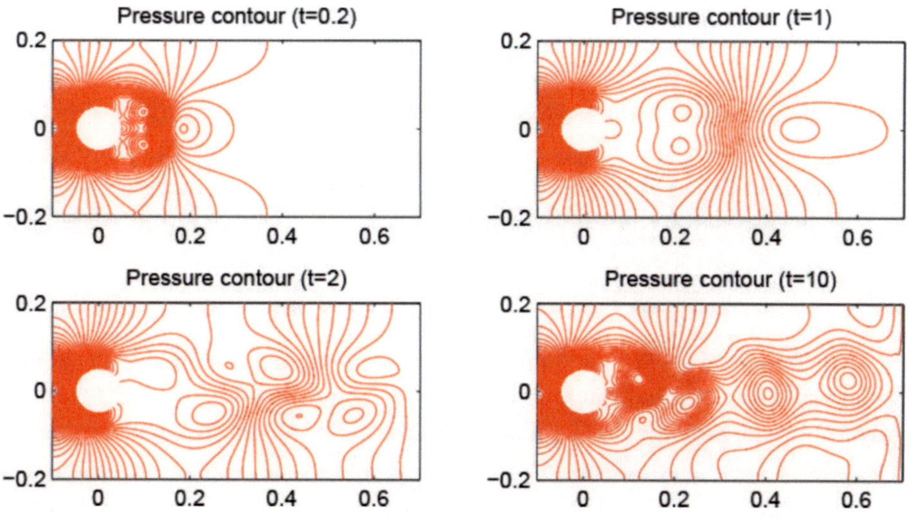

Figure 12.
Pressure contours of flow past a cylinder without a wall at different time.

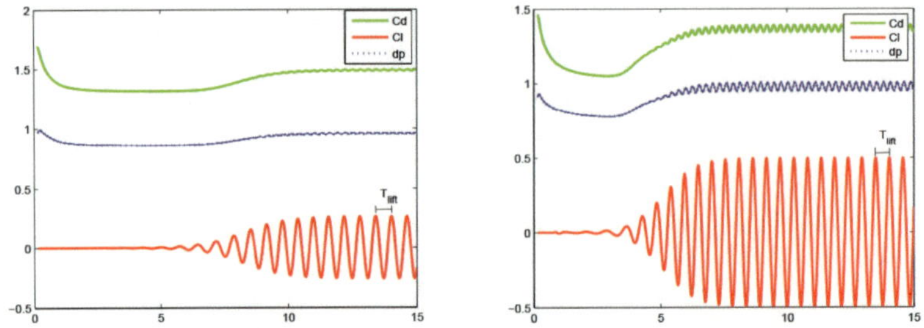

Figure 13.
Drag, lift coefficients and the variation of pressure with time, Re = 80 (left), Re = 140 (right).

aft portion of the cylinder. Given a low total simulation time, the flow field resembles the symmetrical low Reynolds number steady flow. As time progresses however, instabilities are magnified and the upper-lower symmetry increases. Given a total time of beyond $t = 10$, an autonomous and phased locked set of street vortices are generated. Surface pressures (**Figure 13** at $Re = 80, 140$) generated reflect the process, including a steady startup portion and the eventual vortex shedding frequency. Validation at $Re > 41$ requires sufficient time to obtain the unsteady behavior.

To reduce the effect of the boundary layers along the walls, the coefficients of drag and lift as well as the difference of pressure between the leading edge and the trailing edge on the cylinder shall be computed in a larger domain. Then in a domain $\Omega := [-1, 5] \times [-1, 1]$, higher order DG finite elements have been investigated. Based on the velocity $\mathbf{u}_\infty$ and the diameter of the cylinder $D = 0.1$, we will chose different viscosity coefficients $\nu = 1e - 3, 5e - 4, 2.5e - 4$ etc. to simulate flow with different Reynolds numbers, that is, the cases $Re = 100, 200, 400$, respectively. Our interest is the drag coefficient C_d, the lift coefficient C_l on the cylinder and the difference of the pressure between the front and the back of the cylinder

$$d_p = p(t; -0.05, 0) - p(t; 0.05, 0).$$

We use the definition of C_d and C_l given in [4] as follows:

$$C_d = \frac{2}{\rho D u_{\max}^2} \int_S \left(\rho\nu \frac{\partial \mathbf{u}_{\mathbf{t}_S}(t)}{\partial \mathbf{n}} n_y - p(t) n_x \right) ds,$$
$$\text{and} \quad C_l = -\frac{2}{\rho D u_{\max}^2} \int_S \left(\rho\nu \frac{\partial \mathbf{u}_{\mathbf{t}_S}(t)}{\partial \mathbf{n}} n_x + p(t) n_y \right) ds,$$

where $\mathbf{n} = (n_x, n_y)^T$ is the normal vector on the cylinder boundary S directing into Ω, $\mathbf{t}_S = (n_y, -n_x)^T$ the tangential vector and $\mathbf{u}_{\mathbf{t}_S}$ the tangential velocity along S. In the literature, Fey et al. in [27] propose the Strouhal number represented by piecewise linear relationships of the form

$$St = ST^* + \frac{m}{\sqrt{Re}}$$

with different values St^* and m in different shedding regimes of the 3D circular cylinder wake. We originally define the periodic T_{lift} of the lift coefficients (see **Figure 13**) by the periodic $T := \frac{1}{f}$ appearing in the definition of Strouhal number, i.e.,

$$St(Re) = \frac{D}{T_{lift} \mathbf{u}_\infty}, \tag{22}$$

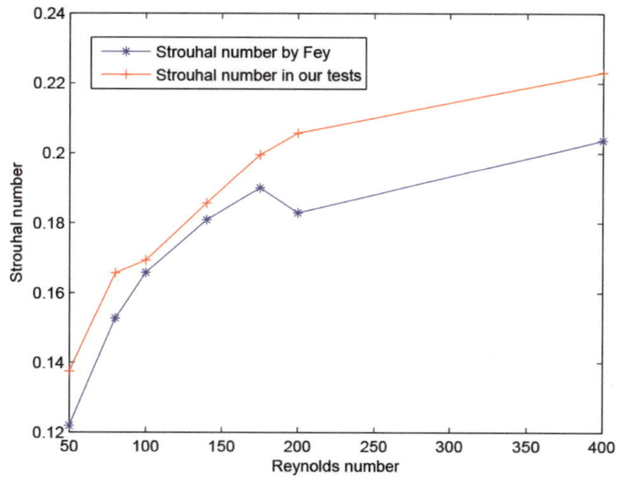

Figure 14.
A comparison of Strouhal-Reynolds-number between our estimate and the linear fit in [27] for $50 \leq Re \leq 400$.

which comes from the classical definition (5). From the evolution of C_d, C_l and d_p as in **Figure 13**, we may find a period T_{lift} of the lift coefficients for different values of Re to calculate Strouhal number by (22). In **Figure 14**, a comparison of Strouhal numbers between the experimental estimates in Fey etc. [5] and our estimates from (22) indicates a behavioral match between the unsteady onset at approximately $Re = 50$ and the beginning of the transition to turbulence at $Re = 180$. Beyond $Re = 180$, the onset of turbulence changes the flow physics by drastically increasing the energy spectrum of the shed vorticity. The transition appears as a marked decrease in the Strouhal number prior to $Re = 200$. As the present SIPG solver does not include a 3D turbulence model, our estimates follow the laminar results into the actual turbulent region. Moreover, **Figure 14** verifies availability of the periodic of vortex street replaced by the periodic of the lift coefficients. There are many results for flow past a cylinder using Reynolds Averaged Navier–Stokes (RANS) method. For large Re up to 2000, the readers are referred to the reference [28] for details, for example, Strouhal number of a cylinder flow could experience a slight fall.

5. Conclusions

A SIPG solver is developed for the incompressible Navier Stokes equations of fluid flow. Two testcases are presented: a lid-driven cavity and a cylinder flow. The DG method produces stable discretizations of the convective operator for high order discretizations on unstructured meshes of simplices, as a requirement for real-world complex geometries. There are still some open problems, such as how the strouhal number of the von Kármán vortex street changes against the Reynolds number after oscillations or noises are added to the incident flow.

Acknowledgements

The author would like to thank INTECHOPEN for their fully sponsor the publication of this chapter and waive the Open Access Publication Fee completely.

Conflict of interest

The author declares no conflict of interest.

Abbreviations

IPDG	Interior penalty discontinuous Galerkin methods
SIPG	Symmetric interior penalty Galerkin method
NIPG	Non-symmetric interior penalty Galerkin method
CFL	Courant–Friedrichs–Lewy

Author details

Lunji Song
School of Mathematics and Statistics, Lanzhou University, Lanzhou, P.R. China

*Address all correspondence to: song@lzu.edu.cn

References

[1] Shen J. On error estimates of the projection methods for the Navier-Stokes equations: Second-order schemes. Math. Comp. 1996; 65(215): 1039–1065.

[2] Temam R. Navier-Stokes Equations: Theory and Numerical Analysis. Providence: AMS Chelsea publishing; 2001.

[3] Castillo P, Cockburn B, Perugia I, and Schötzau D, An a priori error analysis of the local discontinuous Galerkin method for elliptic problems. SIAM J. Numer. Anal. 2000; 38: 1676–1706.

[4] Karniadakis G, Sherwin SJ. Spectral/hp element methods for CFD. New York: Oxford University Press; 2005.

[5] Hesthaven JS. From electrostatics to almost optimal nodal sets for polynomial interpolation in a simplex. SIAM J. Numer. Anal. 1998; 35(2): 655–676.

[6] Hesthaven JS, Teng CH. Stable spectral methods on tetrahedral elements. SIAM J. Sci. Comput. 2000; 21: 2352–2380.

[7] Barlow JB, Rae WH, Pope A. Low-Speed Wind Tunnel Testing. John Wiley; 1999.

[8] Fey U, König M, Eckelmann H. A new Strouhal-Reynolds-number relationship for the circular cylinder in the range $47 \leq Re \leq 2 \times 10^5$. Physics of Fluids. 1998; 10(7): 1547–1549.

[9] Ghia U, Ghia KN, Shin CT. High-Re solutions for incompressible flow using the Navier-Stokes equations and a multigrid method. J. Comput. Phys. 1982; 48: 387–411.

[10] Kazim MNFM, Rasani R, Nuawi MZ, Harun Z, Hau YK, Majid MSA. Analysis of wake region behind bluff body for piezoelectric energy harvester. J. Adv. Res. Fluid Mechan. Therm. Sci. 2019; 55: 249–263.

[11] Song L, O'Neill C. A high-order symmetric interior penalty discontinuous Galerkin scheme to simulate vortex dominated incompressible fluid flow, AIMS Mathematics, 2016; 1(1): 43–63.

[12] Epshteyn Y, Rivière B. Estimation of penalty parameters for symmetric interior penalty Galerkin methods. J. Comput. Appl. Math. 2007; 206: 843–872.

[13] Temam R. Navier-Stokes Equations and Nonlinear Functional Analysis, Volume 66 of CBMS-NSF Regional Conference Series in Applied Mathematics. Philadelphia: SIAM, 2nd edition, 1995.

[14] Drela M. Flight Vehicle Aerodynamics. Boston: MIT Press; 2014.

[15] Goyon O. High-Reynolds number solutions of Navier-Stokes equations using incremental unknowns. Comput. Method. Appl. M. 1996; 130: 319–335.

[16] Schäfer M, Turek S. The benchmark problem 'flow around a cylinder', In Flow Simulation with High-Performance Computers II, Hirschel, E. H.(editor). Notes on Numerical Fluid Mechanics, vol. 52, Vieweg: Braunschweig; 1996. p. 547–566.

[17] Hoerner SF. Fluid-Dynamic Drag, Bakersfield: Hoerner Fluid Dynamics; 1965.

[18] Girault V, Rivière B, Wheeler MF. A splitting method using discontinuous Galerkin for the transient incompressible Navier-Stokes equations. ESAIM: Mathematical

Modelling and Numerical Analysis. 2005; 39(6): 1115–1147.

[19] Girault V, Rivière B, Wheeler MF. A discontinuous Galerkin method with non-overlapping domain decomposition for the Stokes and Navier-Stokes problems. Math. Comp. 2005; 74: 53–84.

[20] Wheeler MF. An elliptic collocation-finite element method with interior penalties. SIAM J. Numer. Anal. 1978; 15: 152–161.

[21] Rivière B, Wheeler MF, Girault V. Improved energy estimates for interior penalty, constrained and discontinuous Galerkin methods for elliptic problems. Part I. Comput. Geosci. 1999; 3: 337–360.

[22] Chorin AJ. On the convergence of discrete approximations to the Navier-Stokes equations. Math. Comp. 1969; 23: 341–353.

[23] Song L, Zhang Z. Polynomial preserving recovery of an over-penalized symmetric interior penalty Galerkin method for elliptic problems. Discrete Contin. Dyn. Syst. – Ser. B 2015; 20(5): 1405–1426.

[24] Song L, Zhang Z. Superconvergence property of an over-penalized discontinuous Galerkin finite element gradient recovery method. J. Comput. Phys. 2015; 299: 1004–1020.

[25] Song L, Gie GM, and Shiue MC. Interior penalty discontinuous Galerkin methods with implicit time-integration techniques for nonlinear parabolic equations. Numer. Methods Partial Differential Equations. 2013; 29(4): 1341–1366.

[26] Foias C, Manley O, Rosa R and Temam R. Turbulence and Navier-Stokes equations. Cambridge University Press; 2001.

[27] Shen J. Hopf bifurcation of the unsteady regularized driven cavity flow. J. Comput. Phys. 1991; 95: 228–245.

[28] Kaya S, Rivière B. A discontinuous subgrid eddy viscosity method for the time-dependent Navier-Stokes equations. SIAM J. Numer. Anal. 2005; 43(4): 1572–1595.

Chapter 5

Discrete Vortex Cylinders Method for Calculating the Helicopter Rotor-Induced Velocity

Evgeny Nikolaev and Maria Nikolaeva

Abstract

A new vortex model of a helicopter rotor with an infinite number of blades is proposed, based on Shaidakov's linear disk theory for calculating inductive speeds at any point in space in the helicopter area. It is proposed to consider the helicopter rotor and the behind vortex column as a system of discrete vortex cylinders. This allows building a matrix of the influence of the vortex system under consideration on any set of points, for example, the calculated points on the rotor itself, on the tail rotor, etc. The model allows calculating inductive velocities at any point near the helicopter using matrix multiplication operation. It is shown that the classical results for the momentum theory remain constant even in the discrete simulation of the helicopter rotor vortex system. The structure of the air flow behind the rotor and the simulation results obtained by the proposed method is compared with the structure of the tip vortices and the results of the blade vortex theory. In addition, the experimental data were compared with the simulation results to verify the correctness of the model under real operating conditions by the helicopter trimming.

Keywords: induced velocity, intensity of circulation, discrete vortex cylinder, influence matrix, Shaidakov's linear disk theory

1. Introduction

The character of the load distribution on the disk rotor vortex theory affects induced velocity. In turn, the inductive flow is the most important factor affecting the value of the inductive losses, as well as forces and moments acting on the helicopter's rotor. Therefore, the efforts of many authors are aimed for creating theories and methods for the simplest way to calculate the induced-velocity field, without which it is impossible to calculate the determination of the aerodynamic characteristics of the rotor.

In conditions of low velocities, induced-velocity field is particularly irregular. This leads to significant changes of aerodynamic forces acting on the blade. The blades begin to oscillate with higher amplitudes, causing significant variable tensions inside blades.

IntechOpen

1.1 Coordinate systems of the vortex cylinder

The properties of a cylindrical vortex surface are considered in detail by Shaidakov [1]. He analytically investigated the properties of a vortex surface that completely covers a beveled vortex surface. Shaidakov studied a vortex surface that starts from a disk plane, has an arbitrary shape in section, and pointed with its one end to infinity. Vortices on the surface are parallel to the base of the cylinder, which lies in the plane of the beginning of the vortex cylinder. When applied to the rotor disk of a helicopter, it is more appropriate to consider not an arbitrary shape of the vortex surface, but a very specific shape, the cross section of which is shown in **Figure 1**. The disk plane is filled with closed contours formed on two sides by arcs of circles and on the other two sides by radial segments. The number of closed contours depends on the number of calculated points along the blade and the number of points along the azimuth. However, the shape of closed contours remains the same.

To study the velocity field caused by a discrete vortex cylinder, we will select two typical sections of it. One section is located at the beginning plane of the vortex cylinder (lies in the plane of the screw disk); the other is parallel to it and intersects the cylinder at an infinite distance from the first plane. The sections of the cylinder with these planes are conventionally designated 1-1 and 2-2, respectively.

It is convenient to calculate induced velocities in the coordinate systems $Oxyz$ ($Ox_1y_1z_1$) shown in **Figure 2**. The vortex cylinder is tilted from the axis $Oz(Oz_1)$ by the inclination angle of the vortex cylinder δ. It is easy to see that the induced-velocity calculated point is always located in a plane parallel to the disk plane, and the projection of the vortex cylinders on this plane is a disk with a radius equal to the radius of the rotor. The origin of the coordinate system $Oxyz$ ($Ox_1y_1z_1$) is always located in the center of this disk.

The right rectangular coordinate system $Oxyz$ is used to record the components of the induced velocity v_x, v_y, v_z. The right-linked coordinate system $Ox_1y_1z_1$ with guide orts $\bar{e}_1, \bar{e}_2, \bar{e}_3$ is used to calculate the components of the induced velocity $v_{x_1}, v_{y_1}, v_{z_1}$. The axis Ox_1 and Oz_1 coincides with the axes Ox and Oz. The axis Oy_1 is

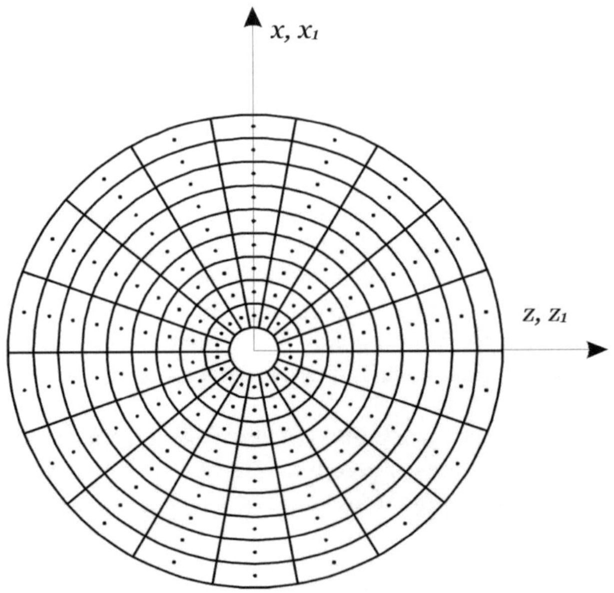

Figure 1.
Scheme for splitting the disk into discrete vortex cylinders.

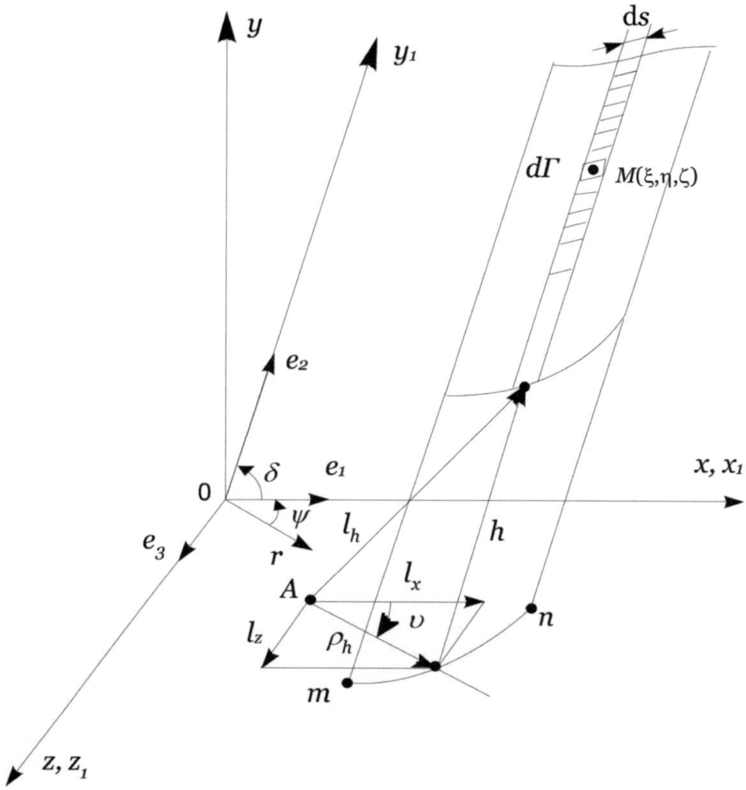

Figure 2.
Coordinate system of the vortex cylinder with arbitrary form [2].

the axis of the cylinder and is inclined to the axis Ox_1 at an angle δ. In both systems, the base plane Oxz (Ox_1z_1) is parallel to the plane of the vortex cylinder and contains the induced-velocity calculated point A. The left skew coordinate $Ory\psi$ system is derived from the $Ox_1y_1z_1$. The left oblique cylindrical coordinate system $Ory\psi$ is obtained by moving the origin to a point A. It is used to bring contour integrals to a form that is convenient for integration.

The last two axis systems are most convenient for deriving equations of the vortex cylinder surface. It is enough to know the distance h along the axis Oy_1 from the base plane of the vortex cylinder and the equation of the projection of the cylinder base on the base plane.

The ratio of induced-velocity's components in systems $Oxyz$ and $Ox_1y_1z_1$ is determined by the following dependencies

$$\begin{aligned} v_x &= v_{x_1} + v_{y_1}\cos\delta; \\ v_y &= v_{y_1}\sin\delta; \\ v_z &= v_{z_1}. \end{aligned} \tag{1}$$

These component interdependences are true for any vector [3].

In the accepted coordinate system $Ox_1y_1z_1$ for an arbitrary pair of vectors $\overline{a}, \overline{b}$ we find a vector product $\overline{c}$ and a scalar product ab

$$c_{x_1} = \frac{1}{\sin\delta}\left(\begin{vmatrix} a_{y_1} & a_{z_1} \\ b_{y_1} & b_{z_1} \end{vmatrix} + \cos\delta\begin{vmatrix} a_{x_1} & a_{z_1} \\ b_{x_1} & b_{z_1} \end{vmatrix}\right); \tag{2}$$

$$c_{y_1} = \frac{1}{\sin\delta}\left(\begin{vmatrix} a_{x_1} & a_{z_1} \\ b_{x_1} & b_{z_1} \end{vmatrix} + \cos\delta\begin{vmatrix} a_{y_1} & a_{z_1} \\ b_{y_1} & b_{z_1} \end{vmatrix}\right); \quad (3)$$

$$c_{z_1} = \sin\delta\begin{vmatrix} a_{x_1} & a_{y_1} \\ b_{x_1} & b_{y_1} \end{vmatrix}; \quad (4)$$

$$ab = a_{x_1}b_{x_1} + a_{y_1}b_{y_1} + a_{z_1}b_{z_1} + \left(a_{x_1}b_{y_1} + a_{y_1}b_{z_1}\right)\cos\delta. \quad (5)$$

The modulus of the vector $\bar{a}$ is calculated by the formula (Eq. (5))

$$a = \sqrt{aa} = \sqrt{a_{x_1}^2 + a_{y_1}^2 + a_{z_1}^2 + 2a_{x_1}a_{y_1}\cos\delta}. \quad (6)$$

The projection of the vector $\bar{c}$ on the axis of the cylinder is denoted c_0. Then, using (Eq. (6)), we will have

$$c_0 = \bar{c}\cdot\bar{e}_2 = c_{y_1} + c_{x_1}\cos\delta. \quad (7)$$

1.2 Components of the induced-velocity vector at any point in area around the rotor

Let us consider a certain part of a cylindrical vortex surface, with the beginning at the plane of the disk, bounded on two sides by two generatrices and leaving the other side to infinity (**Figure 2**). The beginning of the generatrices is denoted by the point's m and n. Select a vortex element ds with circulation $d\Gamma$ at any point $M(\xi,\eta,\zeta)$ in the vortex surface:

$$d\Gamma = \gamma d\eta, \quad (8)$$

where γ is the running circulation in the direction of the cylinder generatrices.

We will calculate the induced velocities from this element at the point $A(x_1, 0, z_1)$ using the formula of Biot-Savart

$$d^2\bar{v} = \frac{d\Gamma}{4\pi}\frac{d\bar{s}\times\bar{l}}{|l|^3}, \quad (9)$$

where $\bar{l}$ is connecting the point A to the point M vector; $d\bar{s}$ is a vortex element represented as a vector. The sign of circulation $d\Gamma$ (or γ) is determined by the direction of the vector $d\bar{s}$. A positive value corresponds to a positive direction in the accepted coordinate system. When calculating the inductive effect from the vortex surface, the contour integral is calculated along the vortex lines. The positive direction of the contour traversal in the right coordinate system corresponds to the right rotation, in the left coordinate system—to the left.

We express the vectors included in the formula (Eq. (8)) in terms of affine coordinates

$$d\bar{s} = d\xi\bar{e}_1 + d\zeta\bar{e}_3; \quad (10)$$

$$\bar{l} = (\xi - x_1)\bar{e}_1 + \eta\bar{e}_2 + (\zeta - z_1)\bar{e}_3. \quad (11)$$

The modulus of the vector $\bar{l}$ is defined by the formula (Eq. (6))

$$l = \sqrt{(\xi - x_1^2) + \eta^2 + (\zeta - z_1)^2 + 2(\xi - x_1)\eta\cos\delta}. \quad (12)$$

Calculate the components of the inductive velocity vectors from the vortex path ds. For this purpose, it is necessary to define expressions for projections of the vector product $d\bar{s} \times \bar{l}$ on the coordinate axis.

$$d\bar{s} \times \bar{l}\Big|_x = \frac{1}{\sin\delta}\left(\begin{vmatrix} 0 & d\zeta \\ \eta & \zeta - z_1 \end{vmatrix} + \cos\delta \begin{vmatrix} d\xi & d\zeta \\ \xi - x_1 & \zeta - z_1 \end{vmatrix}\right)\Bigg|_x; \tag{13}$$

$$d\bar{s} \times \bar{l}\Big|_y = \frac{1}{\sin\delta}\left(\begin{vmatrix} d\xi & d\zeta \\ \xi - x_1 & \zeta - z_1 \end{vmatrix} + \cos\delta \begin{vmatrix} 0 & d\zeta \\ \eta & \zeta - z_1 \end{vmatrix}\right)\Bigg|_y; \tag{14}$$

$$d\bar{s} \times \bar{l}\Big|_z = \sin\delta\left(\begin{vmatrix} d\xi & d\zeta \\ \xi - x_1 & \zeta - z_1 \end{vmatrix}\right)\Bigg|_z. \tag{15}$$

After describing the determinants, we will have the following expressions for induced velocities

$$d^2 v_{x_1} = \frac{\gamma}{4\pi} \frac{d\eta}{|l|^3} \frac{1}{\sin\delta} [\cos\delta(\zeta - z_1)d\xi - \cos\delta(\xi - x_1)d\zeta - \eta d\zeta]; \tag{16}$$

$$d^2 v_{y_1} = \frac{\gamma}{4\pi} \frac{d\eta}{|l|^3} \frac{1}{\sin\delta} [(\zeta - z_1)d\xi - (\xi - x_1)d\zeta - \cos\delta\eta d\zeta]; \tag{17}$$

$$d^2 v_{z_1} = \frac{\gamma}{4\pi} \frac{d\eta}{|l|^3} \eta \sin\delta d\xi. \tag{18}$$

Let us perform integration of formulas (Eqs. (16)–(18)) along the cylinder creators

$$dv_{x_1} = \frac{\gamma}{4\pi} \frac{1}{\sin\delta} \int_h^\infty \frac{d\eta}{|l|^3} [\cos\delta(\zeta - z_1)d\xi - \cos\delta(\xi - x_1)d\zeta - \eta d\zeta]; \tag{19}$$

$$dv_{y_1} = \frac{\gamma}{4\pi} \frac{1}{\sin\delta} \int_h^\infty \frac{d\eta}{|l|^3} [(\zeta - z_1)d\xi - (\xi - x_1)d\zeta - \cos\delta\eta d\zeta]; \tag{20}$$

$$dv_{z_1} = \frac{\gamma}{4\pi} \sin\delta \int_h^\infty \frac{d\eta}{|l|^3} \eta d\xi. \tag{21}$$

After the conversion, we will have

$$dv_{x_1} = \frac{\gamma}{4\pi} \frac{1}{\sin\delta} \left[\cos\delta(\zeta - z_1)d\xi \int_h^\infty \frac{d\eta}{|l|^3} - \cos\delta(\xi - x_1)d\zeta \int_h^\infty \frac{d\eta}{|l|^3} - d\zeta \int_h^\infty \frac{\eta d\eta}{|l|^3}\right] \tag{22}$$

$$dv_{y_1} = \frac{\gamma}{4\pi} \frac{1}{\sin\delta} \left[(\zeta - z_1)d\xi \int_h^\infty \frac{d\eta}{|l|^3} - (\xi - x_1)d\zeta \int_h^\infty \frac{d\eta}{|l|^3} - \cos\delta d\zeta \int_h^\infty \frac{\eta d\eta}{|l|^3}\right] \tag{23}$$

$$dv_{z_1} = \frac{\gamma}{4\pi} \sin\delta d\xi \int_h^\infty \frac{\eta d\eta}{|l|^3} \tag{24}$$

In formulas (Eq. (22)–(24)) we will denote by

$$J_1 = \int_h^\infty \frac{d\eta}{|l|^3} = \int_h^\infty \frac{d\eta}{\left|(\xi - x_1^2) + \eta^2 + (\zeta - z_1)^2 + 2(\xi - x_1)\eta\cos\delta\right|^{\frac{3}{2}}};$$
$$J_2 = \int_h^\infty \frac{\eta d\eta}{|l|^3} = \int_h^\infty \frac{\eta d\eta}{\left|(\xi - x_1^2) + \eta^2 + (\zeta - z_1)^2 + 2(\xi - x_1)\eta\cos\delta\right|^{\frac{3}{2}}} \tag{25}$$

Using the following formulas $\rho_h^2 = (\xi - x_1)^2 + (\zeta - z_1)^2$ and $\xi - x_1 = \rho_h \cos\vartheta$ converting integrals (Eq. (25))

$$J_1 = \int_h^\infty \frac{d\eta}{|l|^3} = \int_h^\infty \frac{d\eta}{\left|\rho_h^2 + \eta^2 + 2\rho_h\cos\vartheta\cos\delta\eta\right|^{\frac{3}{2}}};$$
$$J_2 = \int_h^\infty \frac{\eta d\eta}{|l|^3} = \int_h^\infty \frac{\eta d\eta}{\left|\rho_h^2 + \eta^2 + 2\rho_h\cos\vartheta\cos\delta\eta\right|^{\frac{3}{2}}}. \tag{26}$$

Integrals (Eq. (26)) are table integrals and can be denoted by the following:

$$J_1 = \frac{1}{\rho_h^2(1 - \cos^2\delta\cos^2\vartheta)}\left(1 - \frac{h + \rho_h\cos\delta\cos\vartheta}{l_h}\right); \tag{27}$$

$$J_2 = \frac{1}{\rho_h(1 - \cos^2\delta\cos^2\vartheta)}\left(\frac{\rho_h + h\cos\delta\cos\vartheta}{l_h} - \cos\delta\cos\vartheta\right), \tag{28}$$

where $l_h = \sqrt{\rho_h^2 + h^2 + 2\rho_h h\cos\delta}$.

Using the already known relationships $\cos\vartheta = (\xi - x_1)/\sqrt{(\xi - x_1)^2 + (\zeta - z_1)^2}$ converting the received formulas

$$J_1 = \frac{1}{(\zeta - z_1)^2 + (1 - \cos^2\delta)(\xi - x_1)^2}\left(1 - \frac{h + \cos\delta(\xi - x_1)}{l_h}\right); \tag{29}$$

$$J_2 = \frac{1}{(\zeta - z_1)^2 + (1 - \cos^2\delta)(\xi - x_1)^2} \left(\frac{(\xi - x_1)^2 + (\zeta - z_1)^2 + h\cos\delta(\xi - x_1)}{l_h} - \cos\delta(\xi - x_1)\right), \tag{30}$$

where $l_h = \sqrt{(\xi - x_1)^2 + (\zeta - z_1)^2 + h^2 + 2h\cos\delta\sqrt{(\xi - x_1)^2 + (\zeta - z_1)^2}}$.

Integrals were obtained for the first time by Shaidakov [3] and are referred to in this chapter as Shaidakov's integrals.

Now it is possible to get expressions for differentials of induced velocity's components on the axis of a coordinate system $Ox_1y_1z_1$

$$dv_{x_1} = \frac{\gamma}{4\pi}\frac{1}{\sin\delta}\{(\zeta - z_1)J_1\cos\delta d\xi - [(\xi - x_1)J_1\cos\delta + J_2]d\zeta\} \tag{31}$$

$$dv_{y_1} = \frac{\gamma}{4\pi}\frac{1}{\sin\delta}\{(\zeta - z_1)J_1 d\xi - [(\xi - x_1)J_1 + \cos\delta J_2]d\zeta\} \tag{32}$$

$$dv_{z_1} = \frac{\gamma}{4\pi} \sin \delta J_1 d\xi \tag{33}$$

We get the formula for the component of the induced velocity directed along the vortex cylinder generatrices. We use the expression (Eq. (7))

$$\begin{aligned} dv_0 = &\frac{\gamma}{4\pi} \frac{1}{\sin \delta} [(\zeta - z_1) J_1 d\xi - (\xi - x_1) J_1 d\zeta - \cos \delta J_2 d\zeta] + \\ &+ \frac{\gamma}{4\pi} \frac{1}{\sin \delta} [\cos \delta (\zeta - z_1) J_1 d\xi - \cos \delta (\xi - x_1) J_1 d\zeta - J_2 d\zeta] \cos \delta. \end{aligned} \tag{34}$$

After simple transformations we get

$$dv_0 = \frac{\gamma}{4\pi} \frac{1 - \cos^2 \delta}{\sin \delta} [(\xi - x_1) J_1 d\zeta - (\zeta - z_1) J_1 d\xi]. \tag{35}$$

To calculate the velocity from a limited width vortex cylindrical surface, we need to take the contour integral from the projection of this surface on the base plane (the integral along the length of the arc s).

$$v_{x_1} = \int_s dv_{x_1}; v_{y_1} = \int_s dv_{y_1}; v_{z_1} = \int_s dv_{z_1}; v_0 = \int_s dv_0. \tag{36}$$

2. Discrete vortex cylinders method

The method of discrete vortex cylinders is based on the linear disk theory of Shaidakov described above. It allows you to calculate the induced velocity from the rotor at any point in the space around the rotor. Consider a main rotor with an infinite number of blades [2]. Imagine a vortex system that descends from the rotor in the form of a vortex column, starting at the plane of the disk and going to infinity. The vortex column is supported by a circle with a radius equal to the radius of the rotor. The angle of vortex column inclination to the disk plane depends on the helicopter forward flight speed and the thrust of the rotor. It is calculated using the Shaidakov formula for the angle of inclination of the vortex column.

Each partial volume of the vortex column can be considered as an elementary column of dipoles with a constant density of circulation. Alternatively consider it as an elementary vortex cylinder of arbitrary shape with a linear circulation of closed vortices γ along the generatrix.

In case of the beveled cylinder filling with dipoles, to calculate the inductive velocity from the entire vortex column, it is necessary to make integral sums from the n number final volumes at $n \to \infty$ for the limit case. In this case, the area of the base of the cylinder is divided into n number areas $d\sigma_1, d\sigma_2, \ldots, d\sigma_n$. The area of the cylinder base is divided into finite regions, and the entire vortex column is divided into infinite volumes.

When filling a vortex column with vortex cylinders, the area of the base is filled with closed contours of a specific shape (**Figure 1**), and the column is entirely filled with vortex cylinders of linear circulation γ along the generatrix. The generatrices of the vortex cylinders are parallel to the axis of the vortex column and inclined at an angle δ to the plane of the disk (**Figure 3**).

We propose to consider a vortex column as a collection of a finite number of vortex cylinders resting on the plane of the rotor disk. The plane of the disk is filled with closed vortex contours on two sides by arcs of circles and on the other two sides by radial segments (**Figure 4**). When the disk is split in this way the size of the contours will depend on the number of calculated points along the blade radius and

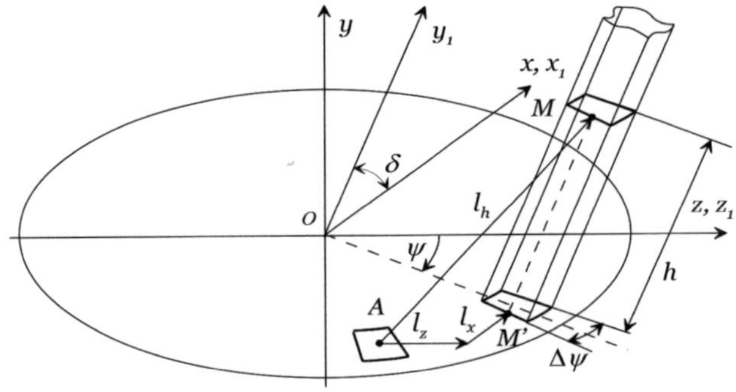

Figure 3.
Coordinate system of a discrete vortex cylinder.

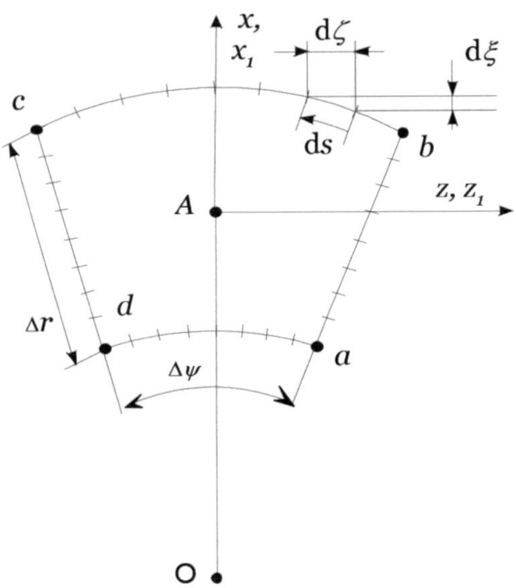

Figure 4.
Scheme for calculating the influence function from a vortex cylinder (the contour integral).

along the circumference of the rotor disk. In this case, the induced velocities will be calculated at points located at the vortex cylinder's axis, and outside the vortex column at any point other than the vortex cylinders surface. This avoids computational difficulties when calculating contour integrals on the surface of cylindrical columns. This point is indicated by a letter A [3]. We will also follow to this designation.

To calculate the components of the inductive velocity at point A by the method of discrete vortex cylinders we will use Shaidakov's formulas (Eqs. (31)–(33), (35)), derived for a discrete vortex cylinder (**Figure 4**).

We assume that the circulation along the contour and along the generatrix of each discrete vortex cylinder is constant. Then in formulas (Eq. (36)) it is possible to take the linear circulation γ as an integral and calculate the induced velocities from the vortex cylinder of the unit circulation, integrating along four segments of the contour (**Figure 4**).

After the obvious transformations, we will have:

$$v_{x1} = \frac{\gamma}{4\pi}\frac{1}{\sin\delta}\int_s\left[\left(l_z\frac{d\xi}{ds} - l_x\frac{d\zeta}{ds}\right)J_1\cos\delta - J_2\frac{d\zeta}{ds}\right]ds \tag{37}$$

$$v_{y1}=\frac{\gamma}{4\pi}\frac{1}{\sin\delta}\int\limits_{s}\left[\left(l_z\frac{d\xi}{ds}-l_x\frac{d\zeta}{ds}\right)J_1-J_2\frac{d\zeta}{ds}\cos\delta\right]ds \tag{38}$$

$$v_{z1}=\frac{\gamma}{4\pi}\sin\delta\int\limits_{s}J_2\frac{d\xi}{ds}ds \tag{39}$$

$$v_0=\frac{\gamma}{4\pi}\frac{1-\cos^2\delta}{\sin\delta}\int\limits_{s}\left(l_x\frac{d\zeta}{ds}-l_z\frac{d\xi}{ds}\right)J_1ds, \tag{40}$$

where $l_z=(\zeta-z_1), l_x=(\xi-x_1)$ (**Figure 2**).

The derivatives $\frac{d\xi}{ds}$, $\frac{d\zeta}{ds}$ are the cosine and sine of the angle of inclination of the arc element ds to the axis Ox_1. Their values do not depend on the position of the point A, but only on the direction of the vector $d\bar{s}$ in the base coordinate system (**Figure 4**). This angle is denoted by

$$\frac{d\xi}{ds}=\cos\tau,\quad \frac{d\zeta}{ds}=\sin\tau \tag{41}$$

Finally, we will have

$$v_{x1}=\frac{\gamma}{4\pi}\frac{1}{\sin\delta}\int\limits_{s}[(l_z\cos\tau-l_x\sin\tau)J_1\cos\tau-J_2\sin\tau]\,ds \tag{42}$$

$$v_{y1}=\frac{\gamma}{4\pi}\frac{1}{\sin\delta}\int\limits_{s}[J_1(l_z\cos\tau-l_x\sin\tau)-J_2\sin\tau\cos\delta]\,ds \tag{43}$$

$$v_{z1}=\frac{\gamma}{4\pi}\sin\delta\int\limits_{s}J_2\cos\tau ds \tag{44}$$

$$v_0=\frac{\gamma}{4\pi}\frac{1-\cos^2\delta}{\sin\delta}\int\limits_{s}J_1(l_x\sin\tau-l_z\cos\tau)\,ds \tag{45}$$

To calculate integrals along the contour, we will use integrating matrices [3]. For convenience, we will calculate the closed loop integral as the sum of integrals $v=\int_{(ab)}dv+\int_{(bc)}dv+\int_{(cd)}dv+\int_{(da)}dv$ over four contours. On segments ab and cd the angle τ is constant along the contour and is equal to the azimuthal angles ψ_{ab} and $\psi_{cd}+\pi$, respectively. On segments bc and da, the angle τ changes along the contour length, equal to the angle of inclination of the tangent in the middle of the arc ds to the axis Ox_1. It depends only on the midpoint of the arc ds azimuthal position.

With the proposed natural partition of the disk into discrete vortex cylinders, the problem of calculating inductive velocities in the disk plane is reduced to multiplying the matrix of influence on the column of linear circulations γ at the calculated points along the rotor disk. Create a matrix $[A]$ of dimension $N\times N$. The elements a_{ij} of matrix $[A]$ are the induced velocities are caused at the i calculated point on the axis of the i vortex cylinder by the unit-strength vortex cylinder j

$$\begin{Bmatrix} v_1\\ v_2\\ \dots\\ v_N\end{Bmatrix}=\begin{bmatrix} a_{11} & a_{12} & \dots & a_{1N}\\ a_{21} & a_{22} & \dots & a_{2N}\\ \dots & \dots & \dots & \dots\\ a_{N1} & a_{N2} & \dots & a_{NN}\end{bmatrix}\times\begin{Bmatrix}\gamma_1\\ \gamma_2\\ \dots\\ \gamma_N\end{Bmatrix} \tag{46}$$

In hovering mode, the main diagonal ($i = j$) of the influence matrix $[A]$ in the disk plane is 0.5, and for the influence matrix in a plane far from the disk plane is 1. Non-diagonal elements of the matrix $[A]$ are close to zero.

3. Helicopter trimming

3.1 The aerodynamic and inertia loads on the blades

The aerodynamic load on the blades is calculated from the known equations of the aerodynamics of a helicopter rotor is the same for main and tail rotors of the helicopter for the spatial movements relative to the longitudinal and transverse axis of the helicopter

$$\overline{U}_x = (\overline{r} - \overline{l}_g)\cos\beta_b + \overline{l}_g + \mu\sin\psi_b + (\overline{r} - \overline{l}_g)\sin\beta_b(\overline{\omega}_x\cos\psi_b - \overline{\omega}_x\sin\psi_b) + \overline{v}_x(r, \psi_b)$$

$$\overline{U}_y = (\overline{V}\sin\alpha - \overline{v}_y(r, \psi_b))\cos\beta_b - (\mu\cos\psi_b + \overline{v}_z(r, \psi_b))\sin\beta_b - (\overline{r} - \overline{l}_g)\frac{d\beta_b}{d\psi_b} + (\overline{r} - \overline{l}_g + \overline{l}_g\cos\beta_b)(\overline{\omega}_x\sin\psi_b + \overline{\omega}_z\cos\psi_b) \tag{47}$$

$$\overline{U}_z = (\mu\cos\psi_b + \overline{v}_z(r, \psi_b))\cos\beta_b + [\overline{v}_y(r, \psi_b) + \overline{l}_g(\overline{\omega}_x\sin\psi_b + \overline{\omega}_z\cos\psi_b)]\sin\beta_b.$$

We add inertial loads [4] to the distributed aerodynamic forces in the blade cross section

$$dt_r = (c_L\overline{U}_x + c_D\overline{U}_y)\overline{U}\overline{b}_r d\overline{r} + dJ_{yb}, dq_r = (c_D\overline{U}_x - c_L\overline{U}_y)\overline{U}\overline{b}_r d\overline{r} + dJ_{xb}. \tag{48}$$

As a result, we obtain the blade flap equations. The integration over the length of the blade gives us the equation

$$M_t - \frac{d^2\beta_b}{dt^2}J_g - (\cos\beta_b J_g + l_g S_g)[\Omega^2\sin\beta_b + 2\Omega(\omega_x\cos\psi_b - \omega_y\sin\psi_b)\cos\beta_b] + \left(\frac{d\omega_x}{dt}\sin\psi_b + \frac{d\omega_z}{dt}\cos\psi_b\right)(J_g + \cos\beta_b l_g S_g) - g\cos\beta_b S_g - K_\beta(\cos\beta_b - \cos\beta_k) = 0 \tag{49}$$

With a trimmed helicopter flight, the flapping of the blades can be represented as a Fourier series. This uniquely determined by the blade flaps angular velocity $d\beta_b/dt$ and angular acceleration $d^2\beta_b/dt^2$. The parameters of the helicopter state $\overline{X} = (V_x, V_y, V_z, \omega_x, \omega_y, \omega_z, \Omega, \gamma_h, \vartheta_h)$ are set by the main and tail rotor control ($\theta_0, \theta_{c1}, \theta_{s1}$ and θ_{tp}) and the load computation is reduced to the computation of the Fourier series coefficients $\beta_0, \beta_{c1}, \beta_{s1}, \ldots, \beta_{cn}, \beta_{sn}$ of the blade flaps β_b. The Fourier coefficients are determined from the equation (Eq. (49)) by Newton's method. The number of coefficients in this case should be equal to the number of rotor azimuth steps.

3.2 The fuselage aerodynamic loads

The fuselage aerodynamic loads depend on the angle of attack or the angle of sliding is evaluated by wind tunnel or calculation by CFD-method.

The angle of attack

$$\begin{aligned} X_f &= C_{Df}(\alpha_f) \cdot \rho V^2 S_f/2 \\ Y_f &= C_{Lf}(\alpha_f) \cdot \rho V^2 S_f/2 \\ M_{zf} &= C_{Qf}(\alpha_f) \cdot \rho V^2 S_f L_f/2. \end{aligned} \tag{50}$$

The angle of sliding

$$\begin{aligned} Z_f &= C_{Zf}\left(\beta_f\right) \cdot \rho V^2 S_f/2 \\ M_{xf} &= C_{Mxf}\left(\beta_f\right) \cdot \rho V^2 S_f L_f/2 \\ M_{yf} &= C_{Myf}\left(\beta_f\right) \cdot \rho V^2 S_f L_f/2. \end{aligned} \tag{51}$$

In this case, the coefficients in equations (Eqs. (50) and (51)) are calculated with CFD-method and corrected by the results of the tests in the wind tunnel.

3.3 Trimming model

The helicopter trimming equations derived from the helicopter dynamics equations. Equating to zero the linear and angular accelerations of the helicopter

$$\begin{aligned} (\overline{\varpi} \times \overline{V})M_h &= \overline{R}_{MR} + \overline{R}_{TR} + \overline{R}_f + \overline{G}_h \\ [\overline{\varpi} \times (J\overline{\varpi})]J_h &= \overline{M}_{MR} + \overline{M}_{TR} + \overline{M}_f \end{aligned} \tag{52}$$

The state vector $\overline{X} = \left(V_x, V_y, V_z, \omega_x, \omega_y, \omega_z, \Omega, \gamma_h, \vartheta_h\right)$ is defined as a data source. As a result of the system of equations (Eq. (52)) solving we obtain the vector $\overline{X} = \left\{\gamma, \vartheta, \theta_0, \theta_{c1}, \theta_{s1}, \theta_{tp}\right\}^T$. This vector is calculated by Newton's method.

4. Results

To illustrate the comparison of result, we distribute the induced velocity in the plane of the rotor disk and at a far distance from it (**Figure 5**). This confirms the results of the Momentum theory: the inductive velocities in the plane of the rotor disk are two times less than the inductive velocities at an infinite distance from it [5].

The tip vortices structure of the rotor shown in **Figure 6**. Comparison of the structure under the main rotor with the blade theory results shows an adequate behavior of the vortex surface in modeling.

Introducing the air flow configuration, we can see which areas of the helicopter are influenced by the induced flow and used in the analysis of information corresponding loads.

Figure 7 shows a comparison of the position of blade vortex theory (disorderly line) and disk theory (black line) for horizontal flight. The tip vortex is shown only from one blade, but the influence of all blades is taken into account.

The results of calculating the normal component of induced velocity were compared with experimental data for forward flight from 75 to 180 km/h. Comparison with the experiment gave good results (**Figure 8**). Experimental and calculated data have got an adequate correlation.

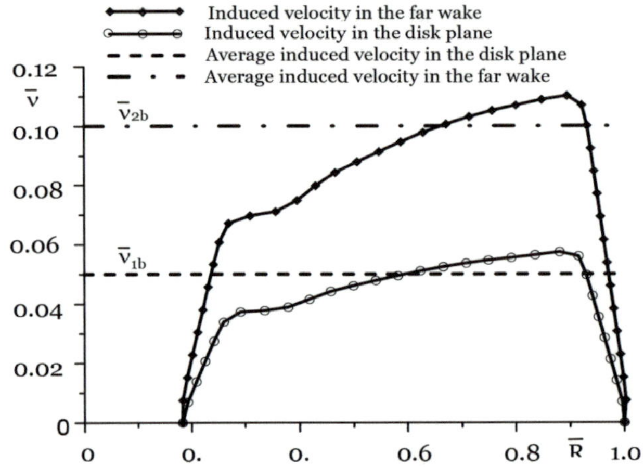

Figure 5.
Normal component of induced velocity on hovering $\alpha = 0°, \mu = 0, C_t = 0.01$.

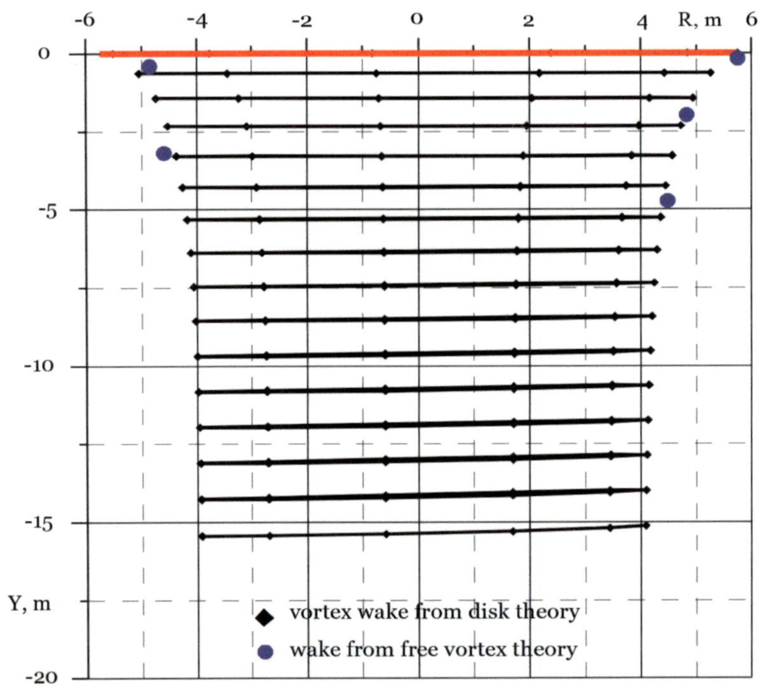

Figure 6.
Tip vortex in hovering.

The application of the described method consists in forming a matrix of influence of the main and tail rotor for any group of points around the rotor. For example, a matrix of influence of the rotor on the fuselage, the matrix of influence of the main rotor on the tail rotor, the matrix of influence of the main rotor on the stabilizer, the matrix effect of the tail rotor on the main rotor, etc., the dimensions of the matrices depend on the selected number of design points on the stabilizer, fuselage, and tail rotor. In this case, these matrices can be formed taking into account the mutual influence of the main and tail rotors.

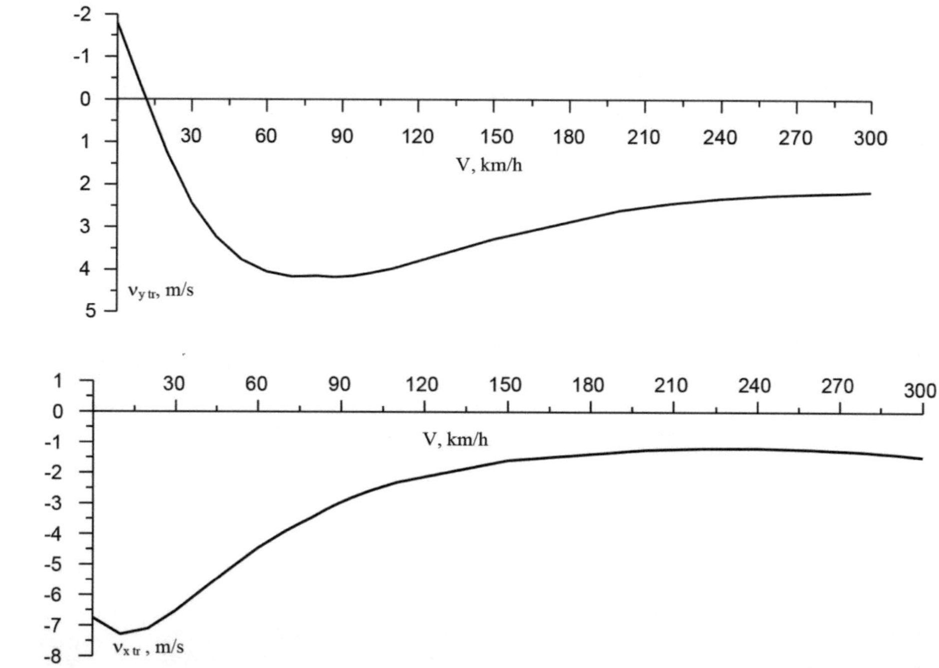

Figure 7.
Induced velocity component in the tail rotor area.

For a single arbitrary point, this will be a matrix-string for the main rotor and a matrix-string for the tail rotor. Multiplying the row by the column of known air circulations on the disk, we calculate the induced velocities at the selected point.

The values of the influence matrix elements depend on only the rotor geometric characteristics and the position of the selected point relative to the rotor and the angle of inclination δ of the vortex cylinder. In the process of calculations, the values of linear circulations and their corresponding induced velocities are refined. Influence matrices must be built in advance for a possible range of angles. During the calculation process, interpolate between the calculated matrices by the necessary values δ.

At different flight speeds, induced velocities have different effects on the tail rotor and stabilizer. In the figures, you can see that induced velocity have a special influence on the low flight speeds mode of $V = 60$ km/h reach 4 m/s (**Figures** 7 and **9**). This is 24% of the flight speed. Therefore it is very important to know and take into account the field of induced velocity's at low flight speeds.

The horizontal and vertical components of induced velocity at a characteristic point in the tail rotor and stabilizer area significantly depend on the flight speed (**Figures** 7 and **9**).

By drawing the configuration of the air flow, we can see which areas of the helicopter are affected by the induced flow. This allows us to correctly configure the loads calculating program.

Figure 10 shows the position of the tip vortex according to the blade theory (disordered line) and the disk theory (black line) for forward flight.

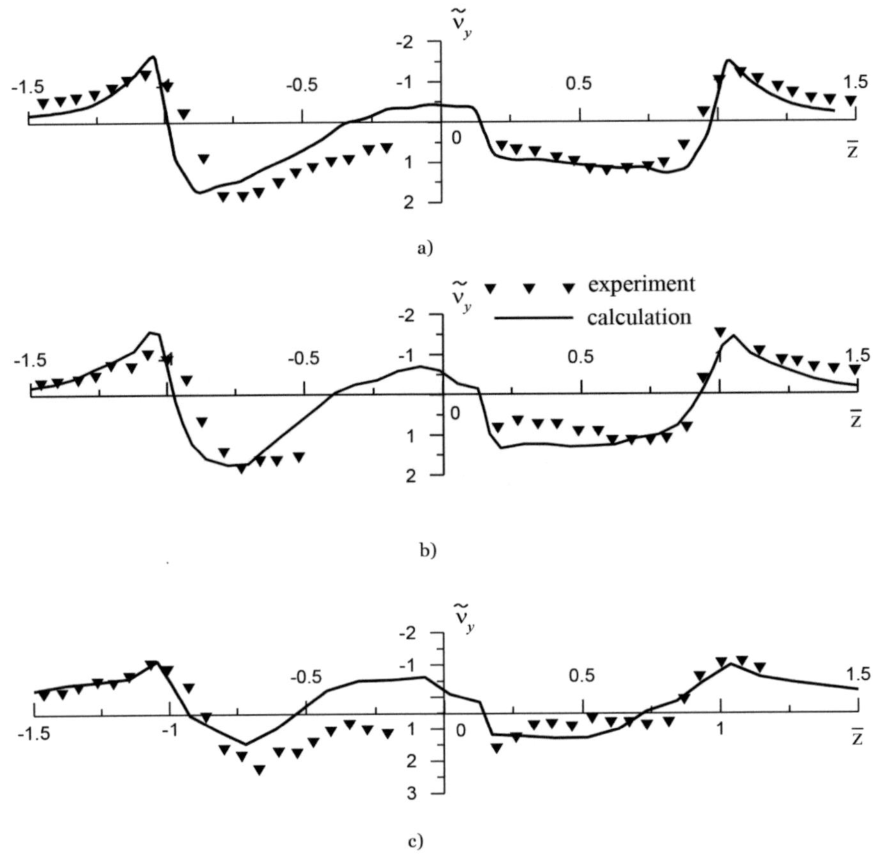

Figure 8.
The normal component of induced velocity in forward flight. (a) $\alpha = -9.2, C_t = 0.01, \mu = 0.095, \bar{x} = 0, \bar{y} = 0.07$, V = 75 km/h. (b) $\alpha = -10.1, C_t = 0.01, \mu = 0.14, \bar{x} = 0, \bar{y} = 0.07$, V = 110 km/h. (c) $\alpha = -9.5, C_t = 0.01, \mu = 0.232, \bar{x} = 0, \bar{y} = 0.07$, V = 185 km/h.

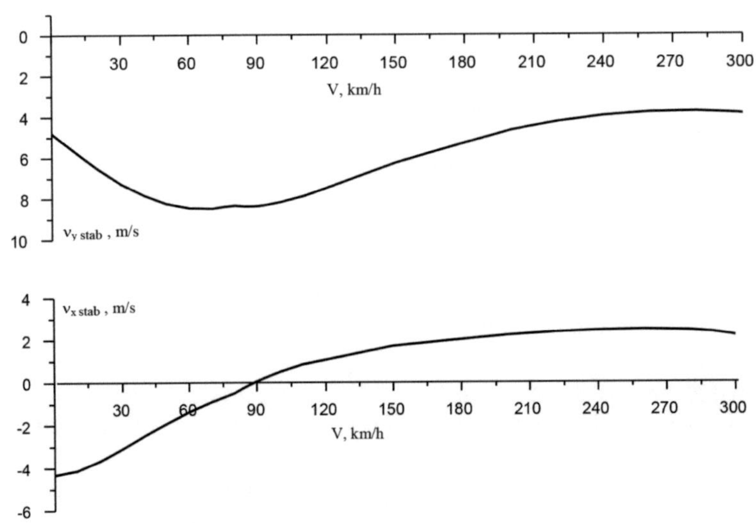

Figure 9.
Components of the induced velocity in the stabilizer area.

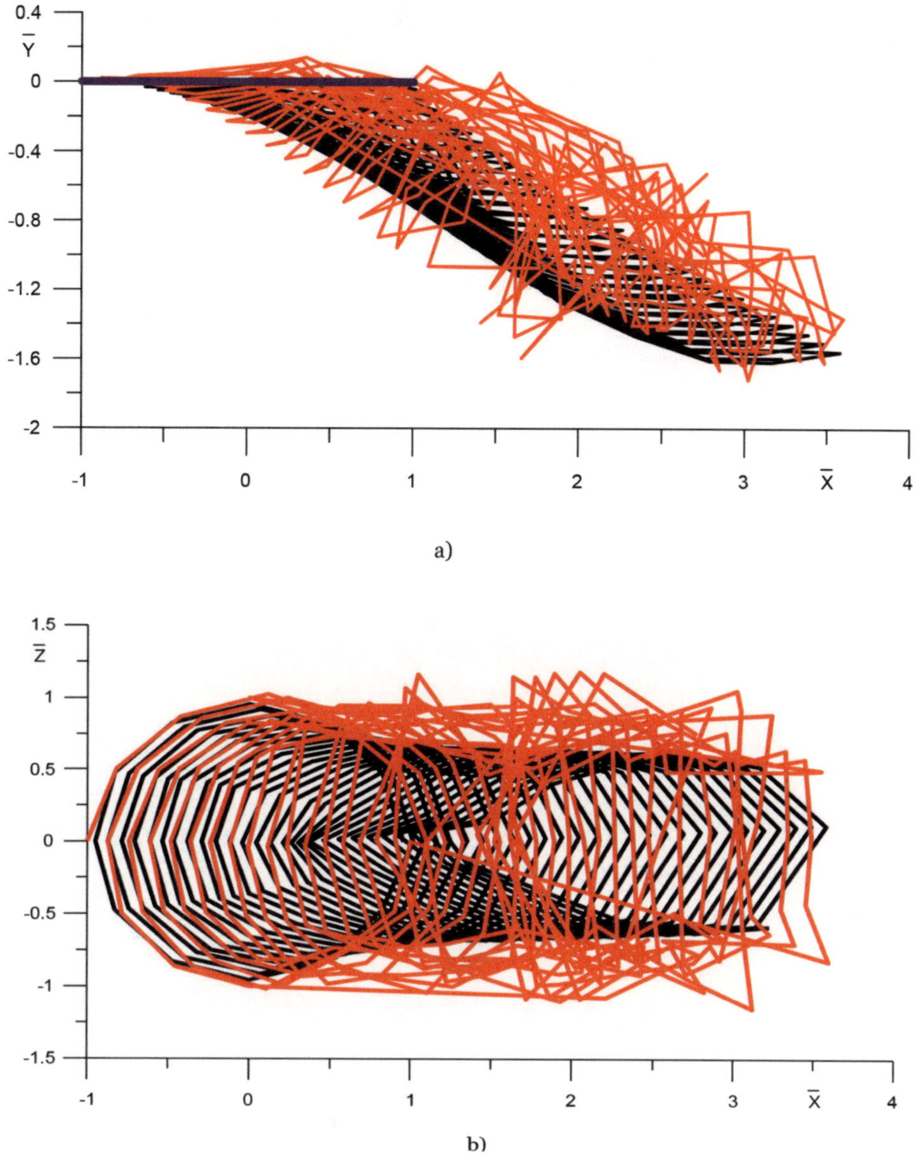

Figure 10.
Tip vortex at low speed (60 km/h, $\mu = 0.076$, $\alpha = -1.23°$, $C_t = 0.01$). (a) side view and (b) overhead view.

Comparison with the blade theory for calculating the position of the tip vortex shows that the method of discrete vortex cylinders gives satisfactory results. The angle of vortex cylinder δ calculated from the disk theory inclination coincides with the angle δ calculated from the blade theory.

As the graphs show, calculating the position of the tip vortex by the method of discrete vortex cylinders gives satisfactory results.

Figure 11 shows the results of comparison of the normal component of the average induced velocity calculated in the cross section along the vane theory [3, 6] and the discrete vortex cylinder method with experiment [2]. Error experiment for flight speed $\mu = 0.095$ is ±15% and the angle of attack $\alpha = -9.2°$.

From these results, we can conclude that the greatest convergence with experimental method of discrete vortex cylinder has got in a cross section $\overline{X} = 0$. In some cases, we have strong disagreement with experiment, but similar to the results obtained by the blade theory.

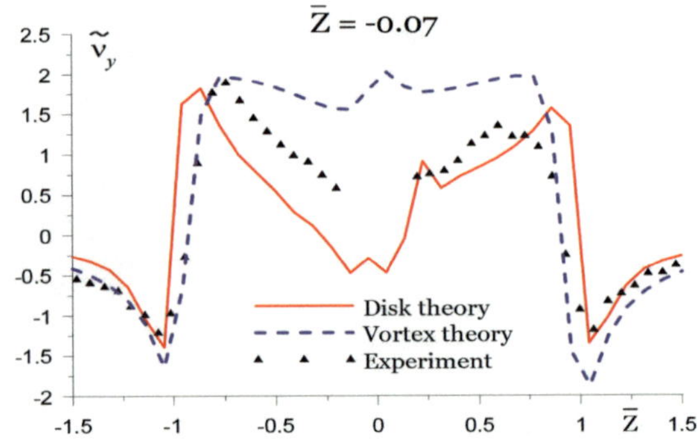

a)

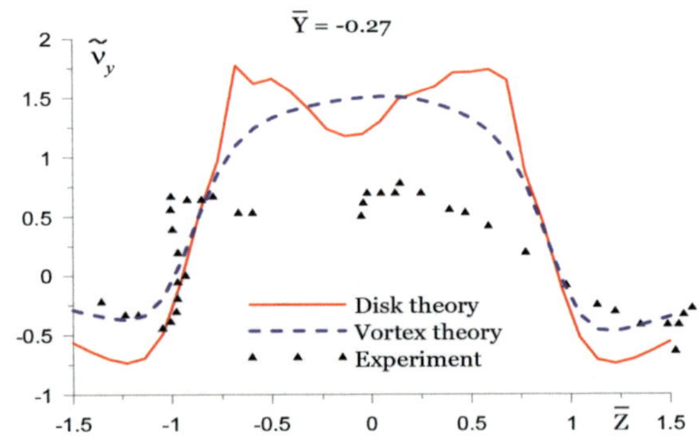

b)

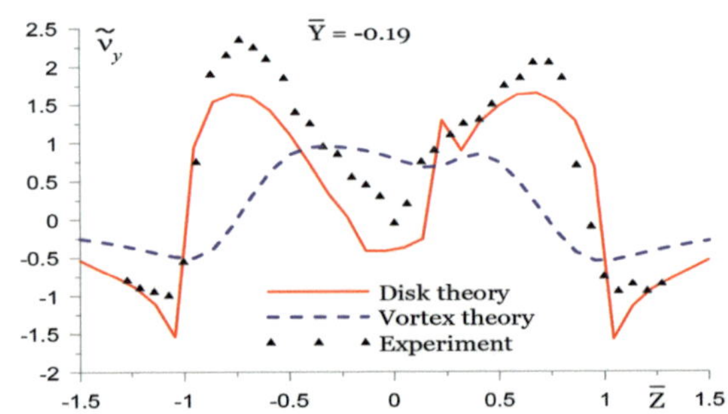

c)

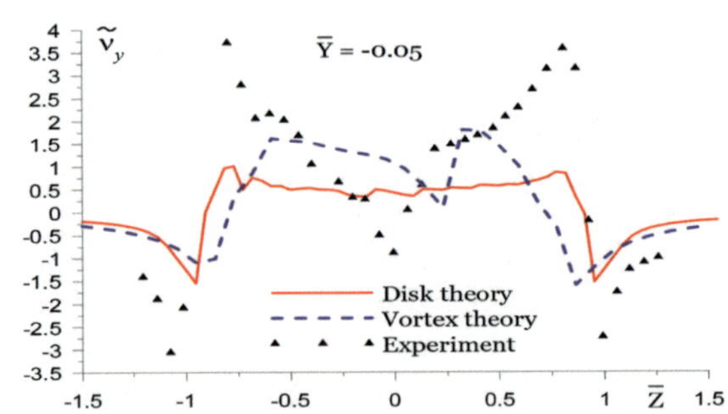

d)

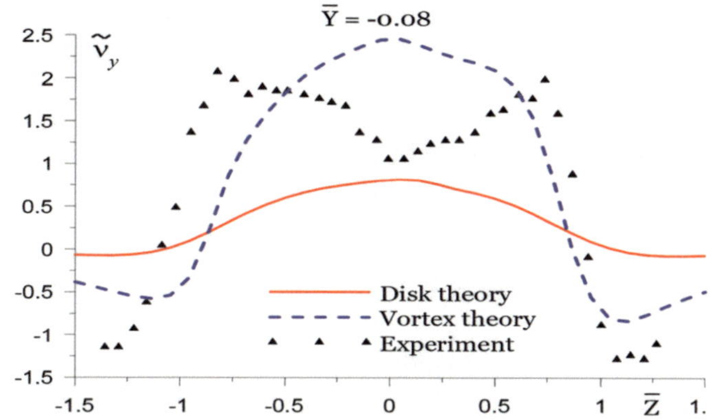

e)

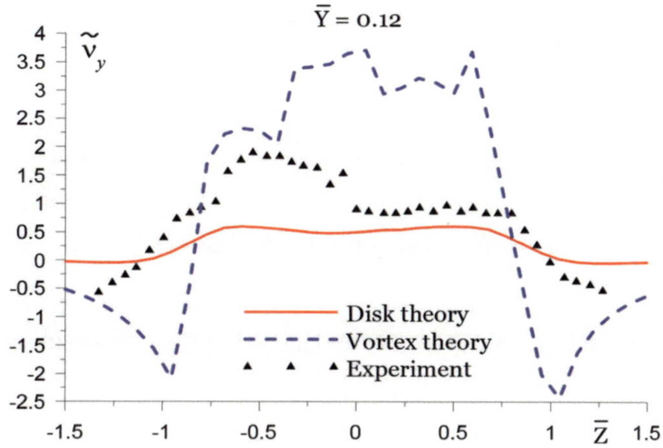

f)

Figure 11.
The normal component of the induced velocity at a distance Y/R under and over of the main rotor ($\mu = 0.095$). (a) $\overline{X} = 0$, (b) $\overline{X} = 0$, (c), $\overline{X} = -0.5$, (d) $\overline{X} = -0.5$, (e) $\overline{X} = 1.7$, and (f) $\overline{X} = -0.5$.

5. Conclusion

In this chapter, the authors have developed a method of discrete vortex cylinders based on the Shaydakov's disk vortex theory. The capabilities of the discrete vortex cylinder method are demonstrated using a helicopter balancing program based on a "semi-rigid" model of the main and tail rotors. Data from numerical calculations are proposed with experimental data from actual flights. Simulations have been conducted. We found the following:

1. The complex procedure for calculating inductive velocities at any point in the space around the rotor was reduced to the procedure of multiplying the row by column. In this case, the row of influence of discrete vortex cylinders coefficients is multiplied by the column of the corresponding vortex cylinders circulations.

2. The use of pre-calculated influence matrices makes it much easier to calculate circulations.

3. The results showed, repeating the results of the classical impulse rotor theory on hover mode in the plane of the rotor and in the wake far behind the rotor.

4. We showed the capabilities of the method in calculating the distribution of inductive speeds in the plane of the rotor and the space outside it, also in calculating the bevel of the flow in the stabilizer and tail rotor regions.

5. We confirmed the fine capabilities of the developed method in calculating inductive speeds at the low-speed horizontal flight mode.

To study the effectiveness of the method of discrete vortex cylinder embedded in the helicopter trimming computer program. The calculations are performed with the original data of the helicopter "ANSAT." The program allows calculating the trimming characteristics of helicopter rotor: aerodynamic performance, the loading of the blades.

The method of discrete vortex cylinders is successfully used for calculating inductive velocities in the aeroelasticity blade model for ANSAT helicopter main rotor loads calculation (for example, see [7]).

6. Future work

1. Modeling of the earth effect using the method of vortices reflected from the earth's surface.

2. Construct a discrete model of the fuselage using the CFD-method to study the effect of main rotor blowing with the tail rotor-induced velocity.

3. Refinement of the discrete cylinders method for calculating the vortex ring regime by the limiting of the vortex cylinder height.

4. Extension of the discrete vortex cylinder method to the nonlinear Shaidakov's vortex theory.

5. Application of the discrete vortex cylinder method to solving helicopter take-off and landing problems.

6. Simulation of vortex formation when landing on a snow-covered and dusty surface.

7. Simulation of helicopter dynamics with rotor blades dynamics using the discrete vortex cylinders method.

Notes

This work was partially supported by the grant in accordance with the Decree of the Government of the Russian Federation No. 220 "on attracting leading scientists to Russian educational institutions" (Agreement No. 11.G34.31.0038).

Notations

$A(x_1, 0, z_1)$	induced-velocity calculated point
a_{ij}	element of influence matrix

b_r	chord at the blade cross section
$C_{Df}, C_{Lf}, C_{Qf}, C_{Zf}, C_{Mxf}, C_{Myf},$	fuselage coefficients
c_L, c_D	section lift and drag coefficients
dJ_{xb}, dJ_{yb}	elementary sectional inertial forces
$d\overline{s}$	vortex element
dt_r, dq_r	elementary sectional aerodynamic forces
h	distance from disk to point A location plane
J_1, J_2	Shaidakov's integrals for a semi-infinite elementary oblique vortex cylinder
J_g	blade rotational inertia, first moment of inertia
J_h	helicopter moment of inertia
M_h, G_h	helicopter mass and weight
$M(\xi, \eta, \zeta)$	vortex elements location point
M_t	aerodynamic flap moment
K_β	flap hinge spring constant
l_g	flap hinge offset
$Oxyz, Ox_1y_1z_1, Ory\psi$	coordinate systems
R	rotor radius
$\overline{R}_{MR}, \overline{R}_{TR}, \overline{R}_f, \overline{M}_{MR}, \overline{M}_{TR}, \overline{M}_f$	main, tile rotors, fuselage forces and moments
r	the cross-sectional radius
S_f, L_f	fuselage area and length
S_g	first moment of inertia
V	flight velocity
v_{1b} , v_{2b}	the induced velocity in hover (at the rotor disk, in the far wake)
$\tilde{V}, \tilde{v}$	related speeds, $\tilde{V} = V/v_{1b}, \tilde{v} = v/v_{1b}$
v_{1b}, v_{2b}	rotor induced velocity in hovering (at the rotor disk, in the far wake)
$\overline{v} : v_x, v_y, v_z$	rotor-induced velocity in $Oxyz$
$\overline{v}_1 : v_{x_1}, v_{y_1}, v_{z_1}$	rotor-induced velocity in $Ox_1y_1z_1$
α_f	fuselage attack angle
$\beta_b, \beta_0, \beta_{c1}, \beta_{s1}, \ldots, \beta_{cn}, \beta_{sn}$	blade flap angle, Fourier series coefficients of β_b
Γ, γ	vortex element circulation, circulation intensity
γ_h, ϑ_h	the helicopter angles of pitch and roll
δ	the inclination angle of the vortex cylinder is calculated by the Shaydakov's transcendental formula ($\delta \neq 0$)

$$\tilde{V} = \frac{2(1 - \sin|\delta|)}{\sqrt{sgn\delta \sin^2\alpha + 2\sin\alpha\cos(\alpha+\delta) + \sin\delta\cos^2(\alpha+\delta)(2 - \sin|\delta|)}}, \; sgn\delta = \frac{\delta}{|\delta|}$$

$\mu = \overline{V}\cos\alpha$	rotor advance ratio
ρ	air density
$\theta_0, \theta_{c1}, \theta_{s1}$	main rotor controls
Ω	rotor rotational speed
ψ	discrete cylinder azimuth
ψ_b	blade azimuth
$\Delta\psi, \Delta r$	discrete vortex dimensions
$\frac{d\xi}{ds} = \cos\tau, \quad \frac{d\zeta}{ds} = \sin\tau$	direction of the vector $d\overline{s}$
$\overline{\omega} : \omega_x, \omega_y, \omega_z$	helicopter angular velocity

Author details

Evgeny Nikolaev* and Maria Nikolaeva
Kazan National Research Technical University named A.N. Tupolev, Kazan City, Russia

*Address all correspondence to: nikjhon@yandex.ru

References

[1] Shaidakov VI. Properties of the oblique cylindrical vortex sheet. In: Helicopter Design. Vol. 381. Moscow: MAI; 1976. pp. 36–57

[2] Nikolaev EI, Nikolaeva MN. Development of the discrete vortex cylinders method for calculating the helicopter rotor induced velocity. Russian Aeronautics. 2016;**59**(1):8–14. DOI: 10.3103/S1068799816010025

[3] Shaidakov VI, Aseev VI. Calculation Algorithms and Programs in Helicopter Design. Moscow: MAI; 1979. p. 48

[4] Braverman AS, Vaintrub AP. Helicopter Dynamics. Limiting Flight Regimes. Moscow: Mashinostroenie; 1988. p. 280

[5] Segalini A, Alfredsson PH. A simplified vortex model of propeller and wind-turbine wakes. Journal of Fluid Mechanics. 2013;**725**:91–116. DOI: 10.1017/jfm.2013.182

[6] Spalart PR. On the simple actuator disk. Journal of Fluid Mechanics. 2003; **494**:399–405. DOI: 10.1017/S0022112003006128

[7] Nikolaev EI. Aeroelasticity of torsional main rotor blades. Russian Aeronautics. 2006;**49**(2):16–23

Chapter 6

Wake-Body Interaction Noise Simulated by the Coupling Method Using CFD and BEM

Masaaki Mori

Abstract

In many engineering applications, obstacles often appear in the wake of obstacles. Vortices shed from an upstream obstacle interact with downstream obstacle and generate noise, for example blades in a turbomachinery, tubes in a heat exchanger, rotating blades like a helicopter and wind turbine and so on. This phenomenon is called wake-body interaction or body-vortex interaction (BVI). The rod-airfoil and airfoil-airfoil configurations are typical models for the wake-body interaction. A rod and an airfoil are immersed upstream of the airfoil. In this chapter, we review the noise mechanism generated by the wake-body interaction and show the numerical results obtained by the coupling method using commercial CFD and acoustic BEM codes. The results show that depending on the spacing between the rod or airfoil and the airfoil, the flow patterns and noise radiation vary. With small spacing, the vortex shedding from the upstream obstacle is suppressed and it results in the suppression of the sound generation. With large spacing, the shear layer or the vortices shed from the upstream obstacle impinge on the downstream obstacle and it results in the large sound generation. The dominant peak frequency of the generated sound varies with increase in the spacing between the two obstacles.

Keywords: rod-airfoil configuration, airfoil-airfoil configuration, wake-body interaction, body-vortex interaction (BVI), coupling method, CFD, BEM

1. Introduction

Flow around a bluff body is one of the basic subjects in fluid mechanics, because it contains not only fundamentally important problems (such as forces acting on the body, transition to turbulence, acoustics, etc.), but also a variety of practical problems (such as structural design of buildings, trains, etc.). In many engineering applications, objects often appear in the wake of an obstacle. When an obstacle is in the wake of another, the flow structure tends to be complex and differs from that of single obstacle.

Flow past two circular or rectangular cylinders in a tandem arrangement is the simplified case of the flow past an array of cylinders and has received increasing attention [1–8]. Liu and Chen [4] and Inoue and Mori [5] showed that two different flow patterns appear depending on the space between the cylinders. When the spacing is small, the shear layer separated from the upstream cylinder does not roll up to form vortices but reattaches to the downstream of the cylinder, and the vortices are shed from the downstream only, and this flow pattern is called *Mode I*. When the spacing is large, the shear layer separated from the upstream cylinder rolls up and forms vortices in front of the downstream of the cylinder, and the roll-up vortices impinge on the downstream cylinder (body-vortex interaction, BVI, wake-body interaction), and this flow pattern is called *Mode II*. Furthermore, Inoue and Mori [5] showed that in *Mode I,* the sound is generated only by the vortex shedding of the downstream cylinder, and in *Mode II,* the sound and strong pressure fluctuation around the downstream cylinder are generated mainly by BVI. The magnitude of the generated sound is much larger in *Mode II* than both *Mode I* and the single cylinder.

Since a *von Karman* vortex street can be regarded as a gust impinging on the obstacle or blade, a rod-airfoil model is another typical one for the study of the body (blade) vortex interaction or wake-body interaction. In this model, a rod is immersed upstream of the blade or airfoil, the wake formed behind the rod interacts with the leading edge of the blade or airfoil. Many studies concerned with the rod-airfoil model have been done both experimentally and numerically [9–21]. Jacob et al. [9] measured the flow field of the rod-NACA0012 airfoil model at the fixed spacing between the rod and the airfoil with varying the inflow velocity, and also measured the far field radiated noise spectra that are generated mainly by the body-wake interaction. They also performed numerical calculations using the Reynolds-averaged Navier-Stokes (RANS) and the large eddy simulation (LES) approaches and compared numerical results with the measured data. Munekata et al. [10–11] measured the flow field of the rod-airfoil model to research the effects of the spacing between the rod (cylinder) and the airfoil and the characteristics of the flow-induced sound generated by the flow around rod-airfoil. They showed that the roll up of the shear layer separated from the upstream rod is suppressed when the spacing between the rod and the airfoil is small, and the interaction between the wake from the rod upstream and the airfoil downstream becomes weak and it results in decreasing the level of the noise radiation. They also showed that the attack angle of the airfoil located downstream affects the characteristics of the flow-induced sound and wake structure at a given spacing between the rod and the airfoil, and the generated sound pressure decreases with the increase of the attack angle of the airfoil. Li et al. [12] performed the experiments of the body-wake interaction noise radiated from the flow around the rod-airfoil model by focusing on the noise control using "air blowing" on the upstream rod and a soft-vane leading edge on the airfoil. Numerical investigations by using the RANS approach have been done by Casalino et al. [13], Jacob et al. [9] and Jiang et al. [14], and those by using the LES approach have been done by Casalino et al. [13], Magagnato et al. [15], Boudet et al. [16], Jacob et al. [9], Greschner et al. [17], Agrawal and Sharma [18], Giret et al. [19], Daude et al. [20], and Jiang et al. [21]. Jiang et al. [21] performed LES simulations of the flow around the rod-airfoil for the inflow velocity U_∞ = 72 m/s and a Reynolds number based on the rod diameter (d) 48,000 (480,000 based on the airfoil chord, c) to clarify the flow patterns, velocity and pressure fluctuations, and noise radiation with varying the spacing between the rod and airfoil. They varied the spacing between the rod and the airfoil, such as L/d = 2, 4, 6, 8, and 10. They showed that when the spacing is small (L/d = 2), the vortex shedding of the rod upstream, the pressure fluctuation, and the noise radiation are suppressed as shown by Munekata et al. [10–11] and when the spacing is large (L/d = 6, 10), the pressure fluctuation, the noise radiation, and the fluid resonant oscillation due to the feedback loop between the rod and the airfoil become stronger.

The airfoil-airfoil model (airfoils in tandem) is also a typical model for the wake-body interaction or body-vortex interaction (BVI). Liu et al. [22–23] performed the measurements to understand the effect of using serration on the aerodynamic and acoustic performance of airfoils in tandem. They studied the wake development, static pressure distributions, and surface pressure fluctuations in detail for a cambered NACA 65-710 airfoil with and without the serration. They showed that the noise is reduced in the case that the upstream airfoil is with the serration.

In this chapter, we simulate the flow around the rod-airfoil model and the noise generated by the wake-body interaction or body-vortex interaction for the cases of L/d = 2 and 10 at a Reynolds number based on the rod diameter (d = 6 mm) 28,800 (288,000 based on the airfoil chord, c) by the coupling method using commercial CFD and acoustic BEM codes in which the acoustic sources are solved in the CFD code and the acoustic field is solved by means of BEM, and compare the results with those obtained by Jacob et al. [9] and Jiang et al. [21]. Then, we simulate the flow around the airfoil-airfoil model (airfoils in tandem) and the noise generation and propagation for the cases of L/c = 0.2, 0.6, and 1 at a Reynolds number based on the airfoil chord (c = 60 mm) 288,000, and compare the results with those for the rod-airfoil model.

2. Analysis model

2.1 Rod-airfoil flow model

A schematic diagram of the flow model is presented in **Figure 1**. The origin is at the leading edge of the airfoil. The coordinates parallel and normal to the free stream are denoted by x and y, respectively. The coordinate in the spanwise direction is denoted by z. The symbol L denotes the spacing between the rod and the airfoil. The lengths are made dimensionless by the rod diameter d and the velocity is scaled by the speed of sound c_∞. The normalized spacing L/d is prescribed to be 2 and 10. The Mach number, M, of a uniform flow is defined by $M = U_\infty/c_\infty$, where U_∞ denotes the velocity of the uniform flow. In this chapter, the Reynolds number is fixed to be Re_d = 28,800 or Re_c = 288,000, and those are based on the rod diameter and the airfoil chord c, respectively. The spanwise length of the rod and the airfoil is $3d$.

2.2 Airfoil-airfoil flow model

A schematic diagram of the flow model is presented in **Figure 2**. The origin is at the leading edge of the downstream airfoil. The normalized spacing L/c is

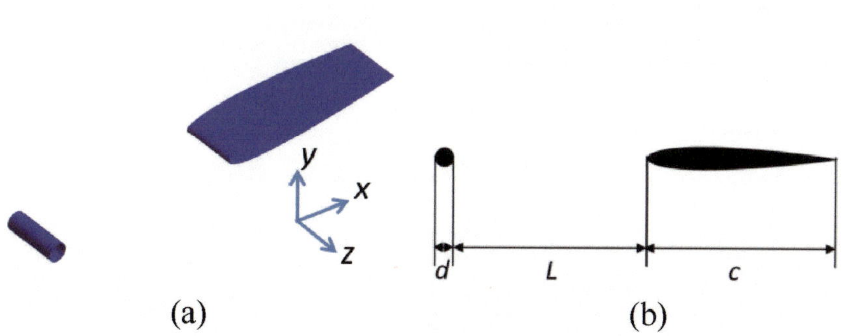

Figure 1.
Schematic diagram of rod-airfoil model. (a) Rod-airfoil model; (b) parameters.

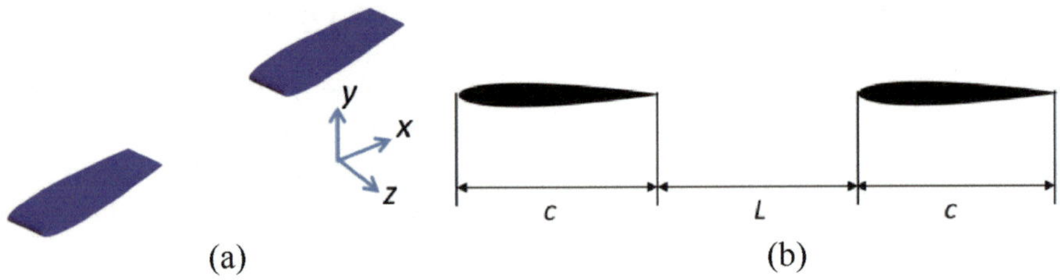

Figure 2.
Schematic diagram of arifoil-airfoil model. (a) Airfoil-airfoil model; (b) parameters.

prescribed to be 0.2, 0.6, and 1.0, and those are equal to L/d = 2.0, 6.0, and 10. In this chapter, the Reynolds number is fixed to be Re_c = 288,000, which is based on the airfoil chord c. The spanwise length of the airfoil and the airfoil is 0.3c, which corresponds to 3d.

3. Numerical procedure

3.1 Transient CFD simulation

Transient flow fields around rod-airfoil and airfoil-airfoil models are simulated at Reynolds number Re_d = 28,800 and Re_c = 288,000; those are based on the rod diameter and the airfoil chord, respectively. The inflow velocity in x direction, temperature, and density are U_∞ = 72 m/s, T_∞ = 300 K, and ρ_∞ = 1.177 kg/m^3, respectively and Mach number is M = 0.207. **Figure 3** shows a computational domain for the rod-airfoil and the airfoil-airfoil models. For these simulations, the three-dimensional computational domain has been applied, as shown in **Figure 3**. Unsteady flow fields are calculated using the commercial CFD code ANSYS Fluent version 2019R1 and its compressible LES (Dynamic Smagorinsky model) calculation features. Steady velocities are imposed on the inflow boundary. Pressure boundary conditions are applied on the top, bottom, and outflow boundaries. Non-reflecting boundary conditions are applied on top, bottom, inflow, and outflow boundaries. No-slip conditions are applied on the walls. At the boundaries at z = 0.009 m and − 0.009 m, the periodic conditions are applied.

The domains of the rod-airfoil simulation model contain 1,169,322 and 1,186,372 hex cells for L/d = 0.2 and 1.0, respectively. The domains of the airfoil-airfoil simulation model contain 1,423,114; 1,435,343; and 1,434,356 hex cells for L/c = 0.2,

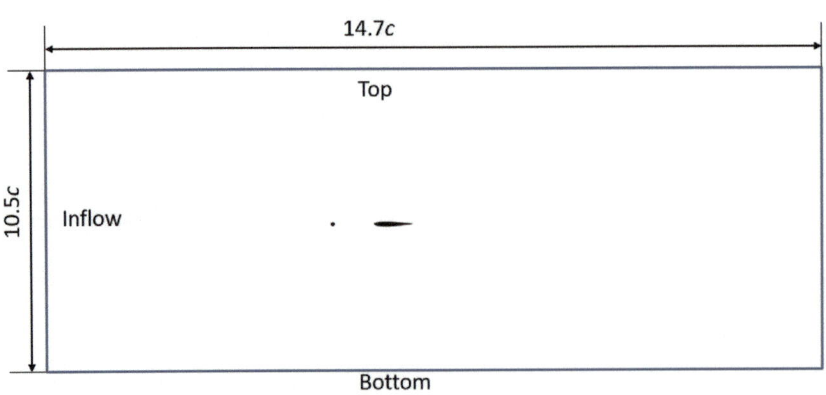

Figure 3.
Computational domain.

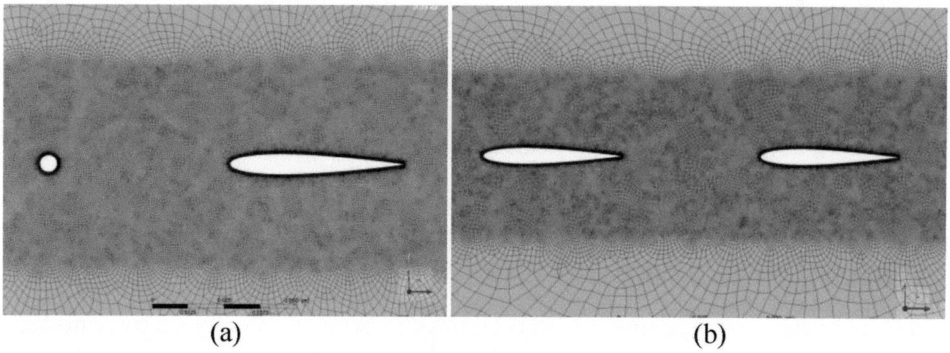

Figure 4.
Computational mesh. (a) Rod-airfoil model; (b) airfoil-airfoil model.

0.6, and 1.0, respectively. **Figure 4(a)** and **(b)** shows the computational meshes near the rod-airfoil and the airfoil-airfoil models, respectively. The cell spacing adjacent to the wall is 0.2 mm (0.033*d*). Steady-state simulations were performed using Spalart-Allmaras (S-A) turbulence model and then used as initial conditions of transient LESs. The transient simulations were performed for 50,000 time steps with a time step size Δt = 2e-6 s.

3.2 Lighthill equation

Lighthill equation [24, 25] in the frequency domain is derived from the equation of continuity and compressible Navier-Stokes equation and as follows:

$$(\nabla^2 + k^2)p = -\frac{\partial^2 T_{lm}}{\partial x_l \partial x_m} \tag{1}$$

where p is the acoustic pressure, k is the wave number, c_∞ is the speed of sound, l and m indicate each direction in the Cartesian coordinates, and v is the flow velocity. T_{lm} is the Lighthill stress tensor and as follows:

$$T_{lm} = \rho v_l v_m + \left(p - c_\infty{}^2 \rho\right)\delta_{lm} - \tau_{lm}, \tag{2}$$

where ρ is the density and is 1.225 kg/m3, δ_{ij} is the Kronecker delta, and τ_{lm} is the viscous stress tensor. For a low-Mach number and high-Reynolds number flow regime, the second and third terms of Eq. (2) are negligible [26–29]. Therefore, the first term is used for the present work.

3.3 Extraction of acoustic source

To convert the acoustic source time histories into the frequency spectra, the discrete Fourier transform (DFT) has been applied. The acoustics sources are extracted from 1250 steps (from t = 0.05 s to 0.1 s) flow field data, the sampling period is 4e-5 s.

3.4 Acoustic simulation

The BEM solver in commercial acoustic simulation package, WAON, is used to solve the acoustic characteristics [30]. In a sound field that satisfies the three-dimensional Helmholtz equation, the Kirchhoff-Helmholtz integral equation [31] for sound pressure is described as follows with respect to a point i and an area S of a surface on a boundary.

$$\frac{1}{2}P(\boldsymbol{r}_i) = \int_{\Gamma}\left(P(\boldsymbol{r}_q)\frac{\partial G(\boldsymbol{r}_i,\boldsymbol{r}_q)}{\partial n_q} - \frac{\partial P(\boldsymbol{r}_q)}{\partial n_q}G(\boldsymbol{r}_i,\boldsymbol{r}_q)\right)dS + p_d(\boldsymbol{r}_q) \tag{3}$$

In this solver, the following simultaneous linear equation is solved:

$$(\boldsymbol{E}+\boldsymbol{B}+\boldsymbol{C})\boldsymbol{p} = j\omega\rho\boldsymbol{A} + \boldsymbol{p_d} \tag{4}$$

Here, p is the acoustic pressure vector, v is the particle velocity vector, and the entries of the influence coefficient matrices are represented as follows:

$$E_{ij} = \frac{1}{2}\delta_{ij}, \tag{5}$$

$$A_{ij} = \int_{\Gamma_1} N_j(\boldsymbol{r}_q)G(\boldsymbol{r}_i,\boldsymbol{r}_q)dS_q, \tag{6}$$

$$B_{ij} = \int_{\Gamma} N_j(\boldsymbol{r}_q)\frac{\partial G(\boldsymbol{r}_i,\boldsymbol{r}_q)}{\partial n_q}dS_q, \tag{7}$$

$$C_{ij} = \frac{jk}{Z_j}\int_{\Gamma_2} N_j(\boldsymbol{r}_q)G(\boldsymbol{r}_i,\boldsymbol{r}_q)dS_q, \tag{8}$$

$$G(\boldsymbol{r}_i,\boldsymbol{r}_q) = \frac{e^{jk|\boldsymbol{r}_p-\boldsymbol{r}_s|}}{4\pi|\boldsymbol{r}_p-\boldsymbol{r}_s|} \tag{9}$$

where δ_{ij} is Kronecker delta, and Γ_1 is a vibration boundary and a part of Γ. Γ is the total boundary. Γ_2 is an absorption boundary and a part of Γ. $\boldsymbol{r_i}$ is the position vector at the node i, $\mathbf{r_q}$ is the position vector of the source point q, and N_j is the interpolation function of the node j. z_j is the acoustic impedance ratio at the node j. G is the fundamental solution of a three-dimensional sound field. With the number of nodes N, the component p of the vector $\boldsymbol{p}$ is expressed as follows:

$$p(\boldsymbol{r}_q) = \sum_{j=1}^{N} N_j(\boldsymbol{r}_q)p_j. \tag{10}$$

The component p_d of the vector $\boldsymbol{p_d}$ is the direct pressure contribution from the acoustic source, which is evaluated by the following equation:

$$p_{\mathrm{d}}(\boldsymbol{r}_{\mathrm{p}}) = \frac{1}{4\pi}\frac{\partial^2}{\partial x_l \partial x_m}\int_{-\infty}^{\infty}\frac{T_{lm}(\boldsymbol{r}_s,\omega)e^{jk|\boldsymbol{r}_p-\boldsymbol{r}_s|}}{|\boldsymbol{r}_p-\boldsymbol{r}_s|}dV \tag{11}$$

where $\partial^2/\partial x_l \partial x_m$ is the directional derivative and V is the volume of the flow field (in this case, the region filled by CFD cell). $\boldsymbol{r_p}$ is a position vector of the monitor point p, and $\boldsymbol{r_s}$ is a position vector of the source point s. There are 8340 boundary elements. The acoustic sources are extracted from CFD results, whose numbers are equivalent to the number of grids of the CFD model. **Figure 5** shows the boundary elements.

3.5 Validation of numerical results

The numerical accuracy of the computation has been examined for the case of L/d = 10 in the rod-airfoil model. In order to validate the present computation, we

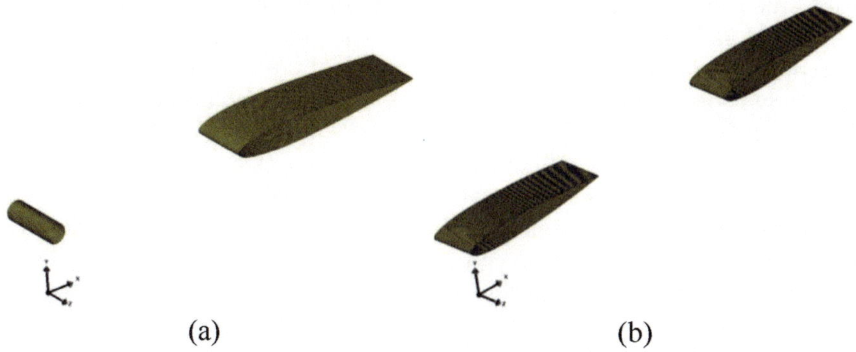

Figure 5.
Boundary elements. (a) Rod-airfoil model; (b) airfoil-airfoil model.

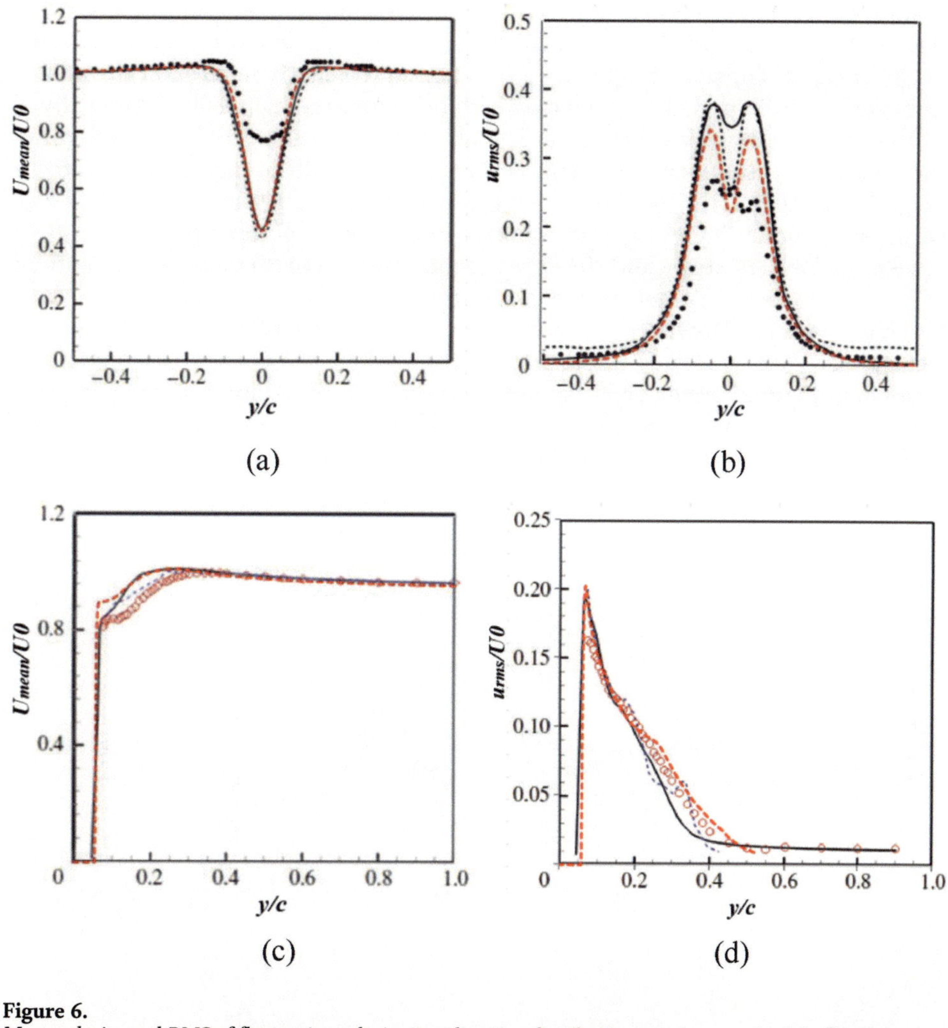

Figure 6.
Mean velocity and RMS of fluctuation velocity in x *direction distributions at* x/c = −0.87 *(a), (b) and* 0.25 *(c), (d). (a) Velocity; (b) RMS value of fluctuation velocity.* •••••• *Measured by Jacob et al. [9]* (Re_d = 48,000); ·········· *Agawal and Sharma [18]* (Re_d = 48,000); ——— *Jiang et al. [21]* (Re_d = 48,000); – – – *present* (Re_d = 28,800). *(c) Velocity; (d) RMS value of fluctuation velocity.* ·········· *Measured by Jacob et al. [9]* (Re_d = 48,000); - - - - *Boudet et al. [16]* (Re_d = 48,000); ——— *Jiang et al. [21]* (Re_d = 48,000); – – – *present* (Re_d = 28,800).

compared the present results with the experimental measurements performed by Jacob et al. [9] and the numerical results performed by Boudet et al. [16], Agawal and Sharma [18], and Jiang et al. [21], respectively. However, it is specified that the present study is performed at Re_d = 28,800, d = 0.06 m, and other experiments and the simulations are performed at Re_d = 48,000, d = 0.1 m. **Figure 6** shows the mean velocity and RMS value of the fluctuation velocity in x (streamwise) direction normalized by the incoming velocity at two locations $x/c = -0.87$ and $x/c = 0.25$. As shown in **Figure 6(a)**, the present calculation predicted similar mean velocity profile in the streamwise direction compared with other numerical results. As shown in **Figure 6(b)**, the RMS value of the fluctuation velocity in the streamwise direction obtained by the present calculation is close to those obtained by the other numerical calculations; however, the difference of the fluctuation velocity near the center-line can be seen. A possible cause for this result is the difference of the methods, meshes and Reynolds number. **Figure 6(c)** and **(d)** shows that the calculated profiles of the mean velocity and the RMS of the fluctuation velocity in the streamwise direction represent a good agreement with those from the experiments and other numerical results.

Figure 7 shows the spectra of the SPL at the location (x = 0.68 m, y = 1.74 m) calculated by the acoustic BEM simulation using the acoustic sources extracted from the CFD results. The peak frequency (St 0.2) and the spectrum around the peak frequency are well predicted compared with the experimental result obtained by Jacob et al. [9]. The peak value of the SPL obtained by the present study is slightly lower than that from the experimental result. There is a difference in the spectrum at high frequencies, $St > 0.6$. A possible cause for these differences of the peak SPL and the spectrum at high frequencies is that the Reynolds number is different between the present study and the experiments, and the mesh and time resolutions for the calculation might not be enough for the accurate prediction of the spectrum at high frequencies. However, the dominant peak SPL and SPLs around the peak frequency are well predicted in the present calculation. The peak frequency St 0.2 as shown in **Figure 7** almost corresponds to the vortex shedding frequency of the cylinder [2], which means that the vortex shedding from the rod plays an important role for the noise generation from the rod-airfoil model.

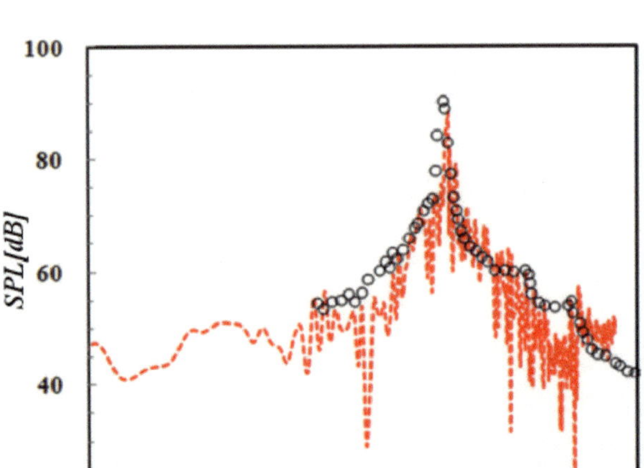

Figure 7.
SPL spectra at the location x = 0.68 *m,* y = 1.74 *m.* • • • • • *Measured by Jacob et al. [9]* (Re_d = 48,000); – – – *present* (Re_d = 28,800).

4. Results

4.1 Rod-airfoil simulation results

4.1.1 Flow patterns

Typical examples of instantaneous vorticity fields are presented in **Figure 8**. In the present calculation, for the case of $L/d = 2$, when the spacing between the rod and the airfoil is small, the boundary layers separated from the rod upstream did not roll up and reattaches to the airfoil downstream. The shear layers rolled up and formed vortices on the downstream airfoil surface, and the vortices were shed and convected downstream. The Kármán-vortex street formed from the rod was suppressed for the short spacing, and this mode is called “non-shedding mode” and “the Kármán-vortex street suppressing mode” as indicated by Munetaka et al. [10] and Jiang et al. [21], respectively. On the other hand, for the case of $L/d = 10$, when the spacing between the rod and the airfoil is large, the boundary layers separated from the rod upstream rolled up and formed vortices in the region between the rod and the airfoil, and the formed vortices shed from the rod interacted with the airfoil and impinged on the leading edge of the airfoil. The impinged vortices were distorted and convected downstream. This mode is called “the Kármán-street shedding mode” as indicated by Jiang et al. [21], and also called wake-body interaction or body-vortex interaction [6, 16, 21]. Similar phenomena such as the

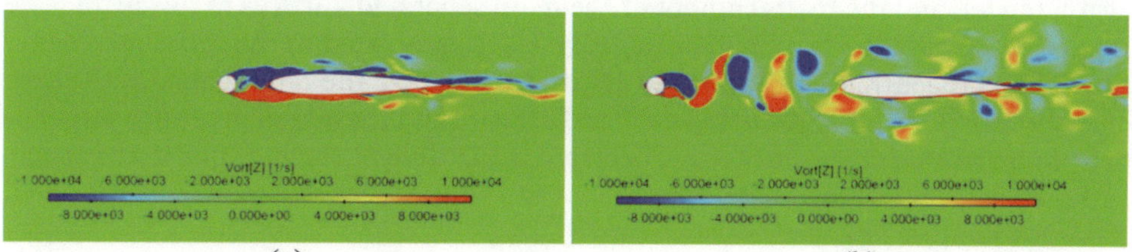

(a) (b)

Figure 8.
Vorticity in the z direction. (a) L/d = 2; *(b)* L/d = 10.

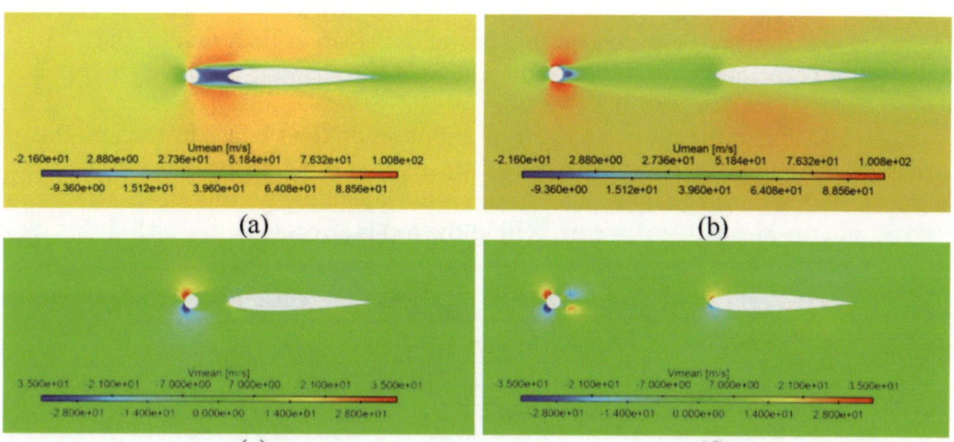

Figure 9.
Time-averaged velocity in the streamwise (x) *and the vertical* (y) *directions. (a), (b) velocity in the streamwise* (x) *direction; (c), (d) velocity in the vertical* (y) *direction. (a)* L/d = 2; *(b)* L/d = 10; *(c)* L/d = 2; *(d)* L/d = 10.

Kármán-street shedding and suppressing modes can be seen in the flow around two obstacles in tandem [2, 4, 6].

Figure 9(a) and **(b)** shows the fields of time-averaged velocity U_{mean} in the x (streamwise) direction. In the space between the rod and the airfoil, **Figure 9(a)** shows that the counter-rotating vortices near the leading edge of the airfoil lead to very small or negative values of U_{mean} for the case of L/d = 2. The reattachments of the counter-rotating vortices to the leading edge of the airfoil due to the approaching rod slow down the flow. By contrast, for the case of L/d = 10, the flow behind the rod was accelerated in front of the leading edge of airfoil as shown in **Figure 9(b)**.

Figure 9(c) and **(d)** shows the fields of time-averaged velocity V_{mean} in the y (vertical) direction. For the case of L/d = 2, the positive V_{mean} appears in the $-y$ region around the leading edge of the airfoil and the negative one appears in the opposite region. This is due to the reattachments of the negative and positive vortices formed from the rod upstream. For the case of L/d = 10, the V_{mean} distribution in the regions around the leading edge of the airfoil is contrary to that for the case of L/d = 2, as shown in **Figure 9(c)**. This phenomenon is indicated by Jiang et al. [21].

Finally, the fields of RMS value of the fluctuation velocity u_{rms} in the streamwise (x) direction are represented in **Figure 10** for L/d = 2 and L/d = 10. As shown in **Figure 10**, for the case of L/d = 2, the u_{rms} values behind the rod and near the central line y = 0 are much lower than those for the case of L/d = 10. This result corresponds to the suppression of the Kármán-vortex street for the case of L/d = 2 in the rod-airfoil model, and the reattachment of the main separated vortices to the airfoil (see **Figure 8**) causes the very small or negative U_{mean} generated in the space between the rod and the airfoil (see **Figure 9(a)**) in this case. On the other hand, in the case of L/d = 10, the u_{rms} values behind the rod and around the airfoil are much larger than those for the case of L/d = 2. Therefore the turbulent fluctuations in the space between the rod and the airfoil for the case of L/d = 10 seem to be much larger than those for the case of L/d = 2.

4.1.2 Near pressure field

Instantaneous snapshots of static pressure and snapshots of mean pressure are represented in **Figure 11**. For the case of L/d = 2, **Figure 11** shows that the value of the static pressure on the upstream side of the rod is larger than that on the downstream side, and that on the leading edge of the airfoil is negative and much lower than that for the case of L/d = 10.

Figure 12 shows the mean pressure distribution on the surface of rod and the airfoil, where x_{rod} is the coordinate in the streamwise direction and the origin of x_{rod} is the located in the stagnation point of the rod. Depending on the spacing between the rod and the airfoil, the pressure behind the rod is affected by the airfoil [21]. For

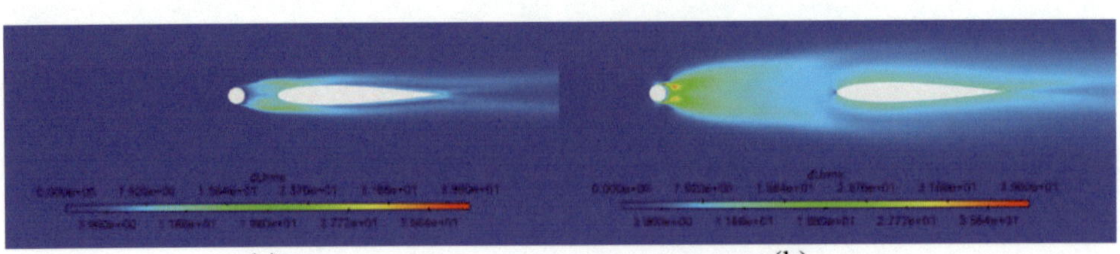

(a) (b)

Figure 10.
RMS value of fluctuation velocity in the streamwise (x) direction. (a) L/d = 2; (b) L/d = 10.

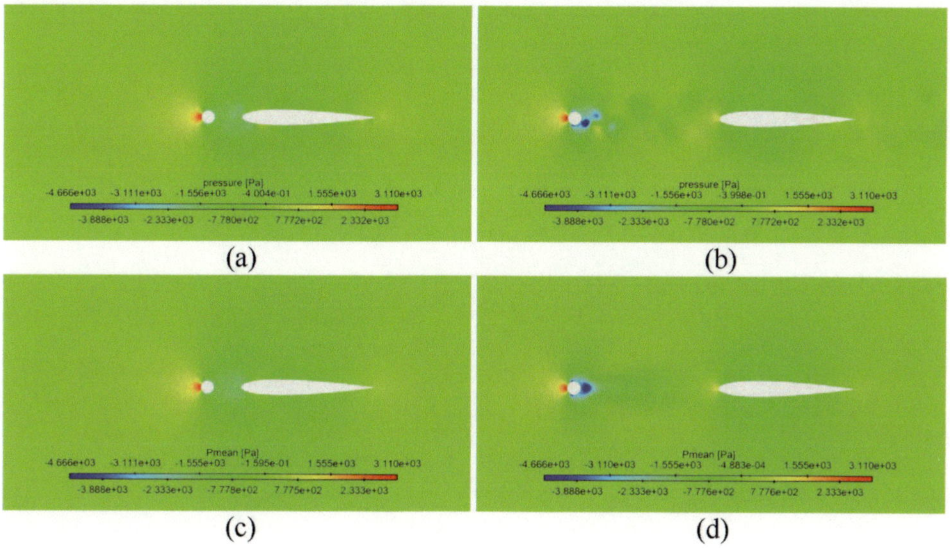

Figure 11.
Static pressure and mean pressure fields. (a) L/d = 2; *(b)* L/d = 10; *(c)* L/d = 2; *(d)* L/d = 10.

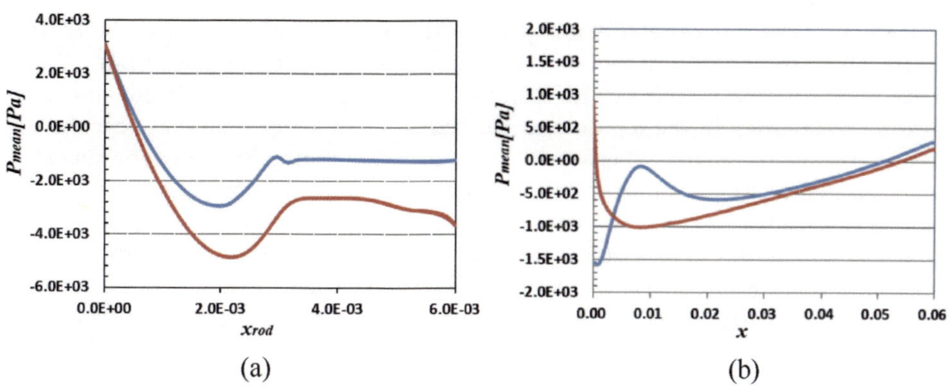

Figure 12.
Mean pressure profile on the rod and the airfoil. (a) Rod. (b) Airfoil. —— L/d = 2; —— L/d = 10.

the case of L/d = 2, as shown in **Figure 12(a)**, the pressure behind the rod is much larger than that for the case of L/d = 10 due to the presence of the airfoil behind the rod. As mentioned in Section 3.1.1, this phenomenon is related to the reattachments of the separated shear layers from the rod to the leading edge of the airfoil, and the Kármán-vortex street formed from the rod is suppressed and the pressure behind the rod increases. **Figure 12(b)** shows that in the case of L/d = 2, the negative pressure occurs around the leading edge of the airfoil, which means that the negative drag force acts on the airfoil, and in the case of L/d = 10, the positive pressure occurs the leading edge of the airfoil and the influence of the rod on the airfoil seems to be small.

Figure 13 shows instantaneous snapshots of a fluctuation pressure (dp) and RMS of the fluctuation pressure (dp_{rms}) fields, and the fluctuation pressure is defined as follows [6].

$$dp = p_s - p_{\mathrm{mean}} \tag{12}$$

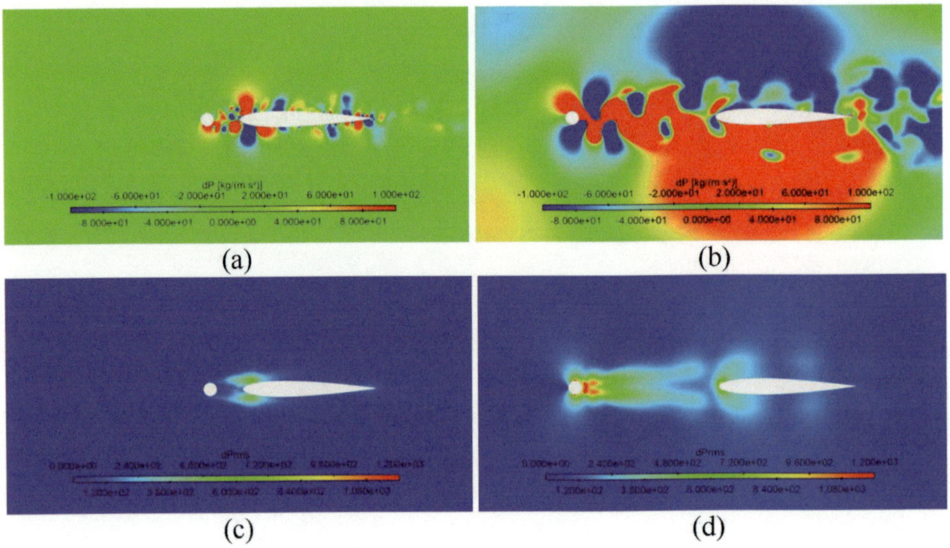

Figure 13.
Fluctuation pressure and RMS of fluctuation pressure fields. (a), (b) Fluctuation pressure, (c), (d) RMS of fluctuation pressure. (a) L/d = 2; (b) L/d = 10; (c) L/d = 2; (d) L/d = 10.

Here, p_s is the static pressure and p_{mean} is the mean pressure as represented in **Figure 11(c)** and **(d)**. As shown in **Figure 13(a)** and **(b)**, in the case of L/d = 2, the pressure fluctuation mainly occurs around the leading edge of the airfoil, and the pressure fluctuation in the region near the rod is much lower than that near the leading edge of the airfoil. This result is related to the reattachment of the shear layers formed from the rod to the airfoil as mentioned in Section 3.1.1. **Figures 8(a), 13(a)** and **(c)** show that the suppression of the vortex shedding from the rod results in the suppression of the noise radiation from the rod-airfoil model, as mentioned by Munetaka et al. [10] and Jiang et al. [21].

On the other hand, in the case of L/d = 10, that is, for large spacing, the pressure fluctuation near both the rod and the leading edge of the airfoil is large, and especially the pressure fluctuation near the rod is larger than that near the leading edge of the airfoil, as shown in **Figure 13(b)** and **(d)**. **Figures 8(b), 13(b)** and **(d)** show that the pressure fluctuation behind the rod is generated by the vortex shedding from the rod. They also show that the pressure fluctuation near the leading edge of the airfoil is generated by the impingement of the vortices shed from the rod onto the leading edge of the airfoil and the distortion of the impinged vortices (wake-body interaction or body-vortex interaction), as mentioned by Boudet et al. [16]. Similar phenomenon is investigated by Inoue and Mori [6] in the simulations of the noise generated by the flow around two square cylinders in tandem. They reported that the distortion of the impinging vortices shed from the upstream cylinder onto the downstream cylinder plays an important role for the noise radiation when the spacing between the two cylinders is large [6].

The distributions of the mean pressure and RMS of the fluctuation pressure in the wake region of the rod along the symmetry line (y = 0) of the rod and the airfoil are represented in **Figure 14**. As shown in **Figure 14(a)**, in the case of L/d = 2, the mean pressure is negative in the wake region between the rod and the leading edge of the airfoil, and it leads to the negative drag of the airfoil as mentioned above. In the case of L/d = 10, the mean pressure behind the rod is negative and much lower than in the case of L/d = 2, and that near the leading edge of the airfoil is positive, which leads to the positive drag of the airfoil. **Figure 14(b)** shows that in the case of

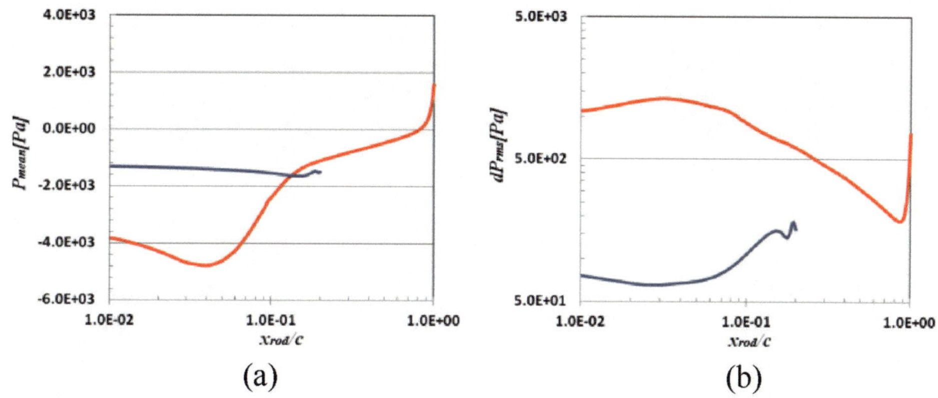

Figure 14.
Distributions of mean pressure and RMS of fluctuation pressure in wake region of rod along symmetry line (y = 0) of rod and airfoil. (a) Mean pressure (b) RMS of fluctuation pressure. —— L/d = 2; —— L/d = 10.

L/d = 2, the RMS of the fluctuation pressure in the region near the rod is much lower than that near the leading edge of the airfoil due to the suppression of the vortex shedding form the rod, and in the case of L/d = 10, the pressure fluctuation near both the rod and the leading edge of the airfoil is much larger than that in the case of L/d = 2 due to the vortex shedding from the rod and the impingement of the vortices on the leading edge of the airfoil, as mentioned in **Figure 13**.

4.1.3 Far acoustic pressure field

Figure 15 shows the spectra of the SPL at the location (x = 0.68 m, y = 1.74 m) calculated by the acoustic BEM simulation using the acoustic sources extracted from the CFD results. The SPL in the case of the single airfoil is also plotted to compare the two cases with it. The SPL in the case of L/d = 10 is much larger than that in the case of L/d = 2, and the peak frequencies for the cases of L/d = 10 and L/d = 2 are 2440 and 1960 Hz, respectively. However, the peak frequency for the case of L/d = 2 does not appear clearly compared to that in the case of L/d = 10, and the

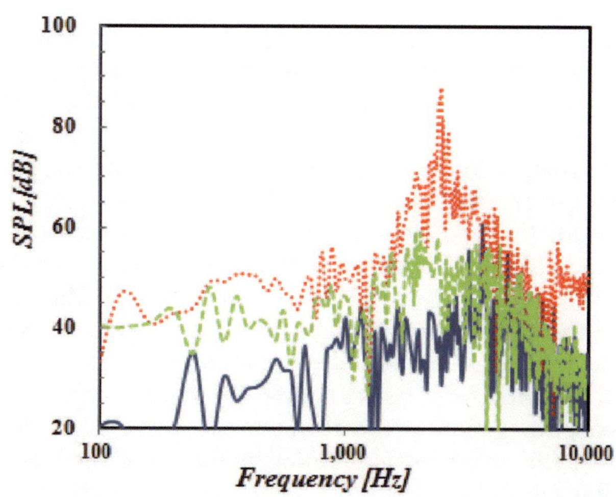

Figure 15.
SPL spectra at the location x = 0.68 *m,* y = 1.74 *m.* ········ L/d = 2; ········ L/d = 10; —— *single airfoil.*

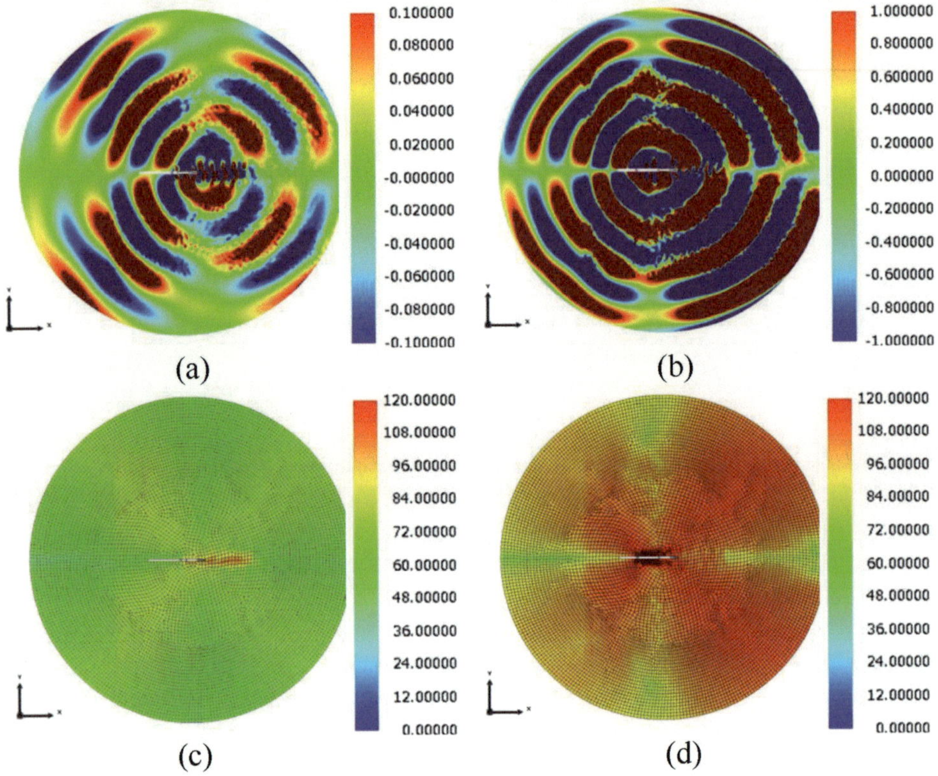

Figure 16.
Far-field sound pressure and SPL fields at the peak frequency. (a), (b) Sound pressure and (c), (d) SPL. (a) L/d *= 2, 1960 Hz; (b)* L/d *= 10, 2440 Hz; (c)* L/d *= 2, 1960 Hz; (d)* L/d *= 10, 2440 Hz.*

peak frequency increases depending on the spacing between the rod and airfoil as reported by Jiang et al. [21]. The peak SPL in the case of single airfoil is larger than that in the case of L/d = 2, and the peak frequency is higher than those in the cases of L/d = 2 and 10. The sound radiation is mainly generated by both the vortex shedding from the rod and the impingement of the vortices shed from the rod onto the leading edge of the airfoil in the case of L/d = 10, as indicated by Boudet et al. [16]. According to Jiang et al. [21], when $L/d \geq 4$, the vortex shedding from the rod and the impingement of the shed vortices onto the leading edge of the airfoil are the main generation sources of the noise radiation. As mentioned in Section 3.1.2, the suppression of the vortex shedding from the upstream rod results in the suppression of the pressure fluctuation in the region between the rod and the airfoil and the noise radiation for the case of L/d = 2.

Figure 16 presents the far-field sound pressure and SPL fields at the peak frequency. The far-field sound pressure and SPL fields show a dipolar nature of the sound radiation, and the lift dipole is dominant in the fields. The magnitude of the generated noise in the case of L/d = 10 is much larger than that in the case of L/d = 2, as indicated above. In the case of the L/d = 2, the noise propagation on the upstream side is larger than that on the downstream side. On the other hand, in the case of the L/d = 10, the noise propagation on the downstream side is larger than that in the upstream direction, which means that impingement of the vortices shed from the rod onto the leading edge of the airfoil, in other words, the wake body interaction is the main generation of the far filed noise radiation.

4.2 Airfoil-airfoil simulation results

4.2.1 Flow patterns

Typical examples of instantaneous vorticity fields are presented in **Figure 17**. In the present calculation, for the case of *L/c* = 0.2, when the spacing between the two airfoils is small, the boundary layers separated from the airfoil upstream did not roll up and reattach to the airfoil downstream. It seems that the reattached shear layers are oscillating at the leading edge of the downstream airfoil. The shear layers rolled up and formed vortices on the downstream airfoil surface, and the vortices were shed and convected downstream. On the other hand, in the cases of *L/c* = 0.6 and 1.0, the shear layers shed from the leading edge rolled up in front of the leading edge of the airfoil and impinged onto the leading edge of the airfoil. In the case of the single, the shear layer and the vortices shed only from the trailing edge of the airfoil.

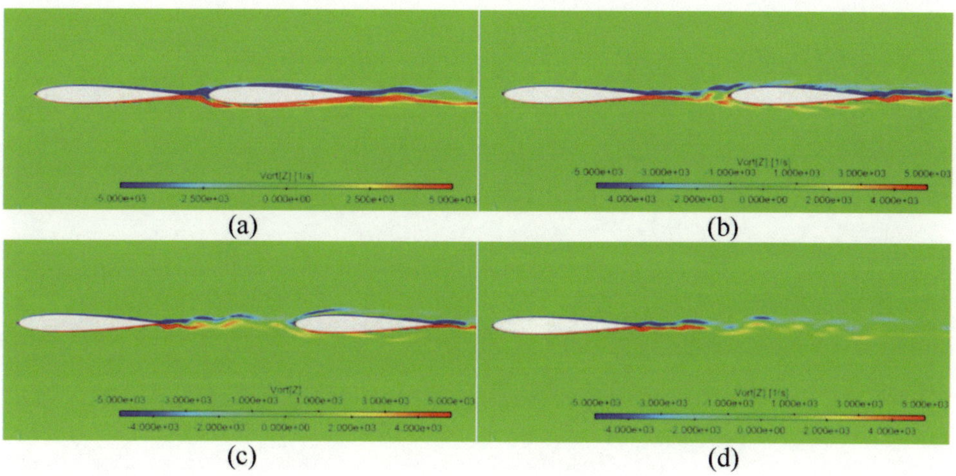

Figure 17.
Vorticity in the z direction. (a) L/c = 0.2; (b) L/c = 0.6; (c) L/c = 1; (d) single airfoil.

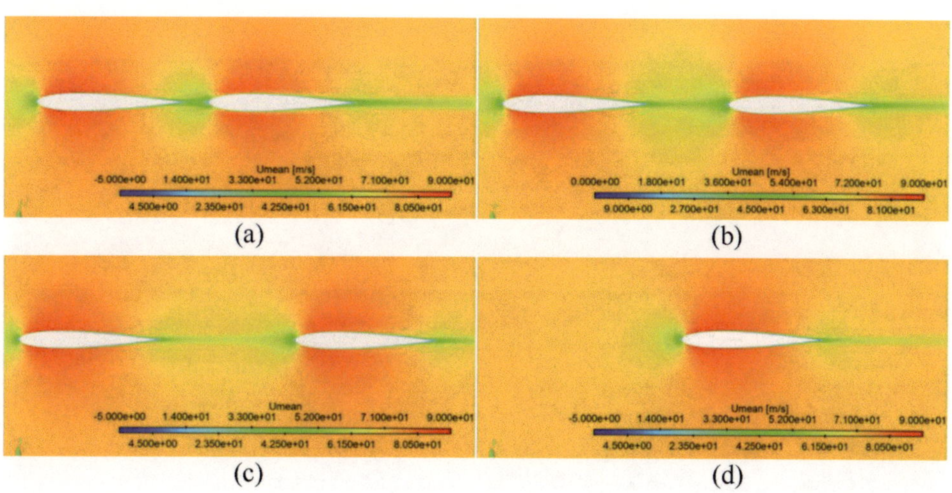

Figure 18.
Time-averaged velocity in the streamwise (x) direction. (a) L/c = 0.2; (b) L/c = 0.6; (c) L/c = 1; (d) single airfoil.

Figure 18(a) and **(b)** show the fields of time-averaged velocity U_{mean} in the x (streamwise) direction. In the spacing between the two airfoils, **Figure 18(a)** shows that the shear layers shed from the upstream airfoil near the leading edge of the airfoil lead to very small or negative values of U_{mean} for the case of L/c = 0.2. The reattachments of the shear layer shed from the upstream airfoil to the leading edge of the airfoil due to the approaching airfoil slow down the flow. By contrast, for the cases of L/c = 0.6 and 1, the flow behind the upstream airfoil was accelerated in front of the leading edge of airfoil as shown in **Figure 18(b)** and **(c)**. These similar phenomena were observed in the rod-airfoil model, as explained in Section 3.1.1. In the case of the single, U_{mean} contour around the airfoil shows similar one of the two airfoils in the case of L/c = 1. The influence of the wake from the upstream airfoil seems to become smaller with the spacing between the two airfoils increasing.

Figure 19 shows the fields of time-averaged velocity V_{mean} in the y (vertical) direction. For the case of L/c = 0.2, the V_{mean} around the leading edge of the

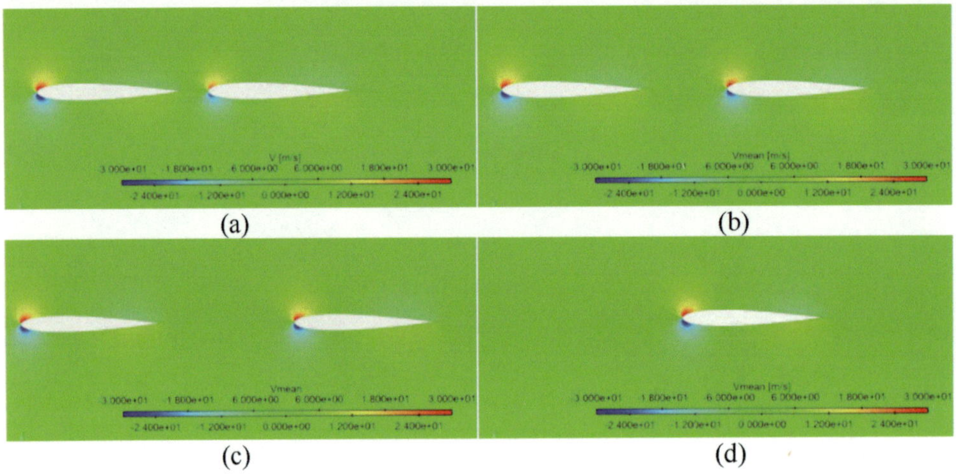

Figure 19.
Time-averaged velocity in the streamwise (x) *direction. (a)* L/c = 0.2; *(b)* L/c = 0.6; *(c)* L/c = 1; *(d) single airfoil.*

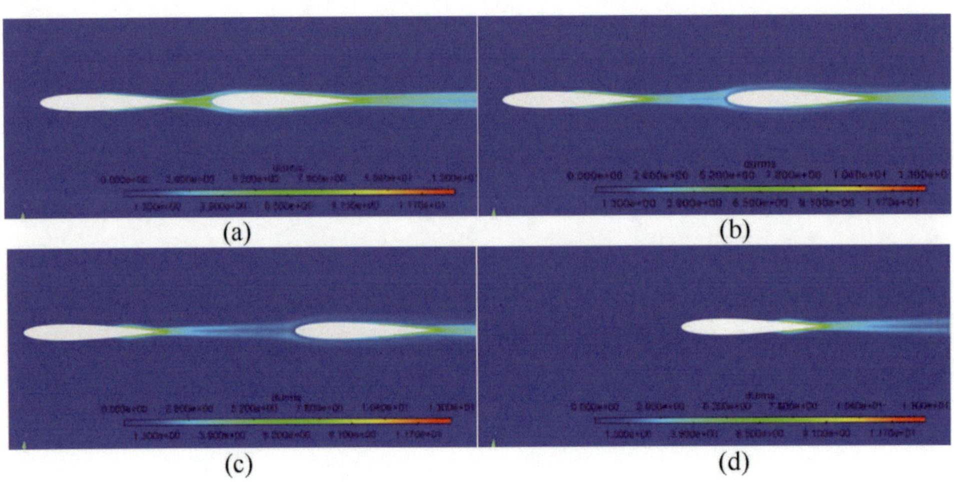

Figure 20.
RMS value of fluctuation velocity in the streamwise (x) *direction. (a)* L/c = 0.2; *(b)* L/c = 0.6; *(c)* L/c = 1; *(d) single airfoil.*

downstream airfoil is lower than that in other cases. This is due to the reattachments of the shear layers shed from the upstream airfoil.

Finally, the fields of RMS value of the fluctuation velocity u_{rms} in the streamwise (x) direction are represented in **Figure 20**. As shown in **Figure 20**, for the case of L/c = 0.2, the u_{rms} values around the trailing edges of the two airfoils are lower than those for the cases of L/c = 0.6, 1.0 and the single airfoil. This result corresponds to the reattachment of the shear layers to the airfoil (see **Figure 17(a)**) in this case. Therefore the turbulent fluctuations around the trailing edges of the two airfoils in the cases of L/c = 0.6 and 1.0 seem to be larger than those for the case of L/c = 0.2.

4.2.2 Near pressure field

Snapshots of mean pressure are represented in **Figure 21**. For the case of L/c = 0.2, **Figure 21** shows that the value of the static pressure on the leading edge of

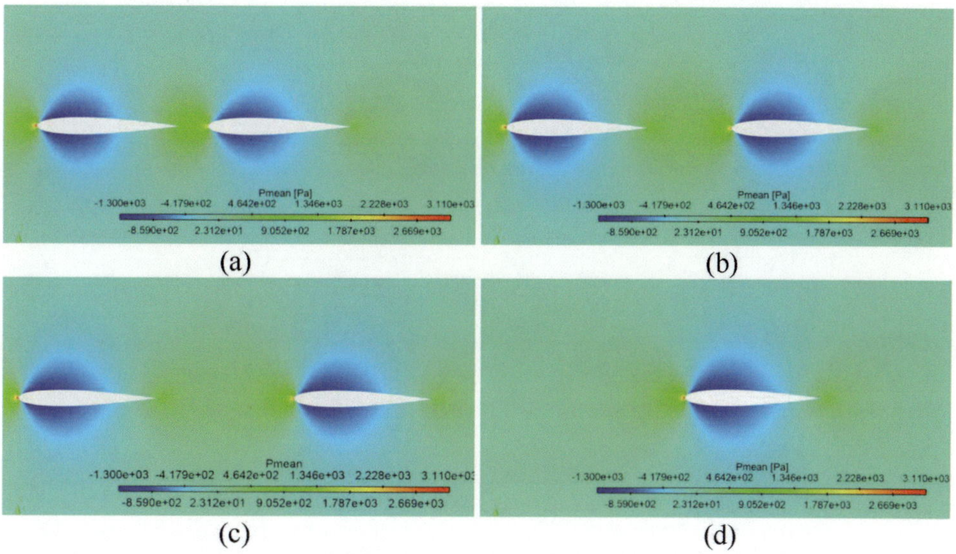

Figure 21.
Mean pressure field. (a) L/c = 0.2; *(b)* L/c = 0.6; *(c)* L/c = 1; *(d) single airfoil.*

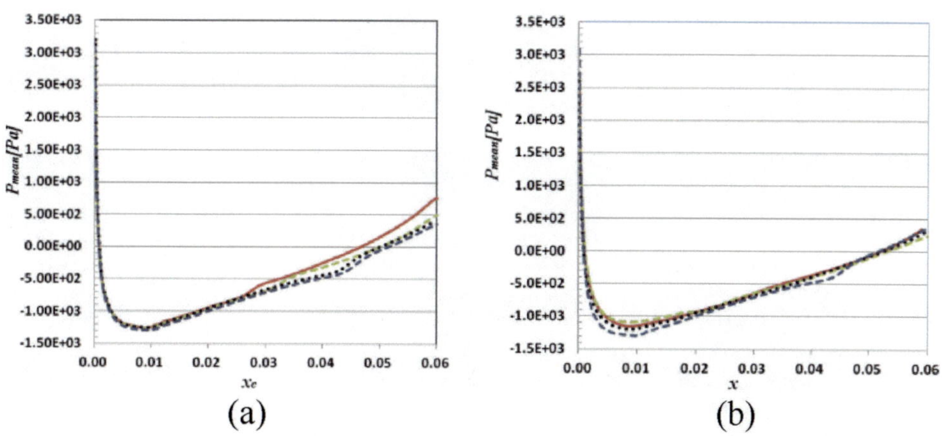

Figure 22.
Mean pressure profile on the upstream and downstream airfoils. (a) Upstream airfoil; (b) downstream airfoil.
—— L/c = 0.2; ······· L/c = 0.6; ········ L/c = 1.0; - - - - *single airfoil.*

the upstream airfoil is much larger than that of the downstream one, and that on the leading edge of the downstream airfoil is much lower than those for the cases of L/c = 0.6, 1.0 and the single airfoil.

Figure 22 shows the mean pressure distribution on the surfaces of the two airfoils, where x_c is the coordinate in the streamwise direction and the origin of x_c is the located in the leading edge of the upstream airfoil. Depending on the spacing between the upstream and downstream airfoils, the pressure behind the upstream airfoil is affected by the downstream one. For the case of L/c = 0.2, as shown in **Figure 22(a)**, the pressure behind the upstream airfoil is larger than that for the cases of L/c = 0.6, 1 and single airfoil due to the presence of the downstream airfoil behind the upstream one. As mentioned in Section 3.2.1, this phenomenon is related to the reattachments of the separated shear layers from the upstream airfoil to the leading edge of the downstream airfoil, and the vortices formed from the upstream

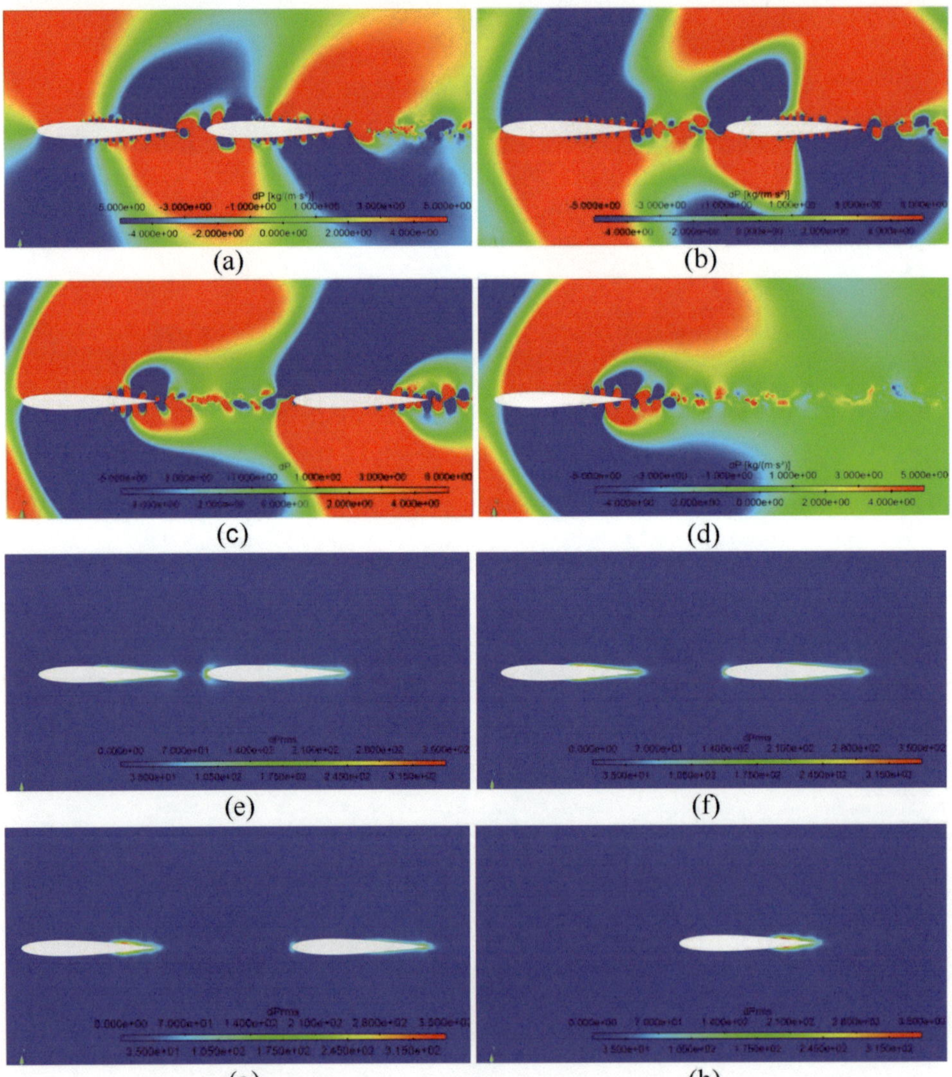

Figure 23.
Fluctuation pressure and RMS of fluctuation pressure fields. (a)–(d) Fluctuation pressure; (e)–(h) RMS of fluctuation pressure. (a) L/c = 0.2; *(b)* L/c = 0.6; *(c)* L/c = 1; *(d) single airfoil; (e)* L/c = 0.2; *(f)* L/c = 0.6; *(g)* L/c = 1.0; *(h) single airfoil.*

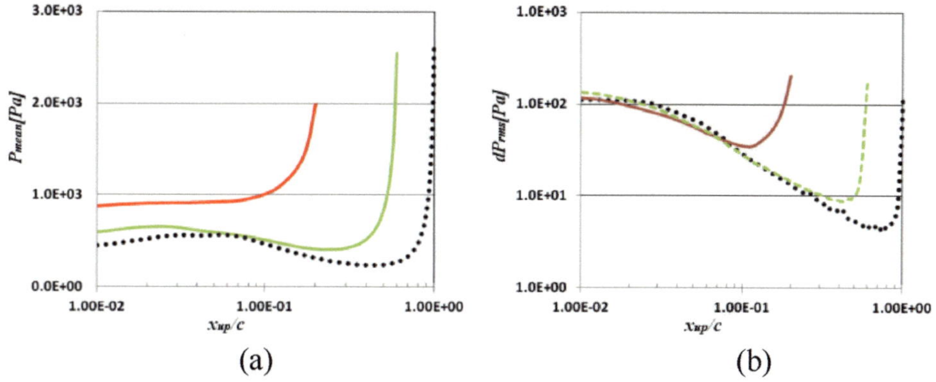

Figure 24.
Distributions of mean pressure and RMS of fluctuation pressure in wake region of upstream airfoil along symmetry line (y = 0) *of two airfoils. (a) Mean pressure; (b) RMS of fluctuation pressure.* —— L/c = 0.2; - - - L/c = 0.6; ·········· L/c = 1.0.

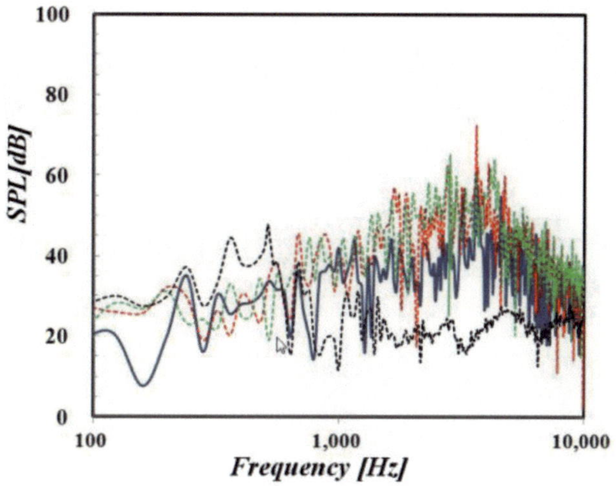

Figure 25.
SPL spectra at the location x = 0.68 *m*, y = 1.74 *m.* ·········· L/c = 0.2; ·········· L/c = 0.6; ·········· L/c = 1.0; —— *single airfoil.*

airfoil is suppressed and the pressure behind the upstream airfoil increases. **Figure 22(b)** shows that in the case of L/c = 0.2, the pressure around the leading edge of the downstream airfoil is lower than that in the cases of L/c = 0.6, 1 and single airfoil, which means that the drag force acting on the downstream airfoil is lower than that for the other cases.

Figure 23 shows instantaneous snapshots of a fluctuation pressure (dp) and snapshots of RMS of the fluctuation pressure (dp_{rms}) fields. As shown in **Figure 23 (a)–(c)**, in the two airfoil cases, the pressure fluctuation occurs around the trailing edge of the upstream airfoil and the leading and trailing edges of the downstream airfoil. In the case of single airfoil, the pressure fluctuation mainly occurs near the trailing edge of the airfoil as shown in **Figure 23(d)**. These figures show the dipolar nature of the fluctuation pressure generated around the trailing edge of the upstream airfoil and the leading and trailing edges of the downstream airfoil.

Figure 23(e)–(g) shows that the pressure fluctuation near the leading edge of the downstream airfoil is generated by the impingement of the shear layers or

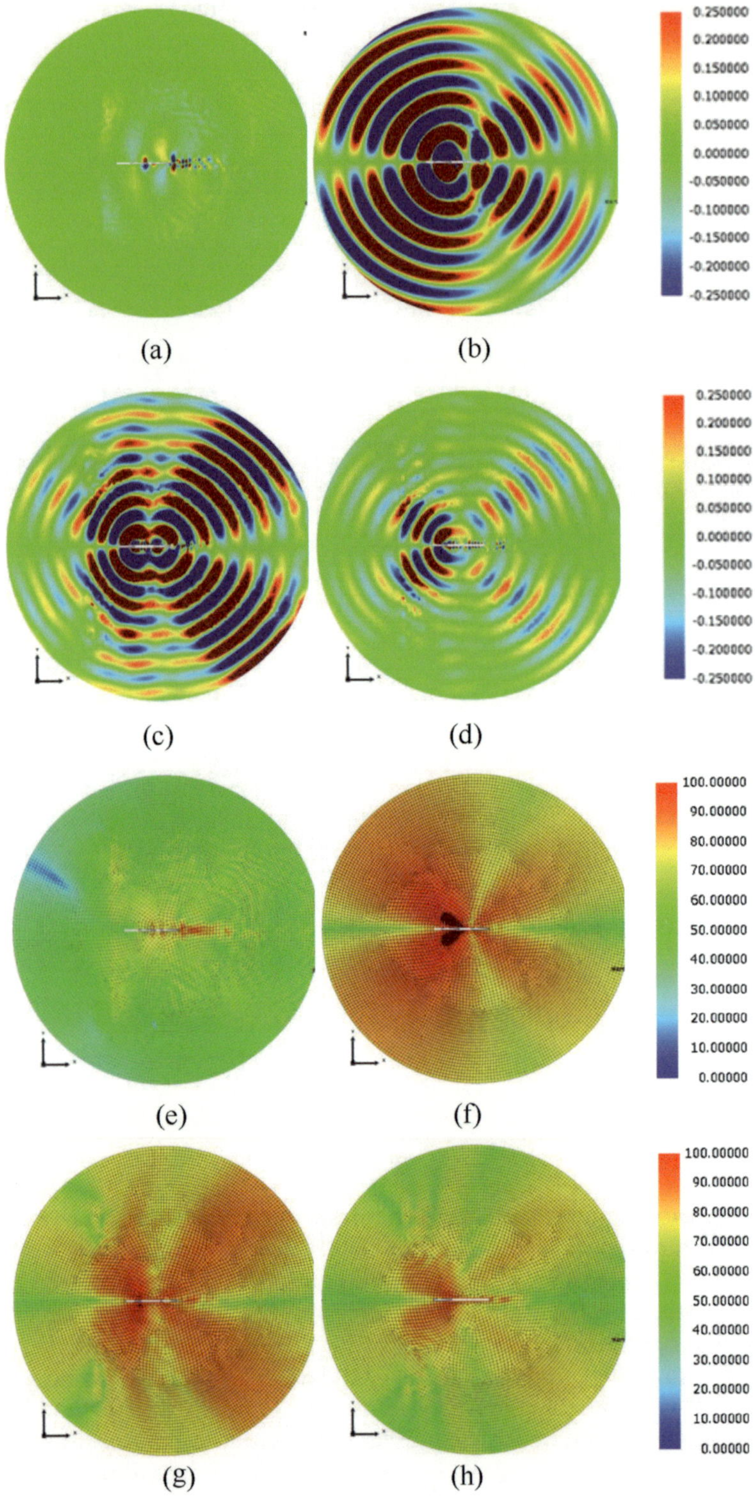

Figure 26.
Far-field sound pressure and SPL fields at the peak frequency. (a)–(d) Sound pressure and (e)–(h) SPL. (a) L/c = 0.2; *(b)* L/c = 0.6; *(c)* L/c = 1; *(d) single airfoil; (e)* L/c = 0.2; *(f)* L/c = 0.6; *(g)* L/c = 1.0; *(h) single airfoil.*

vortices shed from the upstream airfoil onto the leading edge of the downstream airfoil and the distortion of the impinged vortices (wake-body interaction or body-vortex interaction), as mentioned in Section 3.1.2.

The distributions of the mean pressure and RMS of the fluctuation pressure in the wake region of the upstream airfoil along the symmetry line (y = 0) of the two airfoils are represented in **Figure 24**. As shown in **Figure 24(a)**, the mean pressure behind the upstream airfoil is positive and lower than that near the leading edge of the downstream airfoil, and it decreases depending on the spacing between the two airfoils. On the other hand, that near the leading edge of the downstream airfoil increases depending on the spacing between the two airfoils. **Figure 24(b)** shows that the pressure fluctuation near both the trailing edge of the upstream airfoil and the leading edge of the downstream airfoil is large due to the shear layers or vortices shed from the upstream airfoil and the impingement of the shear layers or vortices onto the leading edge of the downstream airfoil, as mentioned in **Figure 23**.

4.2.3 Far acoustic pressure field

Figure 25 shows the spectra of the SPL at the location (x = 0.68 m, y = 1.74 m) calculated by the acoustic BEM simulation using the acoustic sources extracted from the CFD results. The SPL in the case of the single airfoil is also plotted to compare the three cases with it. The SPLs in the cases of L/c = 0.6, 1.0 and single airfoil are much larger than that in the case of L/c = 0.2, and the peak frequencies for the cases of L/c = 0.2, 0.6, 1.0 and single airfoil are 540, 2840, 3640, and 3640 Hz, respectively. **Figure 26** shows the far field sound pressure and SPL fields at the peak frequency. The peak frequency increases depending on the spacing between the two airfoils as reported in the rod-airfoil model. The peak SPL in the case of single airfoil is larger than that in the case of L/c = 0.2. The peak frequency in the case of the single airfoil is higher than those in the cases of L/c = 0.2 and 0.6, and same as that in the case of L/c = 1.0.

In the cases of the two airfoils, the sound radiation is mainly generated by three factors: (1) vortices or shear layers shedding from the upstream airfoil, (2) the impingement of the vortices or shear layers shed from the upstream airfoil onto the leading edge of the downstream airfoil, (3) vortices or shear layers shedding from the downstream airfoil as indicated in Section 3.2.2. As mentioned in Section 3.2.2, the reattachment of the shear layers from the upstream airfoil to the leading edge of the downstream airfoil results in the suppression of the pressure fluctuation in the region between the upstream and downstream airfoils and the noise radiation for the case of L/c = 0.2.

5. Conclusions

In this chapter, we simulated the flow around the rod-airfoil model and the noise generated by the wake-body interaction or body-vortex interaction by the coupling method using commercial CFD and acoustic BEM codes, and compared the results with those obtained by Jacob et al. [9] and Jiang et al. [21]. Then, we simulated the flow around the airfoil-airfoil model (airfoils in tandem) and the noise generation and propagation.

1. In the rod-airfoil model, when the spacing between the rod and the airfoil is small, the shear layers separated from the rod upstream did not roll up and

reattach to the airfoil downstream, and the vortex shedding from the rod is suppressed. It leads to the suppression of the pressure fluctuation near the rod and the airfoil and the noise radiation as reported by Jiang et al. [21].

2. In the rod-airfoil model, when the spacing between the rod and the airfoil is large, the boundary layers separated from the rod upstream rolled up and formed vortices, and the formed vortices shed from the rod impinged on the leading edge of the airfoil. This phenomenon is called the wake-body interaction or body-vortex interaction. It leads to the large pressure fluctuation near both the rod and the leading edge of the airfoil and the large noise radiation.

3. In the airfoil-airfoil model, the flow depends on the spacing between the two airfoils as in the rod-airfoil model. The shear layers shed from the upstream airfoil reattaches to the leading edge of the downstream airfoil with the small spacing between the two airfoils, and the reattached shear layers seem to oscillate at the leading edge of the downstream airfoil. The shear layers shed from the leading edge rolled up in front of the leading edge of the airfoil and impinged onto the leading edge of the airfoil with the large spacing between the two airfoils.

4. In the airfoil-airfoil model, the sound radiation is mainly generated by three factors: (1) vortices or shear layers shedding from the upstream airfoil, (2) the impingement of the vortices or shear layers shed from the upstream airfoil onto the leading edge of the downstream airfoil (the wake-body interaction), (3) vortices or shear layers shedding from the downstream airfoil as in the rod-airfoil model. The SPL and the peak frequency of the generated noise increase with an increase of the spacing between the two airfoils as reported in the rod-airfoil model.

Author details

Masaaki Mori
Cybernet Systems Co., Ltd., Tokyo, Japan

*Address all correspondence to: m-mori@cybernet.co.jp

References

[1] Mahir N, Rockwell D. Vortex shedding from a forced system of two cylinders. Part I: Tandem arrangement. Journal of Fluids and Structures. 1996;**9**: 473-489

[2] Zdravkovich MM. Review of flow interference between two circular cylinders in various arrangements. Journal of Fluids Engineering. 1977;**99**: 618-633

[3] Ljungkrona L, Norberg CH, Sunden B. Free-stream turbulence and tube spacing effects on surface pressure fluctuations for two tubes in an in-line arrangement. Journal of Fluids and Structures. 1991;**5**:701-727

[4] Liu CH, Chen JM. Observations of hysteresis in flow around two square cylinders in a tandem arrangement. Journal of Wind Engineering and Industrial Aerodynamics. 2002;**90**(9): 1019-1050

[5] Fitzpatrick JA. Flow/acoustic interactions of two cylinders in cross-flow. Journal of Fluids and Structures. 2003;**17**:97-113

[6] Inoue O, Mori M, Hatakeyama N. Aeolian tones radiated from flow past two square cylinders in tandem. Physics of Fluids. 2006;**18**:046101

[7] King WFN, Pfizenmaier E. An experimental study of sound generated by flows around cylinders of different cross-section. Journal of Sound and Vibration. 2009;**328**:318-337

[8] Hutcheson FV, Brooks TF. Noise radiation from single and multiple rod configurations. International Journal of Aeroacoustics. 2012;**11**:291-334

[9] Jacob MC, Boudet J, Casalino D, Michard M. A rod–airfoil experiment as benchmark for broadband noise modeling. Theoretical and Computational Fluid Dynamics. 2005; **19**(3):171-196

[10] Munekata M, Kawahara K, Udo T, Yoshikawa H, Ohba H. An experimental study on aerodynamic sound generated from wake interference of circular cylinder and airfoil vane in tandem. Journal of Thermal Science. 2006;**15**(4): 342-348

[11] Munekata M, Koshiishi R, Yoshikawa H, Ohba H. An experimental study on aerodynamic sound generated from wake interaction of circular cylinder and airfoil with attack angle in tandem. Journal of Thermal Science. 2008;**17**(3):212-217

[12] Li Y, Wang XN, Chen ZW, Li ZC. Experimental study of vortex-structure interaction noise radiated from rod–airfoil configurations. Journal of Fluids and Structures. 2014;**51**(3):313-325

[13] Casalino D, Jacob MC, Roger M. Prediction of rod airfoil interaction noise using the FWH analogy. AIAA Journal. 2003;**41**(2):182-191

[14] Jiang M, Li XD, Zhou JJ. Experimental and numerical investigation on sound generation from airfoil-flow interaction. Applied Mathematics and Mechanics. 2011; **32**(6):765-776

[15] Magagnato F, Sorgüven E, Gabi M. Far field noise prediction by large eddy simulation and Ffowcs-Williams Hawkings analogy. AIAA Journal. 2003; **6**:2003-3206

[16] Boudet J, Grosjean N, Jacob MC. Wake-airfoil interaction as broadband noise source: A large-eddy simulation study. International Journal of Aeroacoustics. 2005;**4**(1):93-116

[17] Greschner B, Thiele F, Jacob MC, Casalino D. Prediction of sound

generated by a rod–airfoil configuration using EASM DES and the generalised Lighthill/FW-H analogy. Computers and Fluids. 2008;**37**:402-413

[18] Agrawal BR, Sharma A. Aerodynamic noise prediction for a rod–airfoil configuration using large eddy simulations. AIAA Journal. 2014;**6**: 2014-3295

[19] Giret JC, Sengissen A, Moreau S, Sanjosé M, Jouhaud JC. Noise source analysis of a rod–airfoil configuration using unstructured large eddy simulation. AIAA Journal. 2015;**53**(4): 1062-1077

[20] Daude F, Berland J, Emmert T, Lafon P, Crouzet F, Bailly C. A high-order finite-difference algorithm for direct computation of aerodynamic sound. Computers and Fluids. 2012;**61**: 46-63

[21] Jiang Y, Mao ML, Deng XG, Liu HY. Numerical investigation on body-wake interaction over rod-airfoil configuration. Journal of Fluid Mechanics. 2015;**779**:1-35

[22] Xiao L, Syamir ASA, Mahdi A. Noise control for a tandem airfoil configuration using trailing edge serration. In: Proceedings of the 24th International Congress on Sound and Vibration (ICSV24). London; 2017. pp. 23-27

[23] Xiao L, Syamir ASA, Mahdi A, Yannick M. Aeroacoustic and aerodynamic study of trailing-edge serrated airfoils in tandem configuration. In: Proceedings of the 23rd International Congress on Acoustics (ICA23). Aachen; 2019. pp. 9-13

[24] Lighthill MJ. On sound generated aerodynamically. I. General theory. Proceedings of the Royal Society A: Mathematical, Physical and Engineering Sciences. 1952;**211**(1107):564-587

[25] Lighthill MJ. On sound generated aerodynamically. II. Turbulence as a source of sound. Proceedings of the Royal Society A: Mathematical, Physical and Engineering Sciences. 1954; **222**(1148):1-32

[26] Zhan T, Zhang Y, Ouyang H. Structural vibration and fluid borne noise induced by turbulent flow through a 90° piping elbow with/without a guide vane. International Journal of Pressure Vessels and Piping. 2015;**125**:66-77

[27] Zhang Y, Zhang T, Ouyang H, Li TY. Flow-induced noise analysis for 3D trash rack based on LES/Lighthill hybrid method. Applied Acoustics. 2014;**77**:141-152

[28] Mori M, Masumoto T, Ishihara K. Study on acoustic and flow induced noise characteristics of T-shaped pipe with square cross-section. Advances in Applied Acoustics. 2016;**5**:6-17

[29] Croaker P, Kinns R, Nicole K, Marburg S. Fast low-storage method for evaluating Lighthill's volume Quadrupoles. AIAA Journal. 2013;**51**(4): 867-884

[30] Mori M, Masumoto T, Ishihara K. Study on acoustic, vibration and flow induced noise characteristics of T-shaped pipe with a square cross-section. Advances in Applied Acoustics. 2017;**5**: 10-17

[31] Pierce AD. Acoustics: An Introduction to its Physical Principles and Applications. New York: McGraw-Hill; 1981

Section 3

Experiments and Applications

Chapter 7

Application of Vortex Control Principle at Pump Intake

Zambri Harun, Tajul Ariffin Norizan and Wan Hanna Melini Wan Mohtar

Abstract

Vortex flow in a pump intake could affect a pump operation significantly if not treated appropriately. Many researches have been conducted to determine the best control method for vortex flow in pump sumps so that the pump lifespan can be maximized. In this study, a vortex control principle designed to minimize the impact of submerged vortex flow in pump sump on major pump components is presented. This principle employs a device called the plate type floor splitter which serves the function of eliminating vortices formed on the sump floor and reduces the intensity of swirling motion in the intake flow. A pump sump model was built to carry out the study by installing a floor splitter plate sample under the pump suction inlet and the corresponding parameters used to quantify the swirl intensity known as the swirl angle was measured. Procedures for the measurement were conducted based on ANSI/HI 9.8-2018 standard. A numerical simulation was performed to study the flow in a full-scale pump sump. The results showed that the installation of floor splitter plate can eliminate vortices efficiently and reduce swirl angle significantly. However, optimization of floor splitter design is needed to achieve a reduction effect that can reduce swirl angles to an acceptable value of lower than 5° according to ANSI/HI 9.8-2018 standard.

Keywords: vortex flow, pump sump, anti-vortex device, swirl angle, floor splitter

1. Introduction

Pump intake is the part of a pump that draws fluid from the reservoir called the sump as a result of pressure difference generated by the impeller. In most cases, pumped fluid enters the intake in a swirling motion due to geometric features of the sump [1]. Inappropriate sump design such as abrupt changes in sump boundaries, narrow clearance under the pump inlet and asymmetric orientation of the approach channel to the sump will lead to the formation of swirls and vortices [2]. Strong vortices may cause damages to the pump impeller by channelling air to the impeller surface and initiate adverse effects such as cavitation and vibration [3]. On the other hand, excessive swirls in the intake flow can impose imbalance loading to the impeller and even bring resistance to the impeller rotation by introducing swirl rotation in the opposite direction [4]. Due to site condition and operational restrictions, optimal sump design may not be achieved, and therefore local flow correction devices are used as remedial measures.

These devices which are commonly known as anti-vortex device (AVD) come in different shapes and sizes, depending on its application. The conceptual design of

AVD is outlined in ANSI/HI 9.8-2018 [5] standard which is a guideline to assist engineers and designers in optimal intake sump design. Among the AVD types employed in real applications are floor splitter [6], floor cone [7] and corner fillet [8]. These AVD types serve the purpose of eliminating submerged vortices formed at the sump floor. Floor splitters are the most widely used AVD type due to its effectiveness in eliminating vortices and reducing vorticity in the pump intake flow. There are two versions of floor splitter, namely the prism and the plate types. The use of plate type floor splitter is favourable in many applications due to its fabrication friendly-feature and economic design [9]. However, there are a limited number of articles in the literature which discuss the features of floor splitter plate in detail. In this chapter, the characteristics of swirl angle reduction of a floor splitter plate installed in pump sump are studied.

2. Methodology

The study was carried out by both experimental and numerical approaches. A single intake pump sump model, as shown in **Figure 1**, was utilized for the study in which the sample of a floor splitter was installed beneath an intake suction pipe in the sump model. The layout of the sump model test section and the dimensions of the floor splitter installed is illustrated in **Figure 2(a)** and **(b)**, respectively.

2.1 Swirl angle measurement

The main objective of the study is to evaluate the swirling motion in the intake pipe and associated with submerged vortex without and with the installation of floor splitter plate. Initially, the experiment was conducted without the installation of floor splitter plate to capture the initial conditions of the setup. The measurement of the intensity of swirl in the intake pipe was performed according to the procedure described in ANSI/HI 9.8-2018 standard for pump sump model test. The parameter used to quantify the measurement data is the swirl angle θ which is defined in the following equation:

$$\theta = tan^{-1}\left(\frac{\pi dn}{v}\right) \qquad (1)$$

where d is the inner diameter of the intake pipe, n is the revolution count of the measurement instrument called the swirl metre and a is the average axial velocity at the location of the swirl metre. The swirl metre consists of a shaft with four straight

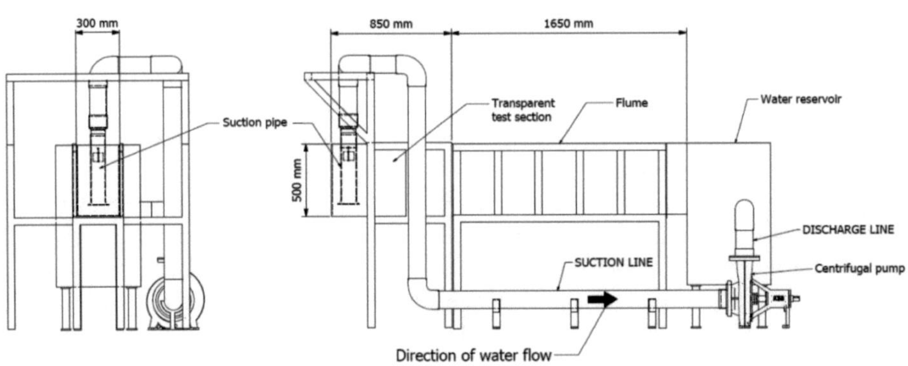

Figure 1.
The experimental rig.

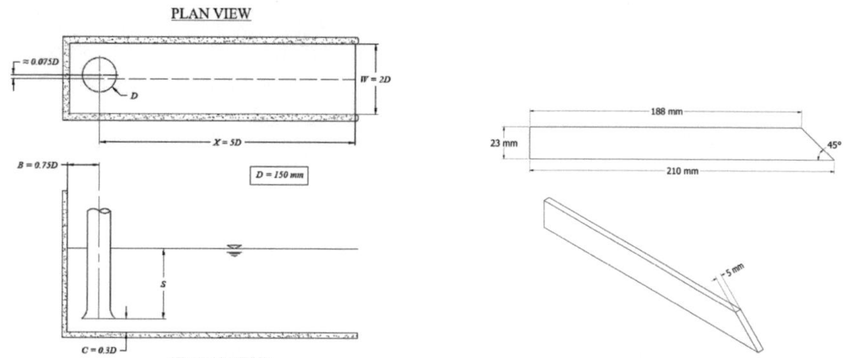

(a) Sump model layout

(b) Dimensions of floor splitter plate used in the study

Figure 2.
Main dimensions of the sump model and splitter.

blades used to capture the swirling motion in the intake pipe, and the revolution count of the swirl metre blade is measured using a tachometer. **Figure 3** shows typical swirl metre installation according to ANSI/HI 9.8-2018 standard. Basically, θ is the convention for describing the ratio between the axial velocity and the tangential velocity of the intake flow which characterizes the intensity of the swirling motion in the fluid. The acceptance criteria according to ANSI/HI 9.8-2018 is that the swirl angle must be lower than 5° to prevent excessive swirl in the intake flow.

In order to generate the submerged vortex, the clearance under the pipe was set to 0.3 times the diameter of the inlet D and two types of flow conditioners

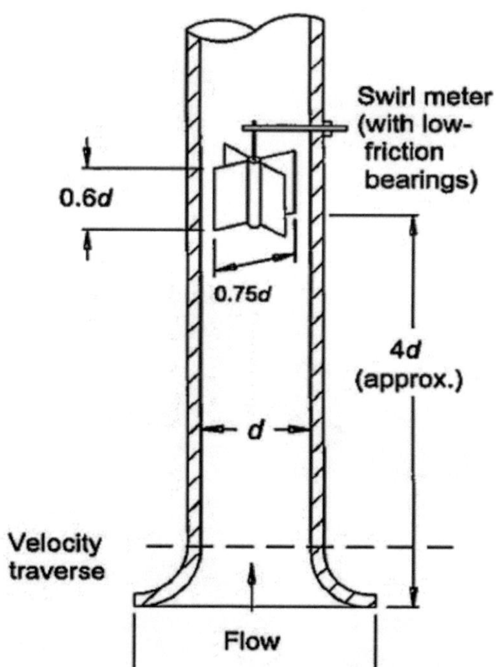

Figure 3.
Swirl metre installation according to ANSI/HI 9.8-2018 standard.

were installed: a sloped floor with an inclination angle of 30° and a sloped wall with the same inclination angle. These flow conditioners were installed at a distance of about 5D from the centre of the intake pipe as shown in **Figure 4(a)** and **(b)**, respectively. The measurement was conducted in a range of pump submergence levels which are normalized by the minimum inlet submergence S_{min}, a threshold value before the occurrence of a surface vortex. S_{min} is calculated by the following equation:

$$S_{min} = D(1 + 2.3Fr_{in}) \tag{2}$$

Fr_{in} is the Froude number at the pipe inlet and is given by:

$$Fr_{in} = \frac{v_{in}}{\sqrt{gD}} \tag{3}$$

where v_{in} is the flow velocity at the inlet and g is the gravitational acceleration. The range of the dimensionless parameter S/S_{min} was set between 0.8 and 1.2.

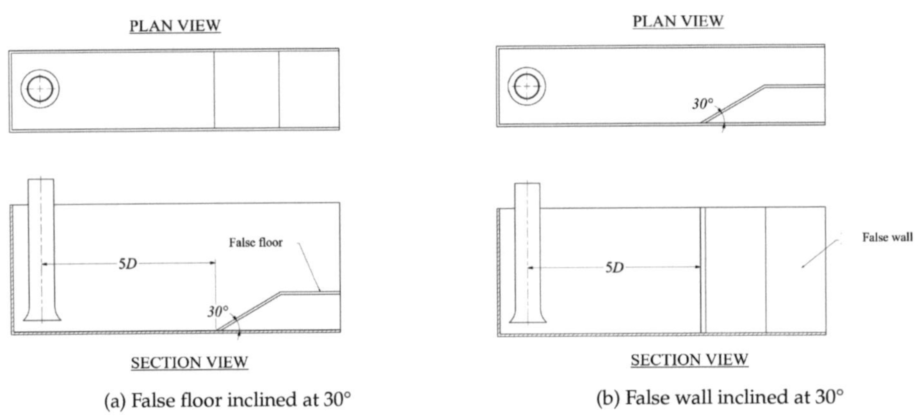

Figure 4.
False floor and false wall arrangements.

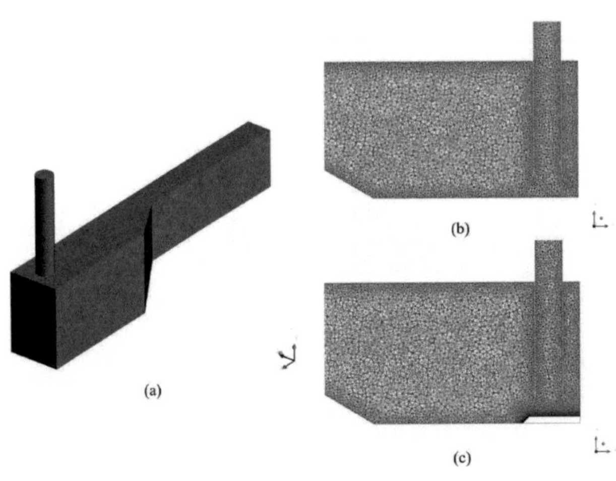

Figure 5.
Numerical model of the full-scale pump sump; (a) the computational domain, (b) model without floor splitter plate, (c) model with floor splitter plate.

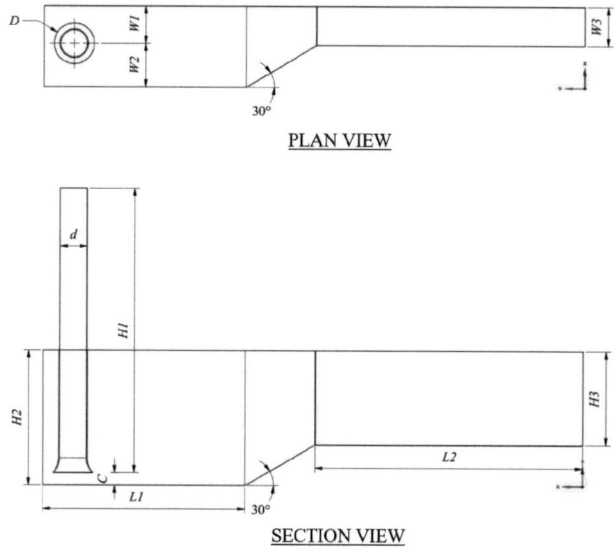

Figure 6.
Dimensions of the full-scale model.

Parameter	Dimension (mm)
Inlet diameter D	1275
Pipe diameter d	850
Right side distance $W1$	1190
Left side distance $W2$	1360
Water entrance width $W3$	1275
Intake pipe height $H1$	9350
Sump height $H2$	4250
Water entrance height $H3$	2975
Floor length $L1$	6375
Water entrance distance from sloped floor $L2$	8417
Clearance C	382.5

Table 1.
Dimension values of the full-scale model shown in Figure 6.

2.2 Numerical simulation of flow in full-scale pump sump

The numerical approach part of the study is set for the simulation of the flow in a full-scale pump sump. As the construction cost for a full-scale pump sump cannot be afforded, a computational fluid dynamics (CFD) simulation was employed as a replacement. The numerical model was validated with experimental data and incorporated with a combined flow conditioner that consists of inclined floor and inclined wall as the ones used in the experiment and built at a scale of 9:1. The flow rate of the pump was set to 2170 l/s, and the pump submergence took the value of S_{min} which is, after the calculation by using Eq. (2), equals to 2678 mm. The mesh structure and the dimensions of the full-scale pump sump are illustrated in **Figures 5** and **6**, respectively, while the values of the model dimensions are listed in **Table 1**.

3. Results and discussion

3.1 Swirl angle evaluation

Figures 7 and **8** show the distribution of swirl angle values at different submergence ratios in the case of false floor and false wall flow conditioner, respectively.

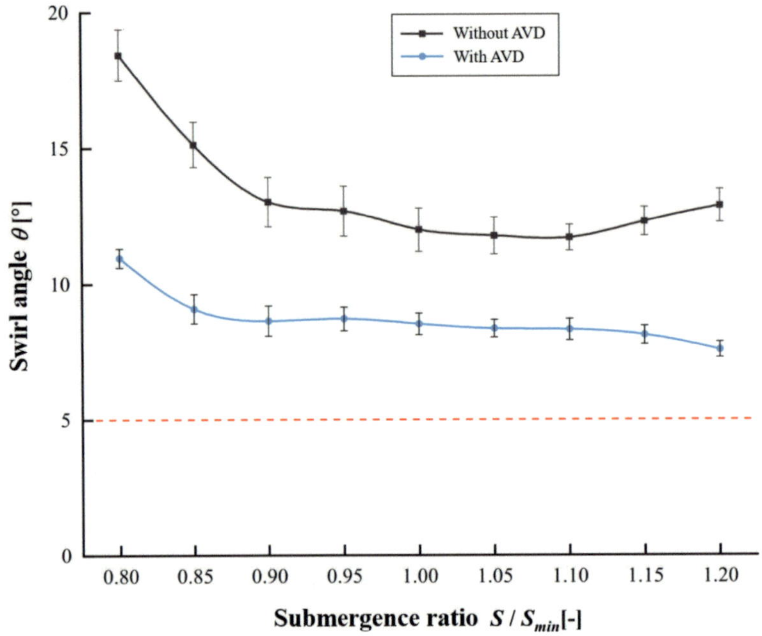

Figure 7.
Swirl angle values at different submergence ratios for the false floor case.

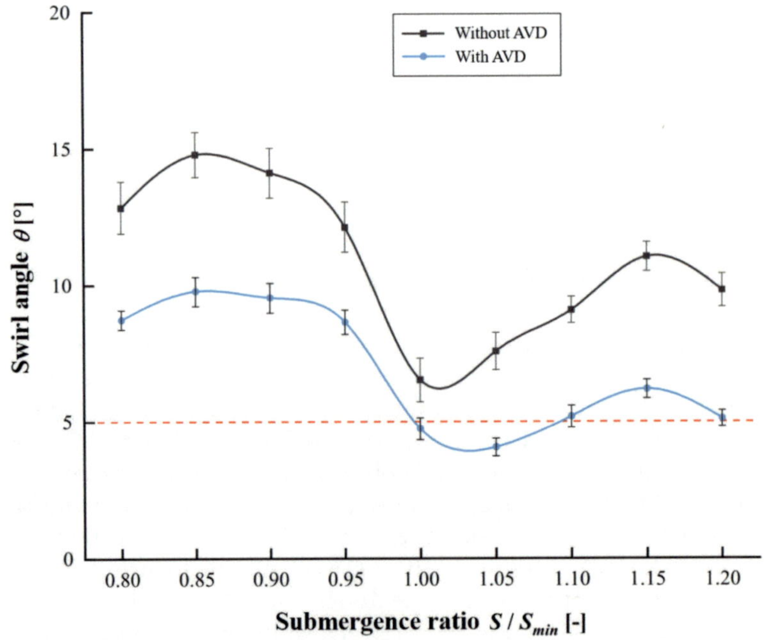

Figure 8.
Swirl angle values at different submergence ratios for the false wall case.

Generally, the installation of floor splitter plate has shown reduction in the swirl angle values. The parameter that can be used to characterize the reduction effect of the floor splitter plate is the swirl angle reduction factor R_θ which is defined as follows:

$$R_\theta = \frac{\theta_{withoutAVD}}{\theta_{withAVD}} \tag{4}$$

In this experiment, the average value of R_θ for the false floor case is 1.53, while in the false wall case, the average value of R_θ is 1.62. In **Figure 7**, the swirl angle values show a decreasing trend with increasing submergence ratio for S/S_{min} greater than 1 when installed with floor splitter plate. This is due to the fact that for S/S_{min} greater than 1, there was only a submerged vortex present in the sump. As the function of floor splitter plate is to eliminate submerged vortices, this result proved that the installation of floor splitter plate has served the purpose. When S/S_{min} is decreased below 1, the swirl angle values increase with decreasing submergence ratio. The inception of free surface vortex at S/S_{min} below 1 has caused bigger fluctuation in swirl angle as can be seen in the larger uncertainties within this region. The higher swirl angle values are contributed by the increase in approach flow velocity at lower water levels. The floor splitter vortex has shown limited swirl angle reduction effect if the submergence ratio is decreased below 1.

In **Figure 8**, the swirl angle values show a sinusoidal trend with increasing submergence ratio for S/S_{min} greater than 1 when installed with floor splitter plate. The trend is contributed by the inception of free surface vortex at S/S_{min} greater than 1. Although the theory behind the minimum inlet submergence S_{min} is that there should be no free surface vortex formed in the sump if the submergence S is greater than S_{min}, this deviation from the theory was contributed by the use of false wall in which the flow has been prerotated at the beginning of the sump. The prerotation has therefore caused the flow to develop a free surface vortex earlier than expected. In the experiment, this situation occurred at S/S_{min} = 1.15. As the swirl angle decreases when S/S_{min} decreases below 1.15, the reduction effect of the floor splitter plate can be observed in the decreasing trend of the swirl angle values. Similar to the case of false floor, the swirl angle increases as the submergence ratio decreases due to the increasing approach flow velocity at low water levels.

Despite the swirl angle reduction effect of floor splitter plate, the fulfilment of the requirement of swirl angle reduction below 5° has not been achieved for most of the cases. In the case of false floor, there is no submergence ratio value at which the swirl angle has been reduced below 5°; however, for the false wall case, the reduction of swirl angle values below 5° can be seen between S/S_{min} = 1.00 and S/S_{min} = 1.05, i.e. the requirement for all submergence ratios when installed with floor splitter plate. This result shows that there is a limiting factor that prevented the swirl angle reduction below 5° and that factor lies on the design of the floor splitter as suggested by Kang et al. [9].

3.2 Simulation of flow in full-scale pump sump model

The first part of the discussion on the result of simulation of flow in full-scale pump sump model is about the vortex elimination by the installation of floor splitter plate. The evaluation is based on the vorticity in the y-axis ω_y due to its influence on the swirling motion of the flow. The value of ω_y is normalized by the ratio of velocity in the suction pipe and the pipe inner diameter ν_d/d. **Figure 9** shows the cross section along x-y plane in which the evaluation of the result in the streamwise direction takes place and its corresponding results which are shown in **Figure 10**.

From **Figure 10**, it can be observed that the core of the vortex, indicated by the high-intensity region extending from the floor towards inside of the pump, has

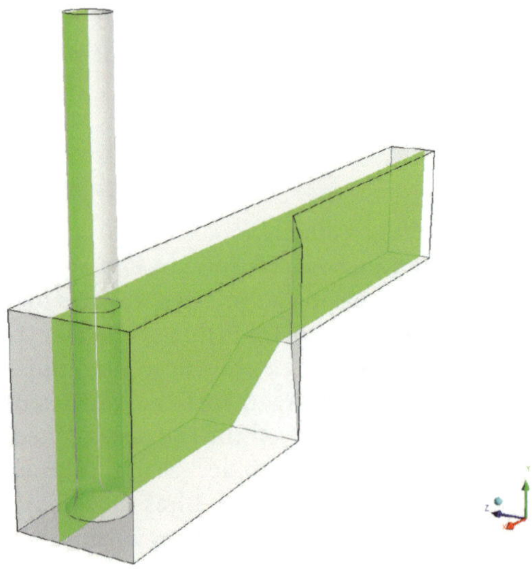

Figure 9.
Evaluation area in x-y plane of the pump sump model (z = 1600 mm).

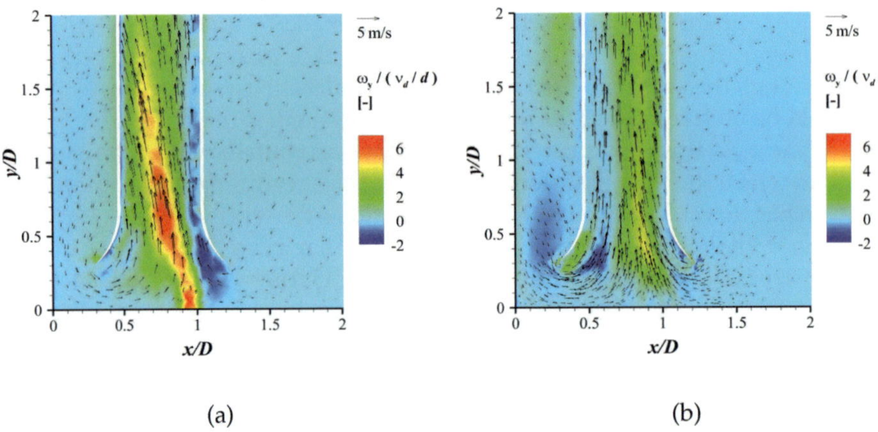

Figure 10.
Contour plot of ωy/(vd/d) at the location of vortex in x-y plane; (a) without floor splitter plate, (b) with floor splitter plate. Dashed circular lines denote pipe diameter.

been eliminated with the installation of floor splitter plate. The vorticity in the pipe has also been reduced which can be seen from the contour colors. The velocity vectors, which appear to point diagonally to the left indicating a strong swirling flow in the pipe, have been straightened in a direction vertically upward towards the direction of suction when installed with floor splitter plate.

When observing the cross section in the spanwise direction (in the plane illustrated in **Figure 11**), similar results are presented. Basically the flow that enters the pump is divided into two regions, namely the right side and the left side flow, due to the geometry of the sump. The flow entrance velocity from the right and the left side of the inlet are nearly the same because of the nearly symmetrical positioning of the pump. The flow entered the pump in a spiral manner without the installation of floor splitter which resulted in vortex formation near to the left side of the pump. When installed with floor splitter plate, the flow is reorganized, and therefore the spiral motion of the flow has been reduced and hence the vortex eliminated. This situation is reflected by the discontinued vortex core shown in **Figure 12** with the installation of floor splitter plate.

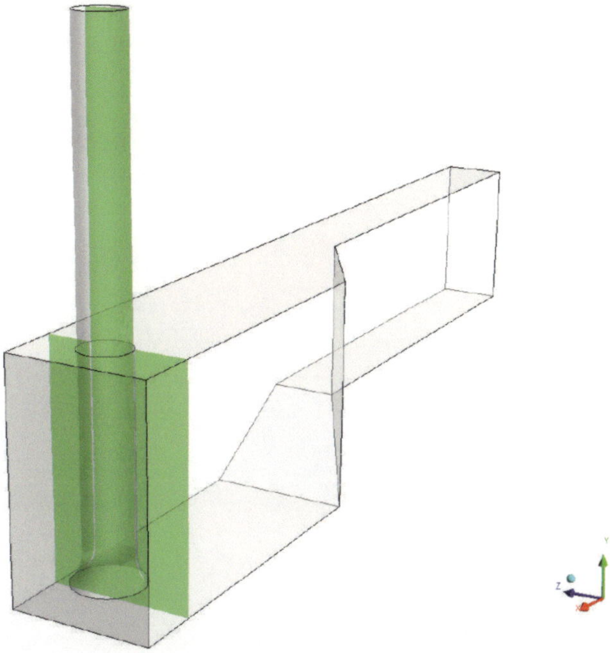

Figure 11.
Evaluation area in y-z plane of the pump sump model (x = 15,860 mm).

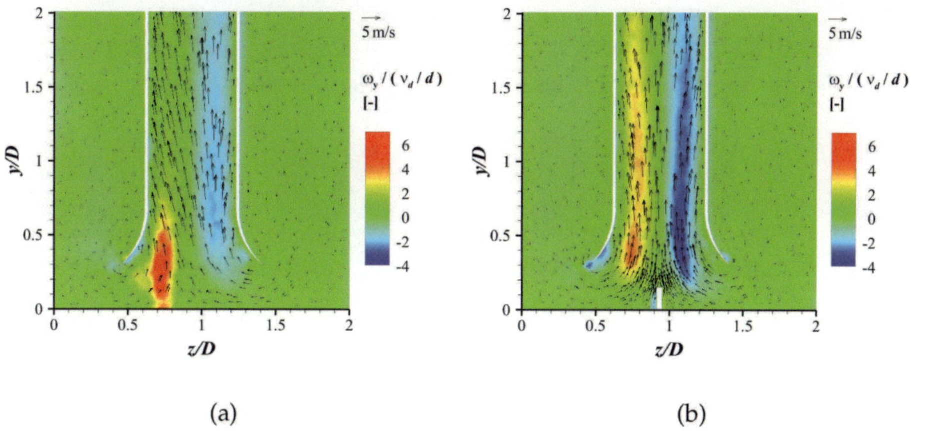

Figure 12.
Contour plot of ωy/(vd/d) at the location of vortex in y-z plane; (a) without floor splitter plate, (b) with floor splitter plate.

As the main function of floor splitter plate is to eliminate vortices formed at the sump floor, an evaluation about the vorticity in the plane at the sump floor is necessary. This location is shown in **Figure 13**. The vortex core is indicated by the spiralling streamline under the pump inlet which can be seen in **Figure 14** in the case without floor splitter plate. As the floor splitter plate was installed, the path of the spiral streamline was interrupted by the plate, and therefore the formation of vortex was prevented. Due to the suction by the pump, a small vortex attached to the side of the floor splitter plate was formed which is inherited from the flow without floor splitter plate as shown in **Figure 14(b)**. However, this vortex constitutes a much smaller vortex core diameter (estimated to be less than 0.1D based on the scale at the x-axis of the graph) and relatively weak compared to the large vortex

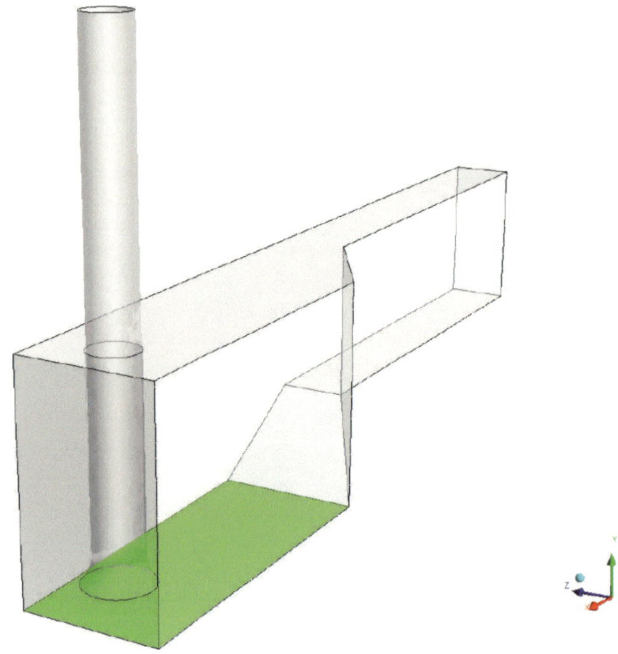

Figure 13.
Evaluation area in z-x plane of the pump sump model (y = 10 mm).

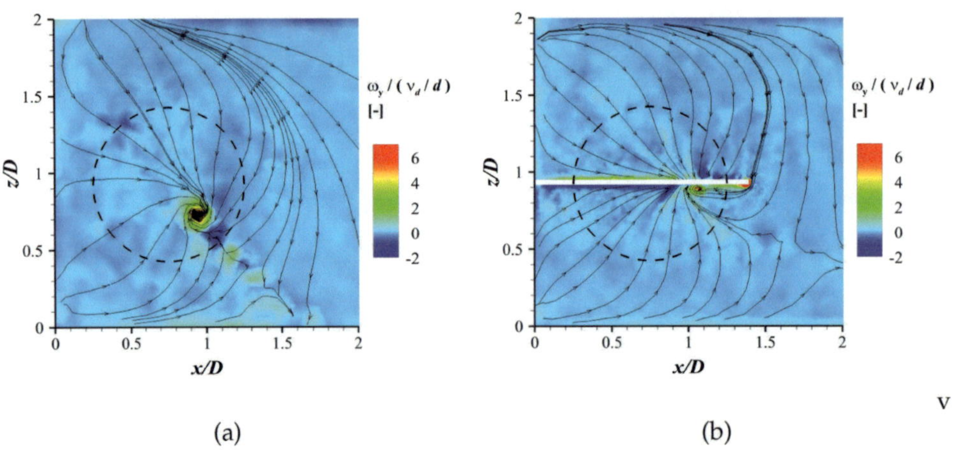

Figure 14.
Contour plot of ωy/(vd/d) at the floor of the sump in z-x plane; (a) without floor splitter plate, (b) with floor splitter plate.

(estimated to be about 0.2D) which can be seen in **Figure 14(a)**, and therefore it can be considered as nondestructive to the pump impeller.

The next part of the evaluation is about the swirl angle reduction characteristics of floor splitter plate installation. For this purpose, an evaluation plane was selected at the position comparable to the installation of swirl metre in the experimental model. The location of the plane is shown in **Figure 15**, and its corresponding results are displayed in in **Figure 16**. The flow at the swirl metre location was rotational with relatively high velocity components as indicated by the velocity vectors. The two visible vorticity regions show the divided inflow field in the pipe as explained in the previous paragraph which is considerably high in reference to the value of ν_d/d as shown in **Figure 16**. With the installation of floor splitter,

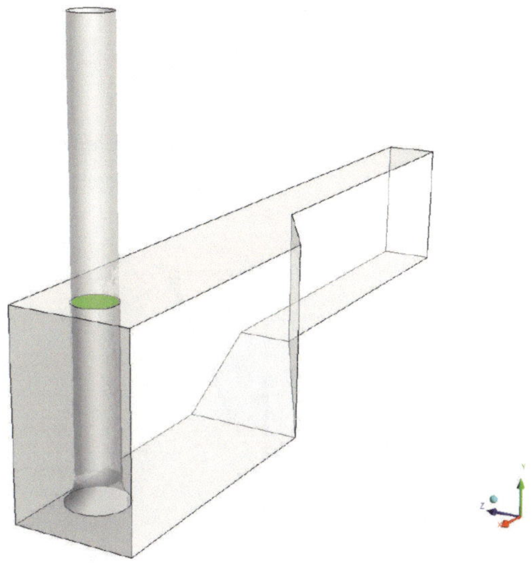

Figure 15.
Evaluation area at the position of swirl metre in z-x plane of the pump sump model (y = 4250 mm).

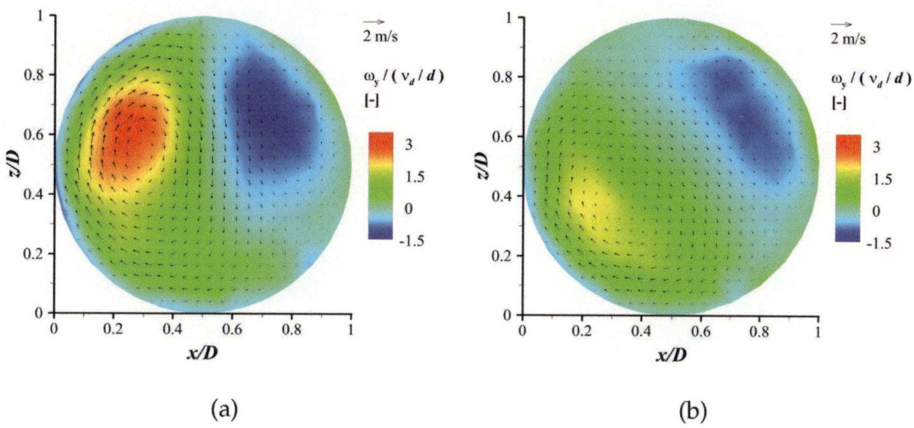

Figure 16.
Contour plot of ωy/(vd/d) at the swirl metre position in z-x plane; (a) without floor splitter plate, (b) with floor splitter plate.

the magnitude of both vorticity regions is significantly reduced and the resulting velocity vectors are also smaller in size compared to the case without floor splitter. This indicates that the spiral flow has been dissolved by the floor splitter plate into a relatively straight flow and the outcome is consistent with the experimental result presented in the previous subsection.

To get a better understanding about the result, a 3D streamline visualization of the intake flow in the sump is illustrated for every case as comparison in **Figure 18**. It can be seen that the intake flow was spiral before the installation of floor splitter plate and as the floor splitter was installed, the spiral motion of the flow was dissolved and went into a relatively straight path. Quantitative values can also be extracted from the result to obtain the associated swirl angle values. The approach for the calculation of swirl angle from the simulation results is based on the principle of Eq. (1) itself where by definition the swirl angle is the angle between the velocity components of the intake flow in the axial and tangential direction. From Eq. (1), the term πdn represents the tangential velocity component, while the term v represents the axial velocity

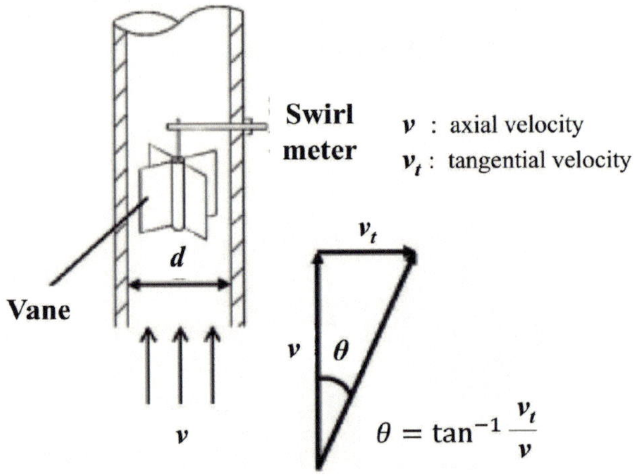

Figure 17.
Swirl angle definition using velocity triangle diagram as shown in Kang et al. [9].

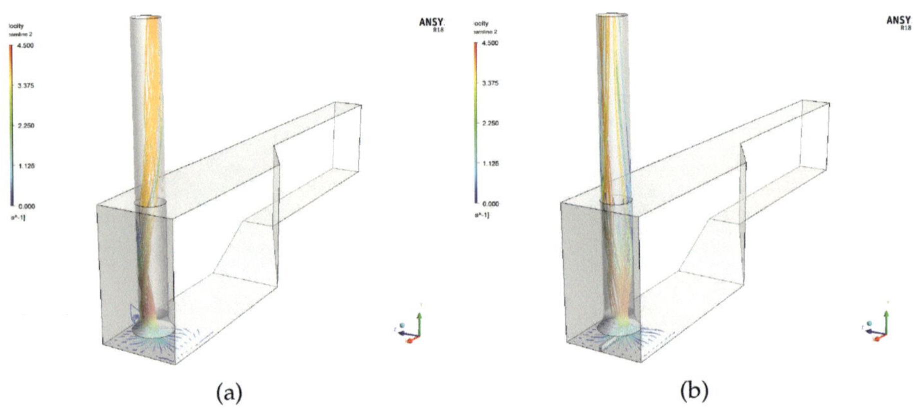

Figure 18.
3D streamline plot showing the intake flow in the sump with the seeding of the flow starts at the floor of the sump; (a) without floor splitter plate, (b) with floor splitter plate.

component; both are at the location of the swirl metre used in the experiment. **Figure 17** shows the velocity triangle diagram which shows the relationship between swirl angle and both of the velocity components in a schematic representation.

Based on this approach in Eq. (1), the velocity components in the axial and tangential direction were derived from the simulation results. As the result was given in vorticity values, the tangential velocity component must be derived from the angular velocity which equals to half of the vorticity [10]. The vorticity of the flow at the position of the swirl metre is calculated by the integration of the vorticity in the plane and divided by the cross section to obtain the vorticity value per unit area. After getting the value of angular velocity, the following correlation is used to calculate the tangential velocity:

$$v_t = r \cdot \omega_y \tag{5}$$

The method to derive the value of axial velocity component from the results was based on the same principle in which the integral value of axial velocity component in the plane was extracted and divided by cross-sectional area of the pipe at the swirl metre location to get the velocity per unit area. The reason of performing integration to find the velocity values is that the swirling motion of the intake flow in the pipe constitutes a solid body rotation and the swirl angle value describes the rotation body as a whole [1], and this is where the integration of the velocity across the cross-sectional area becomes the most practical way of calculating the swirl angle in the simulation. After obtaining both velocity values, the swirl angle was then calculated using the velocity triangle diagram as shown in **Figure 17**.

By following the described procedure, the swirl angle value for the case without floor splitter plate installation is 7.58°, while for the case with floor splitter plate, the swirl angle value is 4.09°. Although these values are based on average velocities as the simulation was conducted in a steady-state simulation and therefore are much smaller than the actual swirl angle values, it can be considered as adequate because they are used for comparison purpose and not for the determination of absolute values. Once again, the results are in agreement with the experimental data. This study complements a previous experimental investigation in which the effects of floor splitter heights have been analysed [11].

4. Conclusions

A study on the application of vortex control principle at pump intake was carried out by using an anti-vortex device type called the floor splitter plate. The device was installed in a pump sump model to eliminate vortices formed at the intake and reduce the swirling motion in the intake pipe as a method to improve pump efficiency in actual applications. Evaluation of the effect was conducted based on experimental and numerical approaches. The experimental part comprised swirl angle measurement which was performed according to ANSI/HI 9.8-2018 standard. To complement the results obtained in the experiment, a numerical simulation of the flow in a full-scale pump sump was conducted. The results showed that the installation of floor splitter plate has successfully eliminated the vortex formed at the sump floor and reduced the swirl angle in the intake flow. However, the reduction effect was not sufficient to achieve the criteria set in the ANSI/HI 9.8-2018 standard which requires the swirl angle to be less than 5°, and therefore optimization of the floor splitter plate design is needed. The simulation of flow in a full-scale pump sump produced similar findings with the experimental results. From the contour and streamline plot, it was found that the immersion of the floor splitter plate has disrupted the vortical flow under the pump inlet and provided a flow straightening effect to eliminate destructive vortices and reduce swirl angle in the pump intake.

Acknowledgements

The research has been funded by the Ministry of Energy, Science, Technology, Environment and Climate Change (MESTECC), Malaysia, under Science Fund grant No. SF1326 and carried out in collaboration with the Department of Irrigation and Drainage (DID), Malaysia.

Author details

Zambri Harun[1*†], Tajul Ariffin Norizan[1†] and Wan Hanna Melini Wan Mohtar[2†]

1 Department of Mechanical and Manufacturing Engineering, Faculty of Engineering and Built Environment, National University of Malaysia, Bangi, Malaysia

2 Department of Civil Engineering, Faculty of Engineering and Built Environment, Universiti Kebangsaan Malaysia, Bangi, Malaysia

*Address all correspondence to: zambri@ukm.edu.my

† These authors contributed equally.

References

[1] Hecker GE. Fundamentals of vortex intake flow. In: Knauss J, editor. Swirling Flow Problems at Intakes. London: Taylor & Francis; 1987. pp. 123-137. DOI: 10.1002/978047974704

[2] Melville BW, Ettema R, Nakato T. Review of Flow Problems at Water Intake Pump Sumps. New York: Electric Power Research Institute; 1994. p. 532. DOI: 10.1002/9780470974704

[3] Padmanabhan M. Design recommendation: Pump sump. In: Knauss J, editor. Swirling Flow Problems at Intakes. London: Taylor & Francis; 1987. pp. 234-249. DOI: 10.1002/978047974704

[4] Cooper P. Centrifugal pumps—Centrifugal pump theory, analysis and performance. In: Karassik I, editor. Pump Handbook. New York: McGraw-Hill; 2008. pp. 478-490. DOI: 10.1002/978047974704

[5] Hydraulic Institute. ANSI/HI 9.8-2018: Rotodynamic Pumps for Pump Intake Design. Parsippany: Hydraulic Institute; 2018. p. 497. DOI: 10.1002/9780470974704

[6] Norizan TA, Reda E, Harun Z. Enhancement of vorticity reduction by floor splitter in pump sump to improve pump efficiency. Sustainable Energy Technologies and Assessments. 2018;**28**:28-36. DOI: 10.1016/s0014-5793(01)03293-8

[7] Yang F, Liu C. Numerical and experimental investigations of vortex flows and vortex suppression schemes in the intake passage of pumping system. Advances in Mechanical Engineering. 2015;**7**(2):547086 (1-9). DOI: 10.1016/s0014-5793(01)03293-8

[8] Hwang KS, Yang CH, Hwung HH. Effects of travelling water screen configurations on pump sump flows. In: Proceedings of the 6th International Conference on Hydrodynamics; 24-26 November 2005. Perth: ICHD; 2005. pp. 451-462

[9] Kang WT, Shin BR, Doh DH. An effective shape of floor splitter for reducing sub-surface vortices in pump sump. Journal of Mechanical Science and Technology. 2014;**28**(1):175-182. DOI: 10.1016/s0014-5793(01)03293-8

[10] Wu JZ, Ma HY, Zhou MD. Vorticity and Vortex Dynamics. Heiderlberg: Springer; 2006. p. 780. DOI: 10.1002/9780470974704

[11] Norizan TA, Harun Z, Abdullah S, Wan Mohtar WHM. Effects of floor splitter height on the effectiveness of swirl angle reduction in pump intake. Journal of Advanced Research in Fluid Mechanics and Thermal Sciences. 2019;**57**(1):32-39

Chapter 8

The Effects of Curved Blade Turbine on the Hydrodynamic Structure of a Stirred Tank

Bilel Ben Amira, Mariem Ammar, Ahmad Kaffel, Zied Driss and Mohamed Salah Abid

Abstract

This work is aimed at studying the hydrodynamic structure in a cylindrical stirred vessel equipped with an eight-curved blade turbine. Flow fields were measured by two-dimensional particle image velocimetry (PIV) to evaluate the effect of the curved blade turbine. Velocity field, axial and radial velocity distribution, root mean square (rms) of the velocity fluctuations, vorticity, and turbulent kinetic energy were presented. Therefore, two recirculation loops were formed close to the free surface and in the bottom of the tank. Moreover, the highest value area of the vorticity is localized in the upper region of the tank which follows the same direction of the first circulation loop. The turbulent kinetic energy is maximum at the blade tip following the trailing vortices.

Keywords: hydrodynamics structure, stirred vessel, curved blade turbine, PIV

1. Introduction

Stirrer geometry and design in a mechanically agitated tank have been studied over a wide range of design aiming at improving the agitation efficiency. Several studies which have already been carried out were interested in enhancing the vessel property, experimentally and numerically. In fact, the experimental study includes the effect of removal of baffles, the impeller geometrical effects, the number of blades on energy efficiency, and the impacts of solid concentration and particle size on power consumption [1]. Mixing time and pattern in the agitated vessel was also experimentally investigated at various mixing Reynolds numbers [2]. Three types of impellers generated in different liquid flows as well as their position were used to investigate their influences on the kinetic parameters in a batch cooling crystallizer [3]. In addition, particle image velocimetry technique (PIV) was used to carry out the turbulent flow inside a cylindrical baffled stirred vessel with a set of speed ranging from 100 to 350 rpm [4]. Ben Amira et al. [5] also studied the hydrodynamic structure of the flow generated by eight concave blade turbines. Furthermore, PIV technique is also used to estimate the turbulence energy dissipation rate in a stirred vessel generated by a Rushton turbine [6]. In addition, both experimental and numerical techniques were developed simultaneously. In fact, Cruz-Díaz et al. [7] modeled the operation of the FM01-LC reactor coupled with a continuous stirred tank (CST) in recirculation mode. Li et al. [8] used the large eddy simulation and the PIV technique to calculate the velocity field generated by a Rushton turbine. In-line high shear mixers (HSMs) with double rows of ultrafine inclined stator teeth were experimentally and numerically investigated under different rotor speeds and flow rates [9]. Furthermore, computational fluid dynamics (CFD) simulations were used to investigate the effects of impeller configuration on fungal physiology and cephalosporin C production by an industrial strain *Acremonium chrysogenum* in a bioreactor equipped with conventional and novel impeller configuration, respectively [10]. Navier-Stokes equation in conjunction with the RNG (renormalization group) of the k-ε turbulent model was used to study the turbulent flow induced by the six flat blade turbines (FBT6), the Rushton turbine (RT6), and the pitched blades turbine (PBT6) in a stirred tanks [11]. Finite volume method was employed to solve the Navier-Stokes equations governing the transport of momentum to compare four different turbulence models used for numerical simulation of the hydrodynamic structure generated by a Rushton turbine in a cylindrical tank [12]. Multiple impellers were used in a stirred vessel to form the micro/nano drug particle in the biopharmaceutical classification system [13].

According to the biography, it is interesting to study the effect of the blade shape in order to improve the hydrodynamic structure. In this paper, we are interested in studying the hydrodynamic structure in a cylindrical stirred vessel equipped by an eight-curved blade turbine.

2. Experimental apparatus

In this study, PIV was used to study the hydrodynamic structure generated with the convex blade turbine in a stirred vessel. For the particle image velocimetry, the flow is illuminated by an Nd-YAG 532 nm green pulsed laser source generated in 2×30 mJ. The acquisition of the two-dimensional image data was taken with a CCD camera with 1600×1200 pixels of resolutions. The results are obtained for 170 images at three different azimuthally planes. The average diameter of the seeding particle was equal to dp = 20 μm with 0.15 g of concentration. The cylindrical vessel was mounted in a squared vessel to decrease reflection which is filled by water. The height of the water is equal to the tank diameter (D = 300 mm). Besides, the turbine diameter is equal to the half of the vessel diameter (T = D/2), and it is placed in the middle of the tank. Four equally spaced baffles which are placed 90° far from one another were used. The rotation speed of the turbine is equal to N = 70 rpm which is equal to a Reynolds of Re = 26,250. The velocity at the tip of the impeller is equal to U_{tip} = 0.55 m/s. The velocity speed is settled by using an electrical motor placed at the top of the tank and controlled automatically by the computer. In addition, the azimuthally plane of the investigation is localized at θ = 10° from the blade. The angular position between two successive blades is settled through a position sensor localized near the electrical motor.

Figure 1 shows the different impellers geometry for the one-staged system namely, flat, concave, and convex blade impeller. **Figure 2** shows the staged system. In fact, the highest impeller is localized at the third position according to the free surface, and the lowest one is localized at the third position according to the bottom. For the first configuration (**Figure 2(a)**), the flat blade impeller is localized at the highest position and the concave blade impeller at the bottom (PD8 h, PI8 concave). For the second configuration, the impeller at the bottom is replaced by the convex impeller (PD8 h, PI8 convex). For the third and the fourth configurations, the flat blade impeller is localized at the bottom, and the highest impeller is occupied by the concave and the convex blade impeller, respectively (PD8 b, PI8 concave; PD8 b, PI8 convex).

a. Flat blade impeller   b. Concave blade impeller   c. Convex blade impeller

Figure 1.
One-staged system.

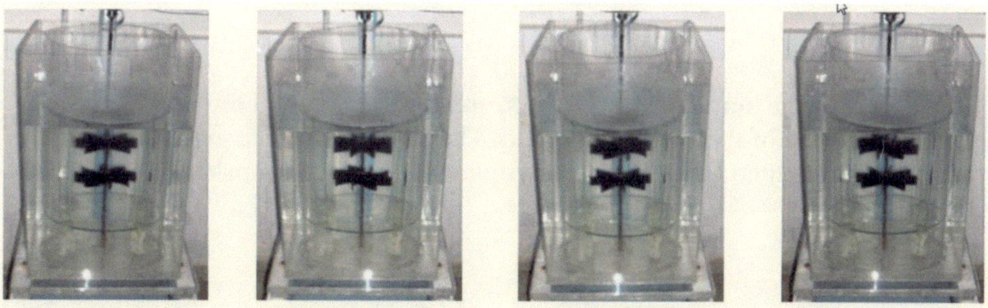

a. PD8 h, PI8 concave   b. PD8 h, PI8 convex   c. PD8 b, PI8 concave   d. PD8 b, PI8 convex

Figure 2.
Staged system.

3. Data analysis

For a given point, the instantaneous velocity components u_{ij} in terms of a mean velocity $\overline{u}_i$ and a temporal fluctuating term u'_{ij} is given as

$$u_{ij} = \overline{u}_i + u'_{ij} \tag{1}$$

where i refers to the velocity component r, z, or θ and j refers to the instant at which the velocity was measured for each acquisition.

The mean velocity is calculated as follows:

$$\overline{u}_i = \frac{1}{Nn}\sum_{j=1}^{Nn} u_{ij} \tag{2}$$

With Nn refers to the snapshot total number. The average of the temporal fluctuating terms which known as a root mean square (rms) value is given by

$$u_{i,rms} = \left(\overline{u'^2_{ij}}\right)^{1/2} = \left(\frac{1}{Nn}\sum_{j=1}^{Nn}\left(u_{ij}-\overline{u}_i\right)^2\right)^{1/2} \tag{3}$$

For two-dimensional velocity data, in the Cartesian coordinate, the turbulent kinetic energy is equal to

$$k = \frac{3}{4}\left(\overline{u'^2} + \overline{v'^2}\right) \tag{4}$$

The proper orthogonal decomposition (POD) is obtained by computing the auto-covariance matrix (R). According to Liné et al. [14], the matrix of instantaneous velocity vector data is calculated as

$$M = \begin{bmatrix} u_1(1) & . & . & u_{Nn}(1) \\ . & & & . \\ . & & & . \\ u_1(n) & . & . & u_{Nn}(n) \\ v_1(1) & . & . & v_{Nn}(1) \\ . & & & . \\ . & & & . \\ v_1(n) & . & . & v_{Nn}(n) \end{bmatrix} \tag{5}$$

Nn refers to the number of snapshots, and n refers to total number of the interrogation area of the whole flow field. The auto-covariance matrix, which is associated to the fluctuating velocity components for each snapshot, is calculated as follows:

$$R = \frac{1}{Nn} M.M^{T} \tag{6}$$

Eigenvalues are calculated by solving the Fredholm integral eigenvalue problem and Karhunen-Loeve analysis:

$$\iint_{\Omega} R(x,z,x',z')\overrightarrow{\phi^{Ki}}(x',z')dx'dz' = \lambda^{ki}\overrightarrow{\phi^{Ki}}(x,z) \tag{7}$$

where Ki refers to the POD mode and Ω refers to the domain of interest.
The eigenfunction is calculated as follows:

$$\phi = \frac{\sum_{n=1}^{Nn} a_n^{Ki} u^n}{\left\| \sum_{n=1}^{Nn} a_n^{Ki} u^n \right\|} \tag{8}$$

4. Experimental results

4.1 Velocity field

Figure 3 shows the velocity field of different types of curved blade turbines for one-staged system. According to these results, two circulation loops were observed, in which the first one is localized in the upper region of the tank near the free surface and the second one is localized in the bottom of the tank. A radial jet is

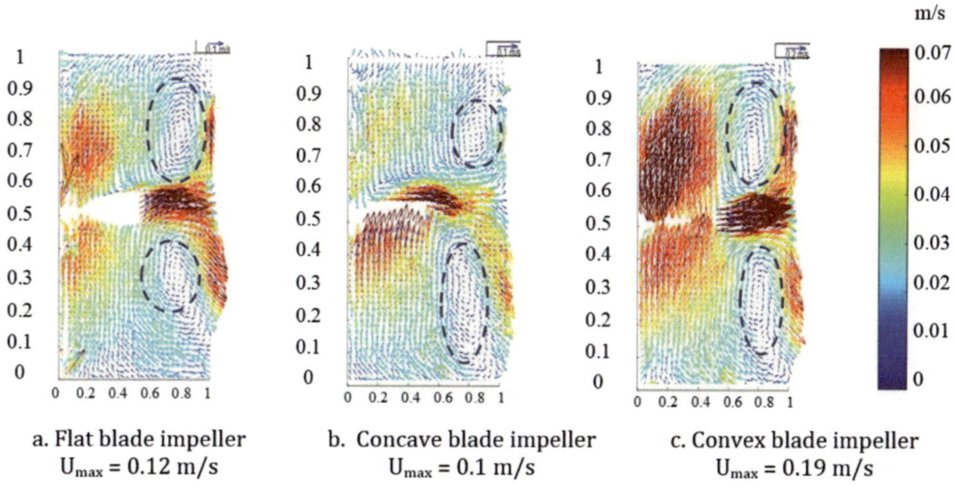

Figure 3.
Velocity field for one-staged system.

described at the blade tip. In fact, the velocity has some deviation in the other azimuthal planes by the effect of the propagation of trailing vortices. The jet flow is more intensive for the convex blade turbine. For the concave blade, a maximum velocity is created at the top edge that deviated to the blade tip. Afterward, near the wall of the vessel, the velocity is divided into the upward and downward flow. Then, it returns to the shaft to create the circulation loop. As a matter of fact, the concave configuration produces a larger lowest loop than the other configurations. Hence, the region located below the impeller (at the bottom of the tank) is more turbulent than the flat and the convex configurations. Moreover, the distribution of the turbulent flow at the upper and the downer regions is more similar at the convex configuration.

The flow distribution of the flat and the curved blade is similar to that presented by Driss et al. [15] for a laminar flow. In fact, a radial jet is created, and then an axial flow is obtained by the wall effect. In addition, the flow slows down significantly far away from the impeller. The shape and the position of the recirculation loops and the trailing vortices are affected significantly by the blade design [15, 16].

The coordination of the center of the highest loops (z_h^*, r_h^*) and the lowest loops (z_b^*, r_b^*) and its radial and the axial extension (z_b, r_b, z_h, r_h) are presented in **Table 1**.

	Flat impeller		Concave impeller		Convex impeller	
Highest loops	z_h^* (z/H)	0.78	z_h^* (z/H)	0.78	z_h^* (z/H)	0.78
	r_h^* (r/R)	0.76	r_h^* (r/R)	0.81	r_h^* (r/R)	0.76
	z_h (z/H)	0.32	z_h (z/H)	0.21	z_h (z/H)	0.32
	r_h (r/R)	0.42	r_h (r/R)	0.33	r_h (r/R)	0.4
Lowest loops	z_b^* (z/H)	0.32	z_b^* (z/H)	0.25	z_b^* (z/H)	0.27
	r_b^* (r/R)	0.72	r_b^* (r/R)	0.75	r_b^* (r/R)	0.8
	z_b (z/H)	0.22	z_b (z/H)	0.37	z_b (z/H)	0.33
	r_b (r/R)	0.4	r_b (r/R)	0.3	r_b (r/R)	0.3

Table 1.
Loop coordination and shape of the one-staged system.

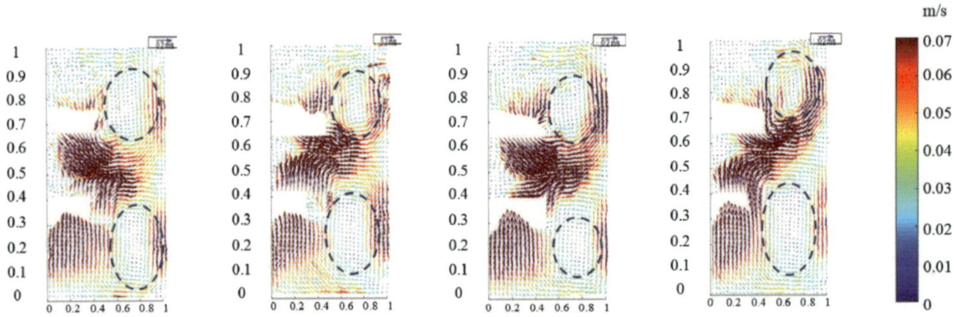

a. PD8 h, PI8 concave U_{max} = 0.19 m/s

b. PD8 h, PI8 convex U_{max} = 0.22 m/s

c. PD8 b, PI8 concave U_{max} = 0.18 m/s

d. PD8 b, PI8 convex U_{max} = 0.18 m/s

Figure 4.
Velocity field for staged system.

	PD8 h, PI8 concave		PD8 h, PI8 convex		PD8 b, PI8 concave		PD8 b, PI8 convex	
Highest loops	z_h^* (z/H)	0.78	z_h^* (z/H)	0.76	z_h^* (z/H)	0.74	z_h^* (z/H)	0.82
	r_h^* (r/R)	0.7	r_h^* (r/R)	0.71	r_h^* (r/R)	0.7	r_h^* (r/R)	0.65
	z_h (z/H)	0.29	z_h (z/H)	0.23	z_h (z/H)	0.26	z_h (z/H)	0.26
	r_h (r/R)	0.46	r_h (r/R)	0.44	r_h (r/R)	0.44	r_h (r/R)	0.46
Lowest loop	z_b^* (z/H)	0.21	z_b^* (z/H)	0.25	z_b^* (z/H)	0.19	z_b^* (z/H)	0.26
	r_b^* (r/R)	0.74	r_b^* (r/R)	0.67	r_b^* (r/R)	0.75	r_b^* (r/R)	0.7
	z_b (z/H)	0.33	z_b (z/H)	0.32	z_b (z/H)	0.24	z_b (z/H)	0.36
	r_b (r/R)	0.46	r_b (r/R)	0.46	r_b (r/R)	0.43	r_b (r/R)	0.47

Table 2.
Loop coordination and shape of the staged system.

In addition, the maximum velocity is greater for the convex blade turbine and weaker for the concave blade turbine due to the interaction between the blade and the flow that is lower for the convex blade and greater for the concave blade (flat turbine U_{max} = 0.22 U_{tip}, concave turbine U_{max} = 0.18 U_{tip}, convex turbine U_{max} = 0.35 U_{tip}).

For the staged system (**Figure 4**), the flow becomes more turbulent, and the loops can reach the free surface as the bottom of the tank. In addition, an oblique flow is created between the two impellers. The highest velocity is produced in the case of the second configuration (U_{max} = 40% U_{tip}) (a, U_{max} = 34.54% U_{tip}; c, U_{max} = 32,72% U_{tip}; d, U_{max} = 32,72% U_{tip}). The position and the shape of the recirculation loops of each configuration are presented at **Table 2**.

The combination of inclined blade turbine and flat turbine shows no great change in terms of acceleration and shape of the recirculation loops. This found locks similar to that developed by Bereksi et al. [17] for the combination between the Rushton and the curved blade. In addition, they proved that the gas holdup is better by the combination between the curved and the Rushton turbine than by the combination between two Rushton turbines.

4.2 Radial and axial velocity

Figure 5 shows the distribution of the radial velocity component of the curved blade turbine. According to these results, the highest value region is localized at the

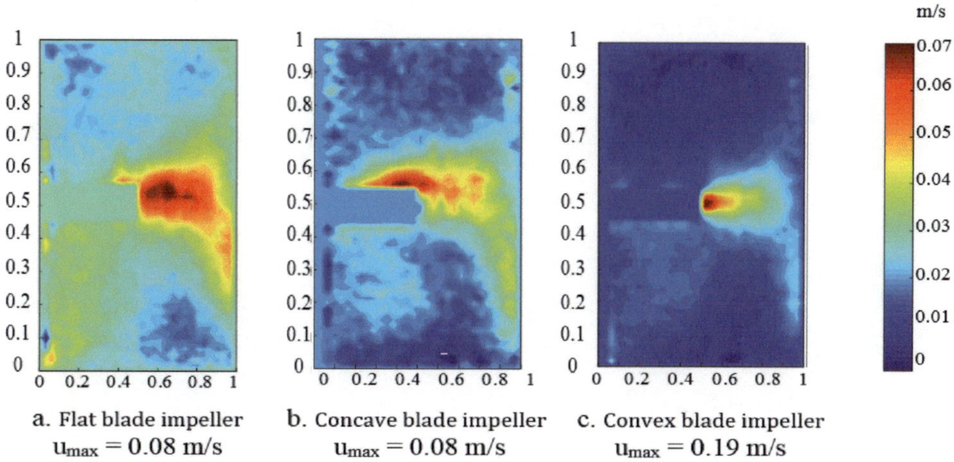

Figure 5.
Radial velocity for one-staged system.

end of the turbine. The bulk region of the tank is described with the lowest value. The flat blade turbine generates a larger radial velocity. Hence, the area of the maximum radial velocity is larger than the other configurations. The maximum radial velocity is spread to reach places farther than the blade, which can be explained by the ability of the blade shape to generate training vortices. The development of the radial velocity component for the concave blade configuration is closer to the axial turbine. In fact, the maximum velocity area is localized at the top edge of the blade and spreads to the same direction as the Von Karman vortex street. The highest value of the radial velocity component is generated by the convex blade configuration that is equal to $u_{max} = 0.35\ U_{tip}$, whereas it is equal to $u_{max} = 0.15\ U_{tip}$ for the other configurations. However, the maximum value remained closer to the blade tip. Therefore, it can be seen that the convex shape of the blade gives the ability to the turbine to move easily within the water and transmit more velocity while not giving it enough capacity to expand much.

For the staged system (**Figure 6**), the maximum value of the radial velocity component is localized between the two blades. This explains the oblique direction of the velocity field at this region. The largest maximum area is defined at the association of the concave and the flat turbines that confirm that the maximum value cannot spread with the convex shape. In addition, the development of the trailing vortices is not as great as the use of the flat and the concave shapes. However, the maximum value of the radial velocity component is found for the second configuration which represents the association of the flat blade at the top and the convex blade at the bottom ($u_{max} = 0.31\ U_{tip}$). The lowest value is found by using the concave blade instead of the convex blade, due to the high strain created by the interaction between the blade and the flow ($u_{max} = 0.22\ U_{tip}$). It can be seen that the blade at the bottom of the tank has the greatest effect on the flow. In fact, the maximum value of the radial velocity is almost similar while we use the flat blade turbine at the bottom ($u_{max} = 0.27\ U_{tip}$).

Figure 7 shows the distribution of the axial velocity component of the curved blade turbine. According to these results, the highest value region is localized at the bottom of the tank close to the blade that represents the suction of the flow of the blade. The second one is localized besides the wall of the tank above the blade at the same direction with the recirculation loops. Then, two lowest value regions were presented. In fact, the largest one is localized close to the free surface, while the

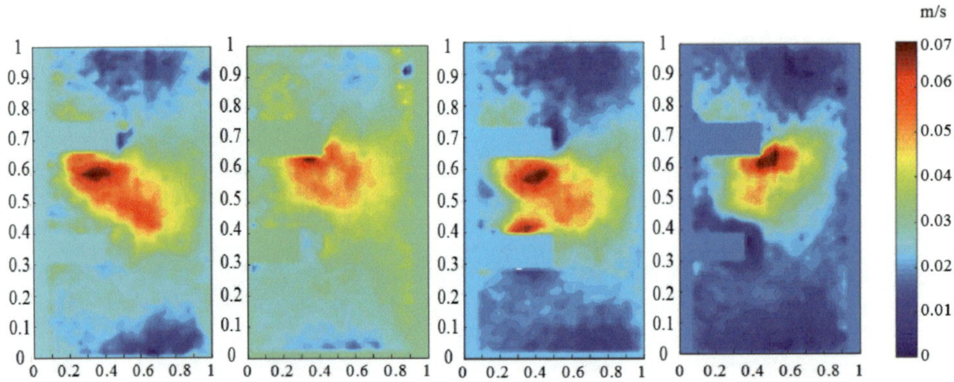

a. PD8 h, PI8 concave $u_{max} = 0.12$ m/s

b. PD8 h, PI8 convex $u_{max} = 0.17$ m/s

c. PD8 b, PI8 concave $u_{max} = 0.15$ m/s

d. PD8 b, PI8 convex $u_{max} = 0.15$ m/s

Figure 6.
Radial velocity for staged system.

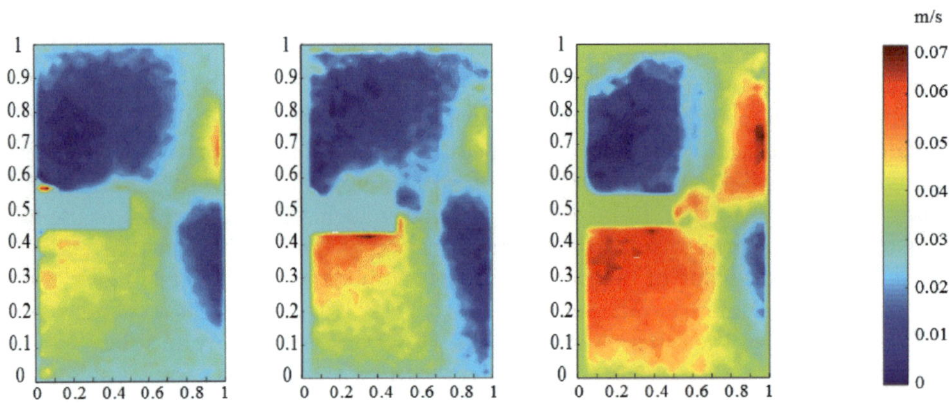

a. Flat blade impeller $v_{max} = 0.12$ m/s

b. Concave blade impeller $v_{max} = 0.09$ m/s

c. Convex blade impeller $v_{max} = 0.08$ m/s

Figure 7.
Axial velocity for one-staged system.

narrowed one is localized besides the wall of the tank at the bottom. The development of the maximum value of the axial velocity component is larger with the convex blade, which confirms that the recirculation loops associated to the convex blade are more important according to the other configurations. According to the velocity field, it can be seen that the recirculation loops associated to the convex blade are larger than the other configurations that are in conjunction with the amelioration of the axial velocity component with the convex blade. The highest maximum axial velocity component is generated by the flat turbine by v_{max} = 0.22 U_{tip} followed by the concave blade by v_{max} = 0.16 U_{tip} and by the convex blade by v_{max} = 0.15 U_{tip}.

For the staged system (**Figure 8**), the maximum value of the axial velocity component is localized at the lowest blade that represents the suction of the flow. The highest maximum value is produced by the second configuration by v_{max} = 0.4 U_{tip}. For the first configuration, the highest value is equal to v_{max} = 0.29 U_{tip}. In addition, the maximum value of the axial velocity is almost similar while we use the flat blade turbine at the bottom (v_{max} = 0.33 U_{tip}).

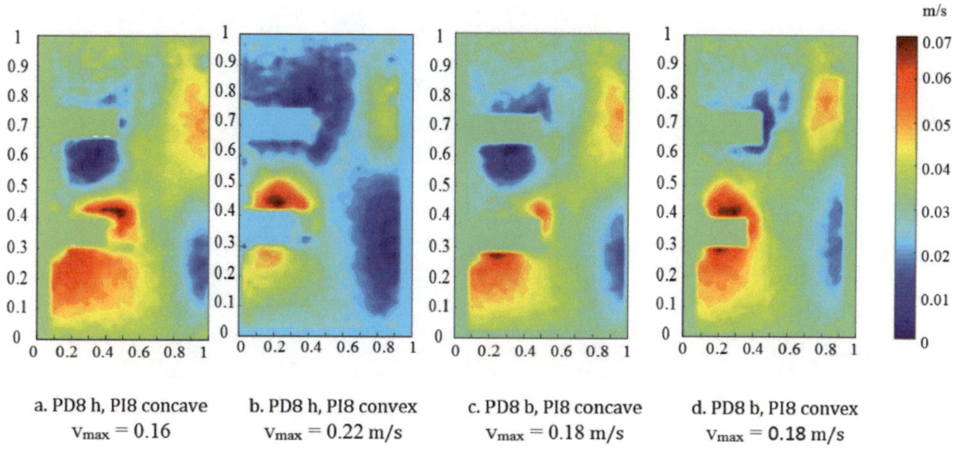

Figure 8.
Axial velocity for staged system.

4.3 Rms of velocity fields

Figure 9 shows the root mean square of the velocity field of the curved blade turbine. The root mean velocity presents the fluctuation of the periodic velocity. According to these results, the highest value region is localized at the same direction of the trailing vortices. In fact, for the convex configuration, the fluctuated velocity is localized close to the blade tip. For the flat blade turbine, the turbulent fluctuation is propagated to the vessel wall. For the concave blade, the fluctuation is localized at the upper and the downer edges of the blade. The maximum variability of the flow occurs due to the trailing vortices following the recirculation loops.

For the staged system (**Figure 10**), the greatest fluctuation is localized at the blade that is placed at the bottom of the tank. These fluctuations can be explained by the suction of the flow from the bottom of the tank. In addition, the maximum values of the turbulent fluctuations are created between the two blades due to the interaction between the blades. The fluctuations generated due to the association of the convex impeller with the flat impeller are narrowed when they are compared to the association of the concave and the flat impeller. This effect reveals that the

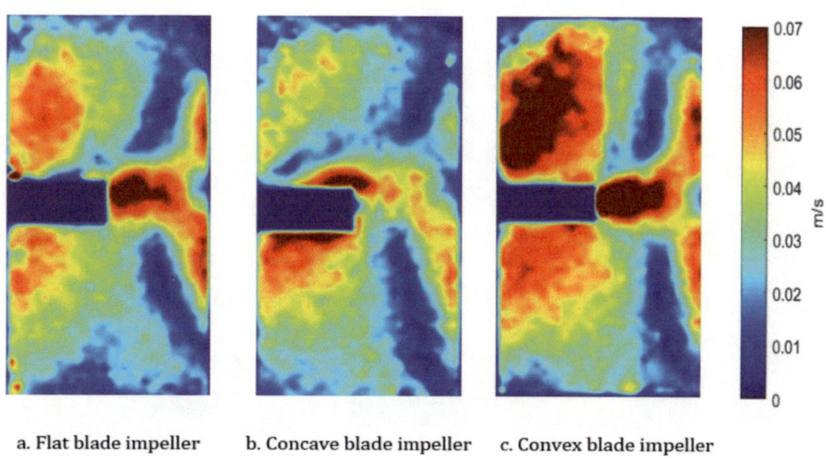

Figure 9.
Rms velocity for the one-staged system.

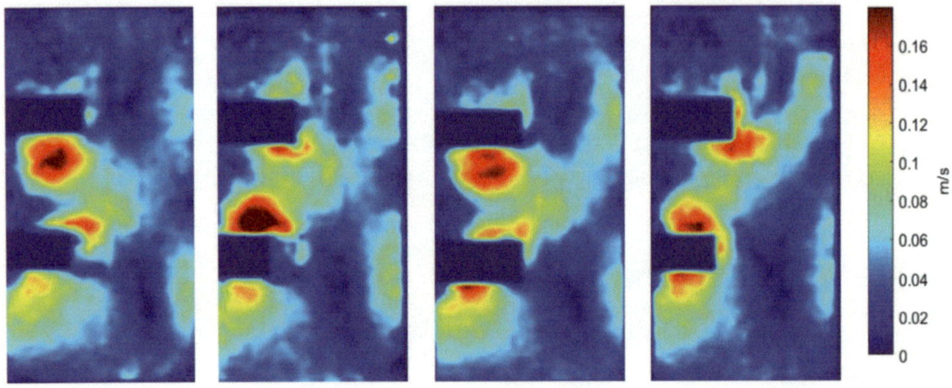

a. PD8 h, PI8 concave b. PD8 h, PI8 convex c. PD8 b, PI8 concave d. PD8 b, PI8 convex

Figure 10.
Rms velocity for the staged system.

convex blade is not able to create a large fluctuation on the turbulent flow and local vortices are created.

4.4 Vorticity

Figure 11 shows the vorticity generated with the curved blade turbines. According to these results, the bulk region of the tank is presented with the medium vorticity value. For the flat and the concave configurations, the propagation of the vorticity is larger than the convex blade. The highest value area is localized in the upper region of the tank which follows the first circulation loop at the same direction of the von Kármán vortex street. The lowest value area is localized in the inferior region of the tank which follows the same direction of the second circulation loop. In fact, it has been noted that the highest recirculation loops are more energetic than the lowest ones. For the convex blade two maximum regions are created presenting the clockwise and the counterclockwise (CW-CCW) vortex pair at the blade tip.

For the staged system (**Figure 12**), the vortical structures are localized at the region between the two blades at the same direction of the discharge flow of each blade, which explains the domination of the trailing vortices at the turbulent flow.

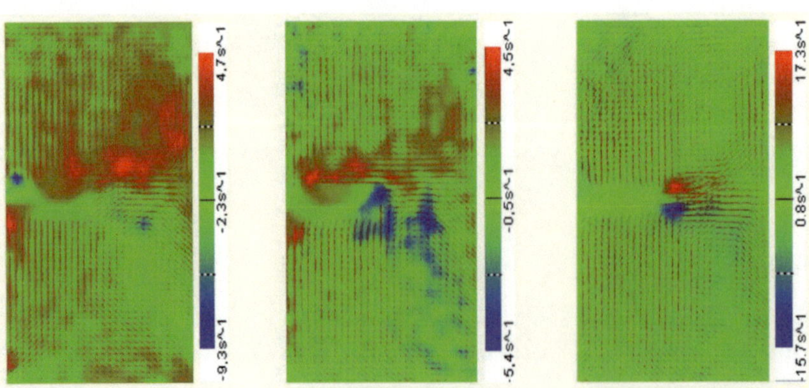

a. Flat blade impeller b. Concave blade impeller c. Convex blade impeller

Figure 11.
Vorticity for the one-staged system.

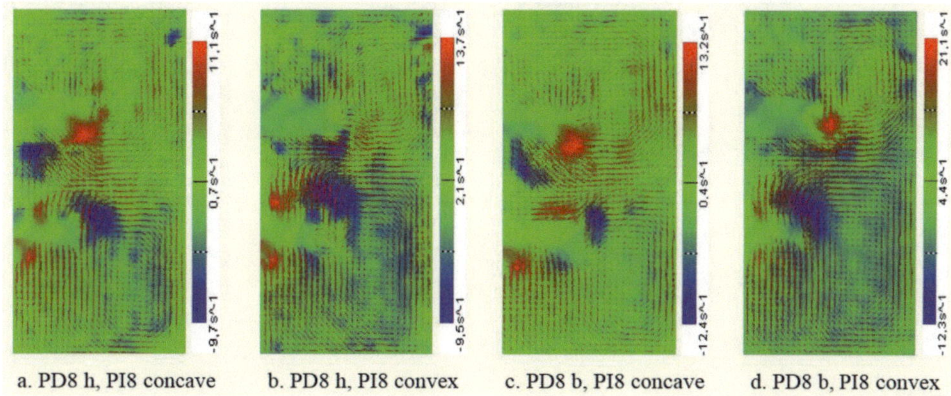

Figure 12.
Vorticity for the staged system.

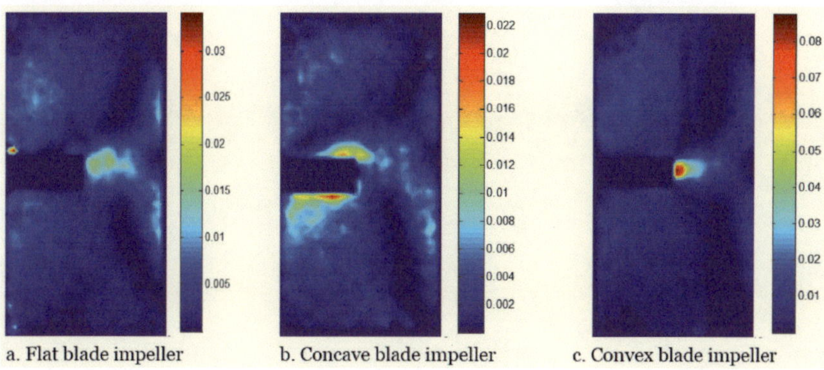

Figure 13.
Dimensionless turbulent kinetic energy distribution of the one-staged system.

4.5 Turbulent kinetic energy

Figure 13 shows the distribution of the turbulent kinetic energy of the curved blade turbine. The turbulent kinetic energy is dimensionless by the square of the tip velocity. According to these results, the turbulent kinetic energy is maximum at the blade tip, and it decreases progressively moving away from the blade. As it was found in the previous sections, the convex blade dissipates the highest energy in the flow. The energy produced by the flat blade decreases by 37% and by 27% for the concave configuration. For the staged system (**Figure 14**), the maximum turbulent kinetic energy is localized between the two blades at the same direction as the trailing vortices. In addition, the turbulent kinetic energy is larger at the association of the concave blade than the convex blade.

4.6 POD analysis

In this section, we used the decomposition of the flow basing on the eigenvalues. This method allows to reveal the smallest vortical structure that cannot be seen by the usual mean flow according to its energetic amount by using the dimensionless eigenfunction. In fact, many vortices are presented with different sizes and shapes. For the one-staged system (**Figures 15–17**), the loop created at the blade tip is the most energetic. The largest one is obtained from the flat turbine. However, the

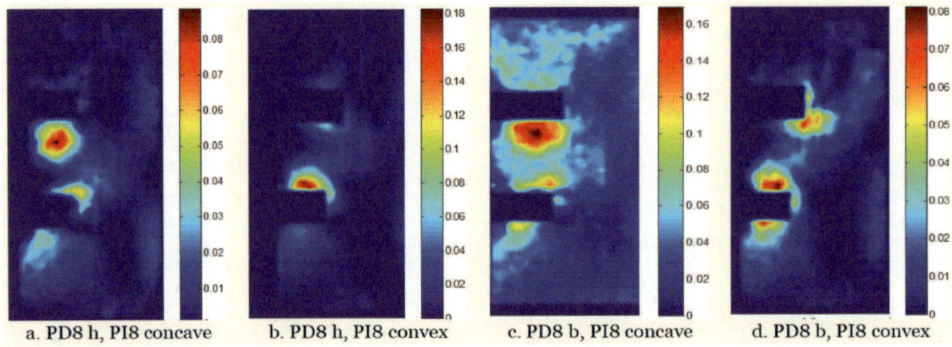

Figure 14.
Dimensionless turbulent kinetic energy distribution of the staged system.

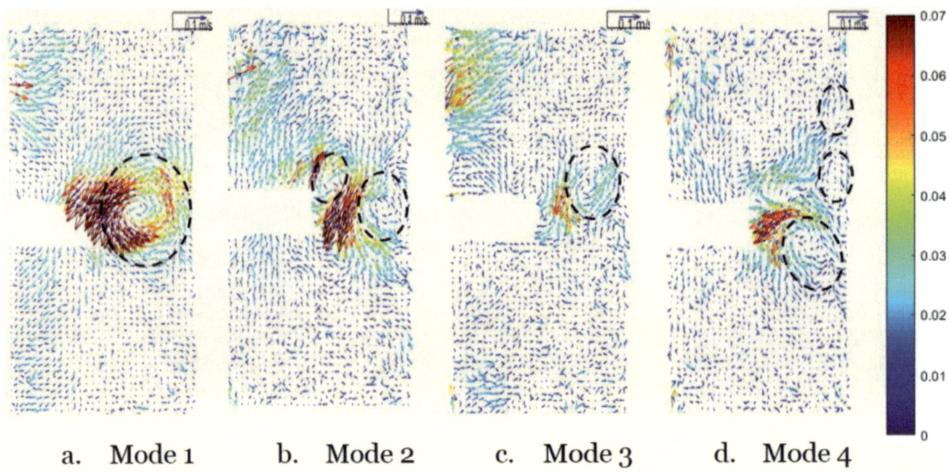

Figure 15.
POD field for the flat blade.

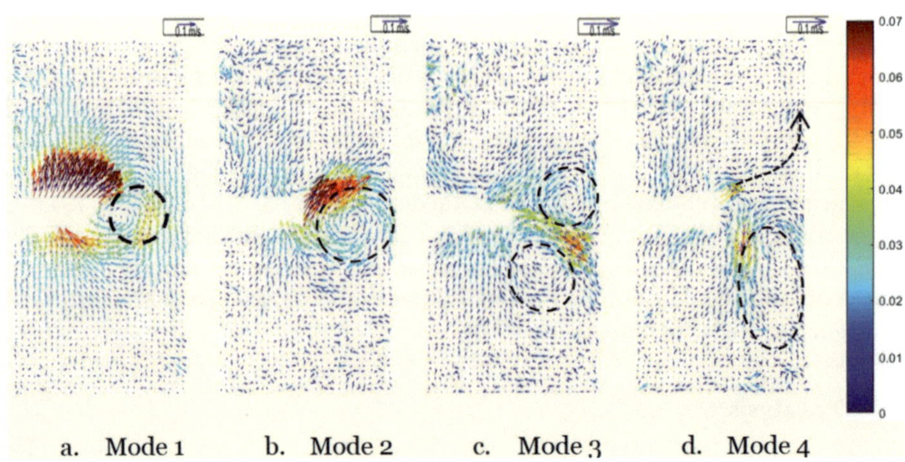

Figure 16.
POD field for the concave blade.

narrowed loop is created by the concave shape, which can be explained by the axial velocity above the impeller. The clockwise and the counter clockwise (CW-CCW) vortex pair at the blade tip can be clearly seen at the highest modes that look similar to the development of the trailing vortices of the Rushton turbine [18, 19]. The trailing vortices are more extended by using the concave blade. However, the flow reaches the bottom of the tank faster by using the convex form.

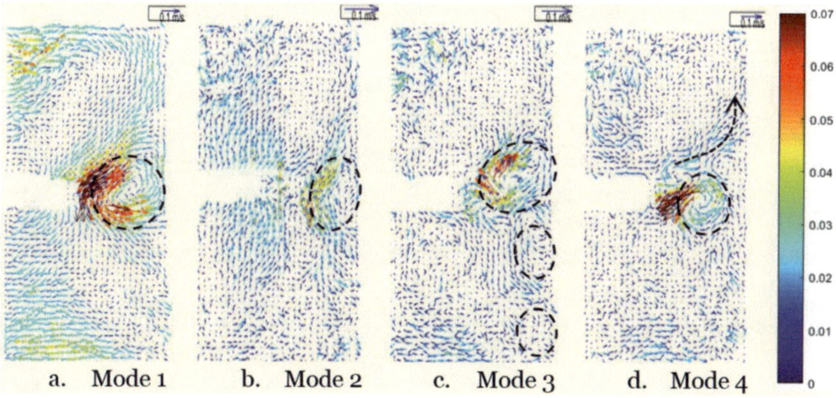

Figure 17.
POD field for the convex blade.

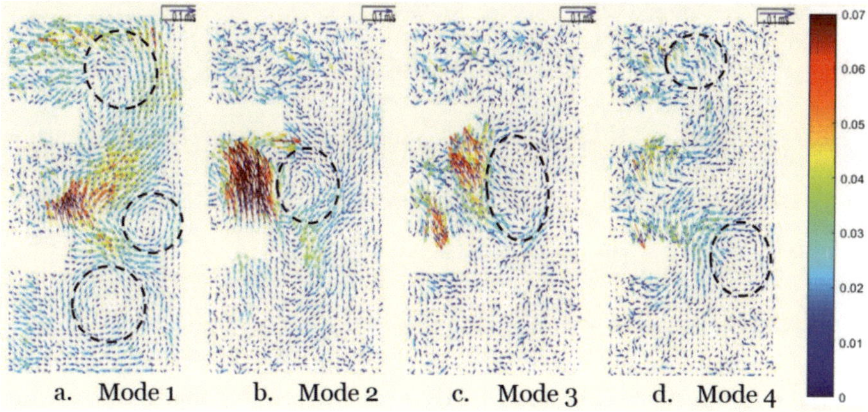

Figure 18.
POD field for the PDh, PI concave.

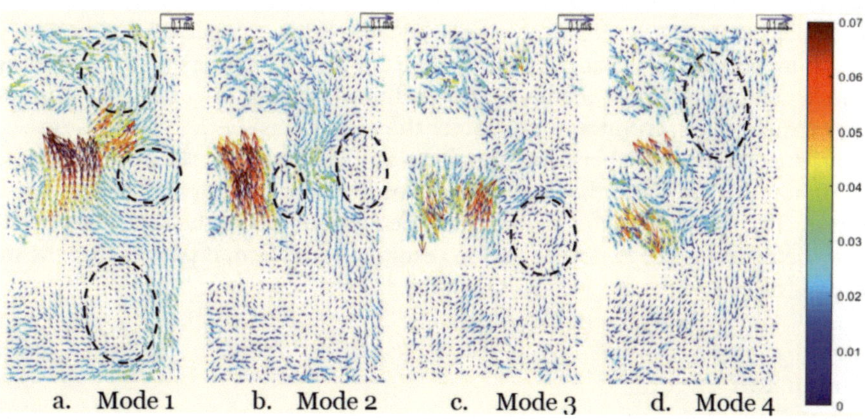

Figure 19.
POD field for the PDh, PI convex.

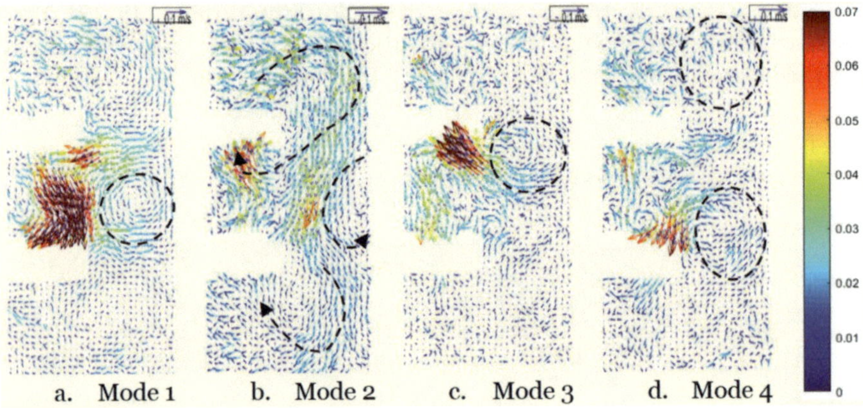

Figure 20.
POD field for the PDb, PI concave.

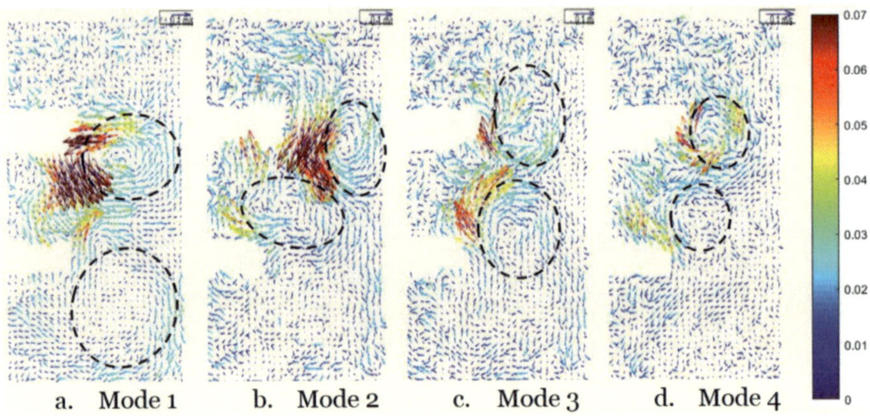

Figure 21.
POD field for the PDb, PI convex.

For the staged systems (**Figures 18–21**), it can be seen that many vortices are created at the region localized between the two blades. This region represents the interaction between the highest and the lowest impeller. Hence, the flow becomes more energetic that explain the cause of the development of these vortices. The flow reach the free surface as well as bottom of the tank faster by mounting the flat turbine at the top of the tank. The trailing vortices become more energetic by using the combination between the flat blade at the bottom and the convex blade at the top.

The development of the different modes shows that the shoes of the combination is extremely important and can affect the mixing inside the vessel. Consequently, the combination between impellers can lead to affect the final product in terms of homogeneity and the cost in terms of the time mixing and power consumption. This found contradicts what has been observed in the study of the mean velocity field, as it gives almost the same results. In addition, it proves that the mean flow is not able to show the real behave of the flow.

5. Conclusion

The objective of this paper is to investigate experimentally the hydrodynamic structure of the curved blade turbine using the particle image velocimetry. Thereby, several results were evaluated which contain velocity field, axial and radial velocity distribution, root mean square velocity, vorticity, and the turbulent kinetics energy. Two circulation loops were presented. The jet flow is more intensive for the convex blade turbine. However, the concave configuration produces a larger lowest loop than the other configurations. Hence, the downer region of the tank is more turbulent than the flat and the convex configurations. The maximum radial velocity generated by the flat blade turbine spreads to reach farther places. This can be explained by the ability of the blade shape to generate training vortices. The convex shape of the blade gives the turbine the ability to move easily within the water and transmit more velocity and energy while not giving it enough capacity to expand much. In fact, it has been noted that the convex blade is not able to create a large fluctuation on the turbulent flow and local vortices are created. In addition, it has been noted that the fluctuation of the flow is dominated by the trailing vortices more than by the recirculation loops. For the staged system, an oblique flow is created between the two impellers, and turbulent fluctuations are greater at this region due to the interaction between the blades.

Author details

Bilel Ben Amira*, Mariem Ammar, Ahmad Kaffel, Zied Driss and Mohamed Salah Abid
Laboratory of Electromechanical Systems (LASEM), National School of Engineers of Sfax, University of Sfax, Sfax, Tunisia

*Address all correspondence to: bba.amira7@gmail.com

References

[1] Wang S, Boger DV, Wud J. Energy efficient solids suspension in an agitated vessel–water slurry. Chemical Engineering Science. 2012;**74**:233-243

[2] Hashimoto S, Chikamochi Y, Inoue Y. Visualization of partially mixed region by use of periodical reaction. Chemical Engineering Science. 2012;**80**: 30-38

[3] Kacunic A, Akrap M, Kuzmanic N. Effect of impeller type and position in a batch cooling crystallizer on the growth of borax decahydrate crystals. Chemical Engineering Research & Design. 2012 0263-8762

[4] Driss Z, Ahmed K, Bilel BA, Ghazi B, Mohamed SA. PIV measurements to study the effect of the Reynolds number on the hydrodynamic structure in a baffled vessel stirred by a Rushton turbine. Science Academy Transactions on Renewable Energy Systems Engineering and Technology (SATRESET). 2012;**2**(4):2046-6404

[5] Ben AB, Driss Z, Abid MS. PIV study of the turbulent flow in a stirred vessel equipped by an eight concave blades turbine. Fluid Mechanics. 2015;**1**(2):5-10

[6] Yianneskis BM. On the quantification of energy dissipation in the impeller stream of a stirred vessel from (fluctuating velocity gradient measurements). Chemical Engineering Science. 2004;**59**:2659-2671

[7] Cruz-Díaz M, Rivera FF, Rivero EP, González I. The FM01-LC reactor modeling using axial dispersion model with a reaction term coupled with a continuous stirred tank (CST). Electrochimica Acta. 2012;**63**:47-54

[8] Li Z, Bao Y, Gao Z. PIV experiments and large eddy simulations of single-loop flow fields in Rushton turbine stirred tanks. Chemical Engineering Science. 2011;**66**:1219-1231

[9] Xu S, Shi J, Cheng Q, Li W, Zhang J. Residence time distributions of in-line high shear mixers with ultrafine teeth. Chemical Engineering Science. 2013;**87**: 111-121

[10] Yang Y, Xia J, Li J, Chu J, Li L, Wang Y, et al. A novel impeller configuration to improve fungal physiology performance and energy conservation for cephalosporin C production. Journal of Biotechnology. 2012;**161**:250-256

[11] Ammar M, Chtourou W, Driss Z, Abid MS. Numerical investigation of turbulent flow generated in baffled stirred vessels equipped with three different turbines in one and two-stage system. Energy. 2011;**36**:5081-5093

[12] Chtourou W, Ammar M, Driss Z, Abid MS. Effect of the turbulence models on Rushton turbine generated flow in a stirred vessel. Central European Journal of Engineering. 2011; **1**(4):380-389

[13] Sanganwar GP, Sathigari S, Babu RJ, Gupta RB. Simultaneous production and co-mixing of microparticles of nevirapine with excipients by supercritical antisolvent method for dissolution enhancement. European Journal of Pharmaceutical Sciences. 2010;**39**:164-174

[14] Liné A, Gabelle JC, Morchain J, Archardf DA, Augier F. On POD analysis of PIV measurements applied to mixing in a stirred vessel with a shear thinning fluid. Chemical Engineering Research and Design. 2013;**91**: 2073-2083

[15] Driss Z, Kchaou H, Baccar M, Abid MS. Simulation numérique de

l'écoulement laminaire dans une cuve agitée par une turbine à pâles incurvées. In: Récents Progrès en Génie des Procédés. Vol. 92. Paris, France: SFGP; 2005) ISBN: 2–910239–66-7

[16] Jing Z, Zhengming G, Yuyun B. Effects of the blade shape on the trailing vortices in liquid flow generated by disc turbines. Fluid Flow and Transport Phenomena, Chinese Journal of Chemical Engineering. 2011;**19**:232-242

[17] Bereksi MS, Kies FK, Bentahar F. Hydrodynamics and bubble size distribution in a stirred reactor. Arabian Journal for Science and Engineering. 2018. DOI: 10.1007/s13369-018-3071-z

[18] Moreau J, Line A. Proper orthogonal decomposition for the study of hydrodynamics in a mixing tank. AIChE Journal. 2006;**52**:2651-2655

[19] Gabelle JC, Morchain J, Archardf DA, Augier F, Liné A. Experimental determination of the shear rate in a stirred tank with a Non-Newtonian fluid: Carbopol, transport phenomena and fluid mechanics. AIChE Journal. 2013;**59**:2251-2266

Chapter 9

Urban Heat Island Effects in Tropical Climate

Luz E. Torres Molina, Sara Morales and Luis F. Carrión

Abstract

This paper reviews some of the characteristics of urban climates and the causes and effects of urban heat island (UHI) issues in the tropical climate. UHI effect is a kind of heat accumulation phenomenon within the urban areas due to urban construction and human activities. It is recognized as the most evident characteristic of urban climate. The increase of land surface temperature in San Juan, Puerto Rico, caused by the UHI effect was influenced by the change of land use and material types in construction. The impacts of daily temperature, surface albedo, evapotranspiration (ET), and anthropogenic heating on the near-surface climate are discussed. Analyzed data and field measurements indicate that increasing albedo and vegetation cover can be effective in reducing the surface and air temperatures near the ground. Some mitigation and prevention measures are proposed for the effects of UHI, such as a flash flood warning system.

Keywords: urban heat island (UHI), albedo effects, evapotranspiration, anthropogenic heat

1. Introduction

When people talk about the climate, the urban heat island (UHI) is not a new terminology. This began to be relevant decades ago when the cities began to be megacities built without any environmental planning and when the trees and grass were converted to concrete and asphalt. According to the United States Environmental Protection Agency, an urban heat island is created in developed areas where the built surfaces absorb and retain radiation from the sun. The impacts of UHI include increased energy consumption, higher concentration of air pollutants, and increase of air temperature level and flash flood.

Temperature is the most significant atmospheric parameter in researches about the effects of heat island. Studies have documented that urban areas have air and surface temperatures that are, on average, 1.8–5.4°F higher than temperatures in surrounding rural areas and there is potential for up to a 22°F difference in more extreme situations [1]. According to the Intergovernmental Panel on Climate Change (IPCC), global average temperatures have risen by 0.6°C (1.1°F) since 1970 and can be expected to rise another 1–4°C (1.8–7.2°F) by the end of the twenty-first century, depending on future societal practices and the amount of greenhouse gas emissions released into the atmosphere. The 2009 US Global Change Research Program report entitled "Global Climate Change Impacts in the United States" illustrated that the average mainland US temperature has increased by 1.1°C (2°F) since about 1960, precipitation has increased by 5%, and the frequency of heavy precipitation has also increased by a factor of two [2].

Another important aspect to be studied is evapotranspiration (ET) because the increase of impervious surfaces prevents the movement of air and water which are key in the cooling processes of evaporation and transpiration and creates a typical "heat urban islands" of warmer temperatures, while the green vegetation in the surrounding rural areas better regulate surface temperatures.

This could be evident with the analysis between cities such as San Juan, Puerto Rico (urban), and Gurabo, Puerto Rico (rural). A decrease of evapotranspiration energy from the rural area to the suburbs and finally to the urban area is expected. This change will coincide with a decrease in vegetation coverage. Without the immediately available energy outlet of evaporation, urban and suburban areas must store more energy during the day. The stored energy is subsequently released to the atmosphere at night, primarily through higher radiant emissions and to a lesser extent via increased convection [3].

The flash flood is a consequence of an urban heat island effects. Most people consider that sudden floods are the product of weather phenomena such as tropical waves through even hurricanes. Previous studies have shown that most of flash floods are caused by anthropogenic behavior generally named anthropogenic heat, which is generated by human activity and comes from many sources, such as buildings, industrial processes, and change in land use from pervious to impervious.

2. Local climate

The climate of the Caribbean is characterized as subtropical with relatively dry winters and wet summers [4]. The dominant large-scale atmospheric influence on the climate is the North Atlantic subtropical high (NAH). The average temperature at the San Juan, Puerto Rico, station last year was 80.74°F. **Figure 1** shows the annual temperature between the years 2000 and 2018 for the months of January and September [5]. On the other hand, the amount of rainfall varies considerably throughout the study area. Most of the rainfall occurs during the month of August with 7.15 inches on average for the last 20 years. The month of February is considered the dry season with 2.32 inches.

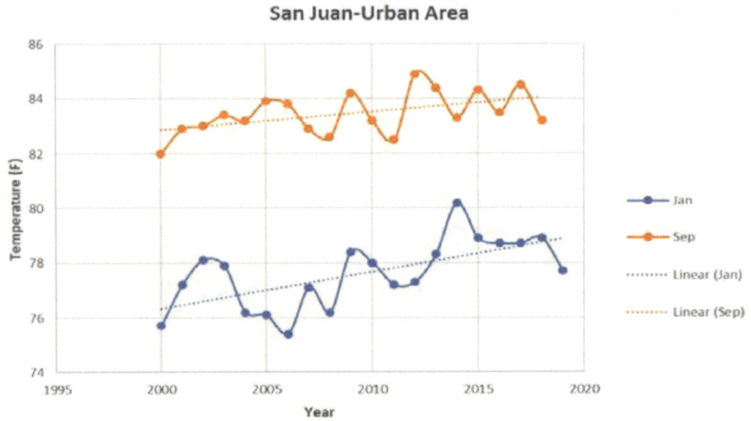

Figure 1.
San Juan, Puerto Rico, annual temperature (°F) pattern for 2000–2019.

3. Study locations

The study UHI is based on the premise that significantly warmer surface temperatures exist in urban settings as opposed to their surrounding rural areas. To make sure that this premise is correct for the Puerto Rico case, seven locations have been selected, which cover the island (see **Figure 2**).

The research began with the use of historical temperature data taken from the southeast regional climate center webpage. The website has the advantage of having historical data such as temperature and precipitation of at least 50 years. For this research, the temperature was the parameter for defining the problem. Data with more than five decades were used to study the changes in temperature between the urban area and the rural area.

The city of San Juan, Puerto Rico (18.44, −66, geographic coordinates in decimal degrees latitude and longitude), is designated as an urban area, and it is the capital city of Puerto Rico. The number of residents in the city is declining from 428,800 in 1957 to 321,000 in 2019. The city designated as a rural area is Gurabo, Puerto Rico (18.25, −66, geographic coordinates in decimal degrees latitude and longitude), located at 20 km from San Juan, Puerto Rico. In 1957 it was estimated that it had 16,600 inhabitants, and in 2019 it has 46,000 residents [6].

Figure 3 shows the difference in temperature between urban and rural areas. The data was taken from 1957 to 1967 at each station and displays a difference in temperature on average of 2.62°F. Furthermore, the same comparative analysis was carried out for the years between 2008 and 2018 (see **Figure 4**). For this case, the difference between temperature increases was around 3.46°F.

Stations located in San Juan, Puerto Rico, and Gurabo, Puerto Rico, show that both period 1957–1967 and period 2008–2018 temperatures are on the rise in heavily urbanized areas where there is a conversion of natural vegetation to urban dwellings.

A NASA study found that in the summer months, the temperature in New York was on average 4°C higher than in the surrounding area. Studies from the 1960s already pointed to the phenomenon of heat island, but the effect is becoming more intense due to climate change.

Another way to verify the existence of heat islands is to make a parallel analysis of urban area temperature behavior as it moves away from its center and approaches a rural area. As shown in **Figure 5**, the first station away from the urban location is

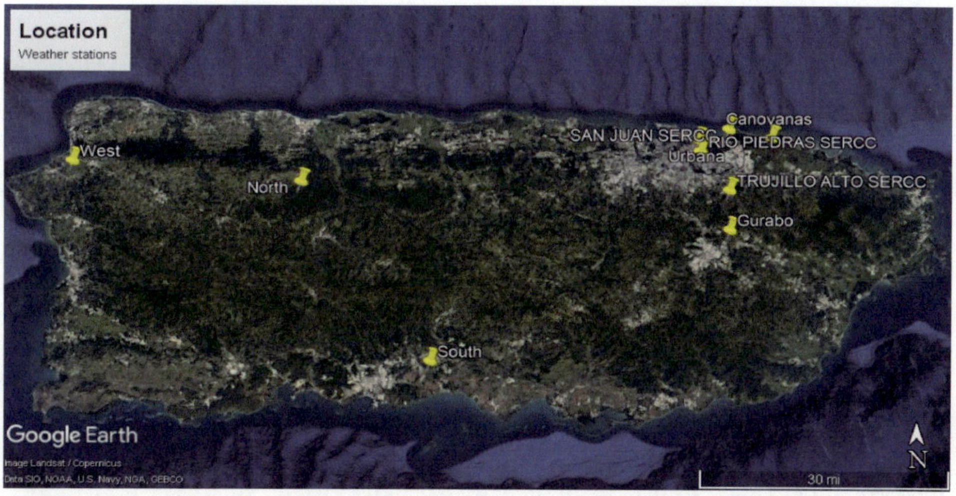

Figure 2.
Study locations, Puerto Rico.

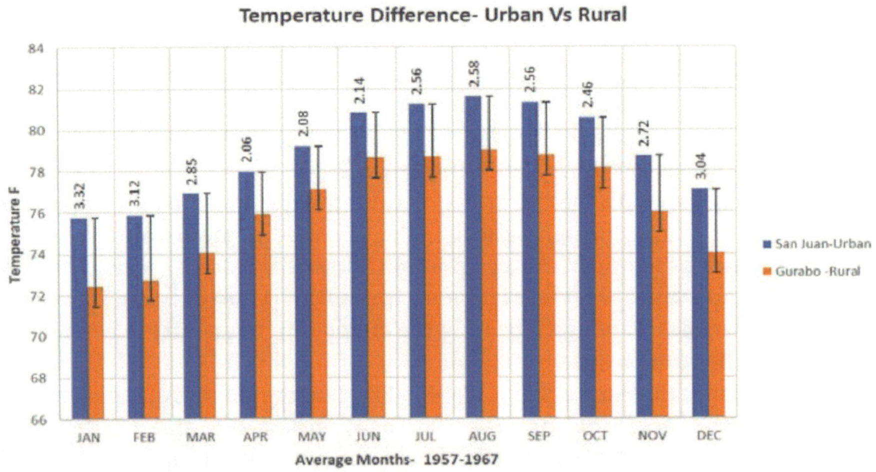

Figure 3.
Difference of temperature (°F) between urban and rural areas (1957–1967).

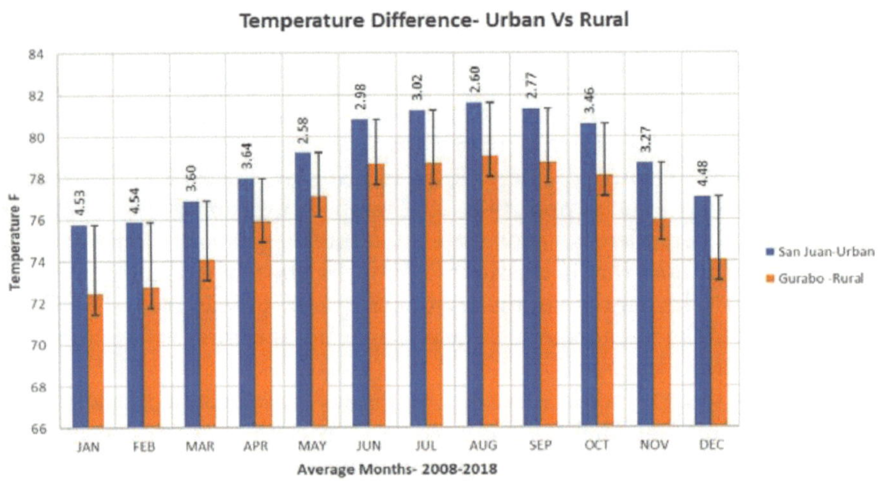

Figure 4.
The difference of temperature (°F) between urban and rural areas (2008–2018).

Rio Piedras, located at 6 km; the next station is Trujillo Alto, located at 12 km from San Juan, Puerto Rico; and the last is Gurabo, Puerto Rico, considered as a rural area located at 20 km from San Juan, Puerto Rico.

Considering the temperature average during the last 55 years in the urban area (San Juan, Puerto Rico) for all months, this shows the highest values compared to the other stations, during the 12 months of the year. It is observed that the coldest months during the period studied are the months of January and February and those with the highest temperatures are August and September. The latter matches with the peak of the hurricane season, where the temperature on the surface of the Atlantic Ocean is at its highest and optimum level for the formation of more powerful hurricanes (see **Figure 6**).

Stations located in San Juan, Puerto Rico, show that both January (minimum) and September (maximum) are on the rise in heavily urbanized areas where there is a conversion of natural vegetation to urban dwellings. This theory was before validated by Gonzalez and Comarazamy in 2009 [7]. Urban data analysis also found

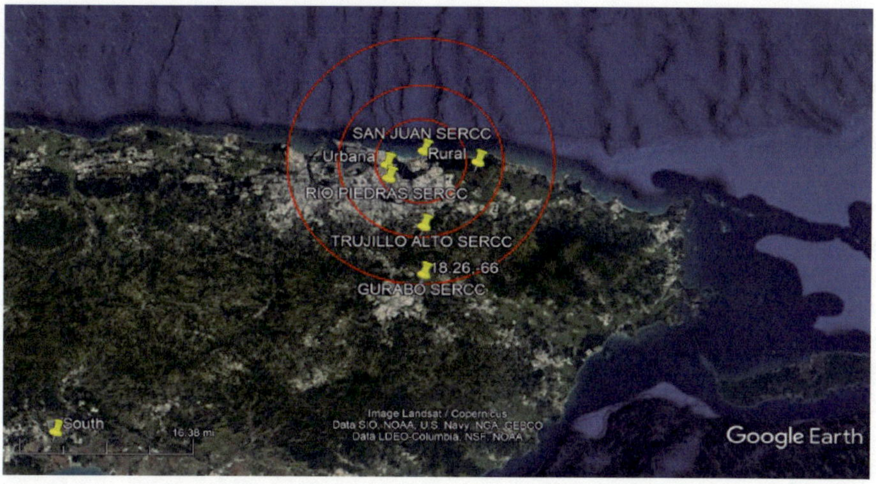

Figure 5.
Location of stations: San Juan, Rio Piedras, Trujillo Alto, and Gurabo, Puerto Rico.

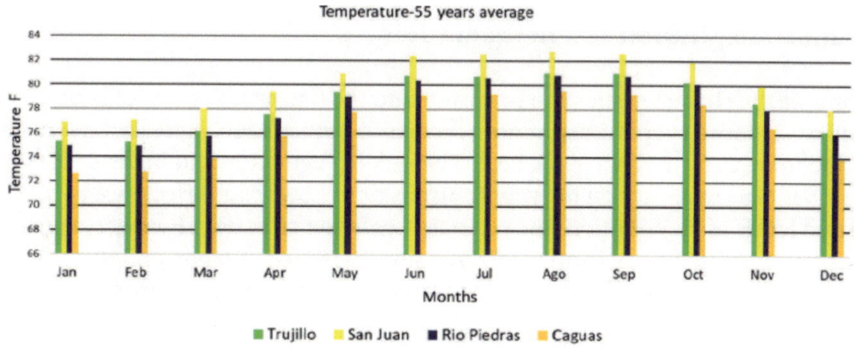

Figure 6.
Average temperature (°F) using data of 55 years in four different stations.

warming in the San Juan, Puerto Rico area, with a trend of 0.09%/year. from 1957 to 2012. This result was obtained when estimating the delta of change of all the months of the year from 1957 to 2012. **Figure 7** shows the projection for the warmer month (September). Using the 0.09% growth (obtained from 1957 to 2012) and the recent data (2000–2019), it is observed that, although there is a difference in some years, after each decade this projection must be more accurate.

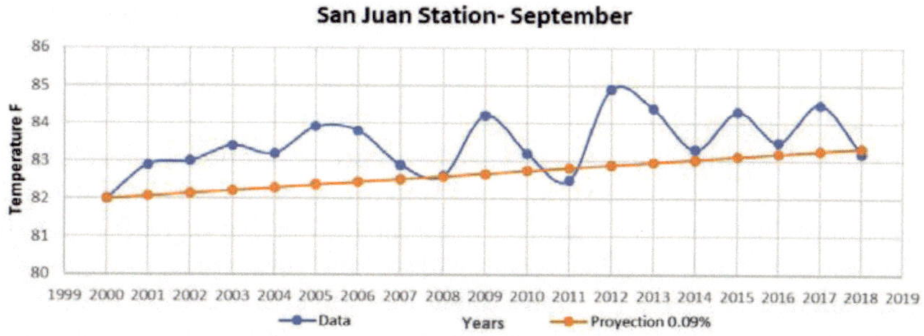

Figure 7.
San Juan data station (°F) and projection data with 0.09% increase yearly.

Rosenzweig et al. [8] analyzed the New York city heat island effect model by taking six case study areas and tested the mitigation strategies. They found that vegetation helps to keep surfaces cool more effectively than increasing the albedo. But they suggested that in order to reduce the temperature in New York city, replacement of low-albedo materials with high-albedo light-colored materials will work great as 64% of the surface area of the city can be replaced easily.

Sailor [9] describes that the urban heat island effect mitigation can be done in two ways. One is by increasing the albedo of the urban surface, and the other is by increasing evapotranspiration. On the other hand, white materials which have albedo greater than 0.60 instead of black materials having albedo of 0.05–0.10 can be used as roofing materials. They found that the roof temperature dropped by 25°C for 0.60 albedo compared to that of 0.20 albedo. More solar radiation could be reflected if the road and highway pavements were of high-albedo materials. White cement mixtures can be made for which the albedo should be higher than the most reflective gray cement mixtures. However, use of high-albedo materials for roads and highway pavement may not be so much effective because of the sky view factor. Even if, it is used, some of the reflection will be intercepted by the buildings surrounding it.

4. Evapotranspiration and net radiation

Some of the characteristics can be sorted into the four main causes of heat island formation: reduced evapotranspiration, increased heat storage, increased net radiation, and increased anthropogenic heat. The lack of vegetation and increase of surface impervious cause a reduction in evapotranspiration. Low solar reflectance and increased levels of air pollution foment increases in net radiation. Evapotranspiration is energy transmitted away from the Earth's surface by water vapor, and this is a process plants use to keep themselves cool, and it increases when there is more moisture available. On the other hand, net radiation encompasses four separate radiation processes taking place at the Earth's surface. Net radiation = Incoming solar – Reflected solar + Atmospheric radiation – Surface radiation. The first term in the equation is related with the amount of energy radiating from the sun. This varies in Puerto Rico mainly according to the time of the day, the amount of cloud cover, and the atmospheric pollution levels. The second term, reflected solar, is directly related with the "albedo," which is the amount of solar energy that bounces off a surface. The higher albedo, the greater is the amount of reflected energy. The third term is heat emitted by particles in the atmosphere, such as clouds, pollution, and Sahara dust and the last term, surface radiation, is heat radiated from a surface itself. A relatively warmer surface radiates more energy to its surroundings, which is the case of San Juan city in north Puerto Rico area. In 2002, the north area city had approximately 70% impervious surface cover, 13% grass area, and 16% tree areas [10].

A comparative study for evapotranspiration and net radiation was successfully accomplished, as part of the evidence to determine the heat island effect. Using GOES satellite data, the ground level, 1 km resolution net radiation, evapotranspiration, and rainfall parameters became available in Puerto Rico in March 2009 [11]. **Figure 8** shows that during the last 10 years, evapotranspiration measures have been higher in the area designated as rural, the difference between rural and urban is very significant, and this evidences a tendency to a separation that increases with time.

In the case of precipitation between the urban and rural areas, using 10-year historical data, it is observed that rainfall has been greater in the urban center (**Figure 9**), under conditions of similar land use. It would be assumed that the amount of precipitation increases evapotranspiration. But this is not the case when there is an UHI phenomenon.

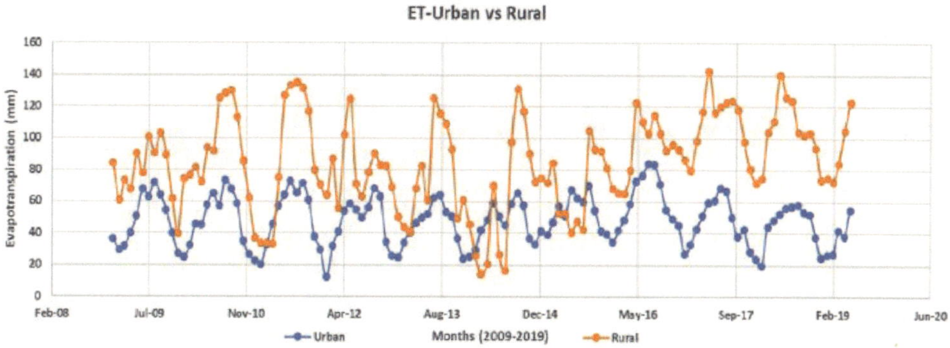

Figure 8.
Evapotranspiration (mm) urban vs. rural data (2009–2019).

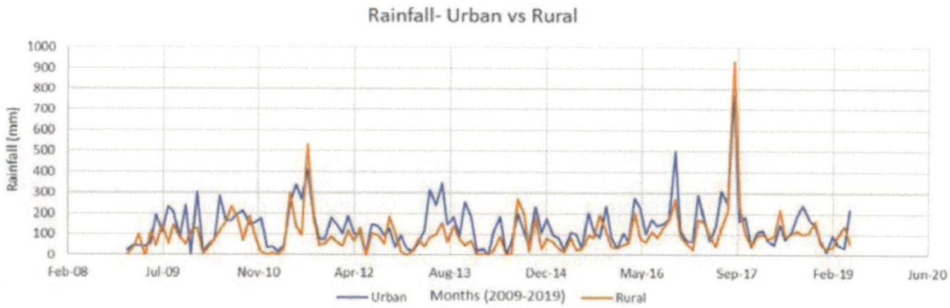

Figure 9.
Rainfall (mm) urban vs. rural data (2009–2019).

The earth in natural conditions absorbs rain, which is evaporated by the sun's rays on hot days and released to the atmosphere, cooling the environment. The lack of vegetation in the cities contributes to the heat island effect. But in the cities, the rain ends up in the sewage systems. The urban centers usually also have few trees. Vegetation plays a crucial role due to evapotranspiration.

Another valid theory is that the net radiation collected by an urban setting is generally greater than that collected in a rural area. The difference is due to many factors, such as, for the Puerto Rico case, lower solar reflectance of urban materials and restrictive urban geometries. Another significant factor, even though it occasionally happens, is the pollutant due to Sahara dust. However, the difference in net radiation between urban and rural areas is not significant. A possible little difference is observed in **Figure 10** where the urban area has higher values. During the 10 years of study, an average of 0.6656 MJ/m2/day represents the difference in net radiation.

Another main characteristic of the climate is wind speed. Physical characteristics of urban climate such as tall buildings, paved streets, and parking lots affect wind flow. The differences of urban climate and rural climate are attributable in large part to the altering of the natural terrain through the construction of artificial structures and surfaces. In theory, the center of a city is warmer than the outlying areas. Monthly minimum temperature readings at related urban and rural sites frequently show that the urban site is 3–4°F warmer than the rural site, for this case. Two primary processes influence the formation of this "heat island." During summer, urban masonry and asphalt absorb, store, and reradiate more solar energy per unit area than vegetation and soil typical of rural areas. Furthermore, a small amount of this energy can be used for evaporation in urban areas, which characteristically exhibit greater precipitation runoff from streets and buildings. At night,

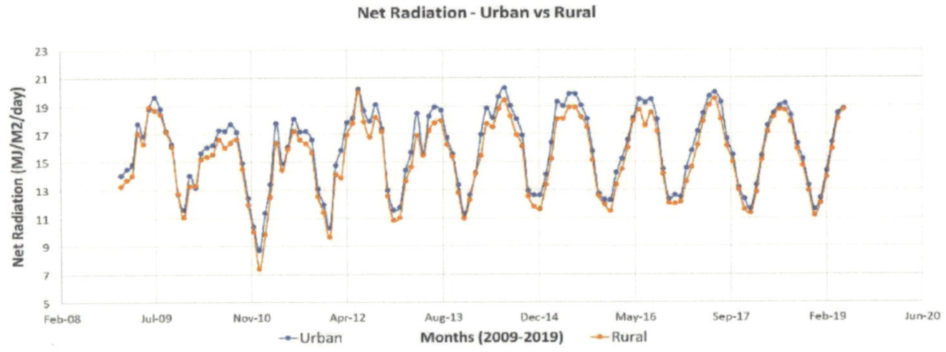

Figure 10.
Net radiation urban vs. rural data (2009–2019).

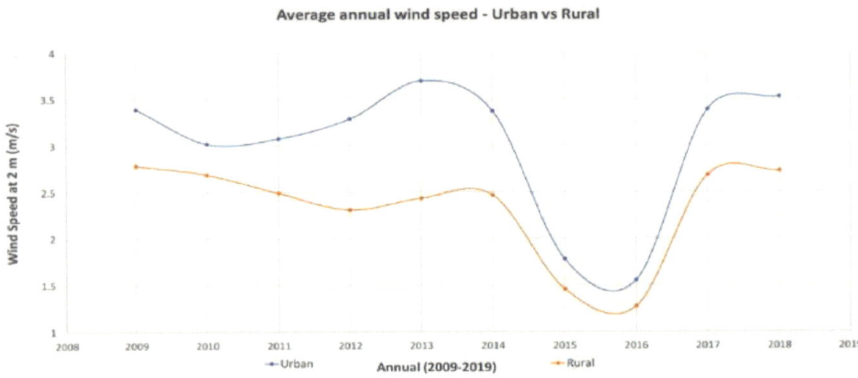

Figure 11.
Average annual wind speed - urban vs rural area.

radiative losses from urban buildings and street materials keep the city's air warmer than that of rural areas. For this case, it is important to consider the location of the urban site and rural site. San Juan is located on the NE coast of the island of Puerto Rico. It is surrounded by waters of the Atlantic Ocean. The climate is tropical marine, slightly modified by insular influence when land breezes blow. San Juan is representative of most of the coastal localities on the island. That proximity to the coast makes wind speeds higher than in the rural area (see **Figure 11**).

5. Surfaces and materials

Buildings' volume, orientation, and the aspect ratio of the spaces between them affect the exposure of urban surfaces to the solar radiation. The concentration of concrete structures without green surfaces between them increases the air temperature of urban areas. In this case study, the concentration of buildings is significantly higher in San Juan than in Gurabo, Puerto Rico (see **Figure 12**). The gray plots indicate the building structures in each city. At a first glance, it is observed that there is a greater amount of building structures in the city of San Juan. The city of San Juan has three times more building structures than the city of Gurabo. The city of San Juan has 127.98 km^2 of the surface area of which 15.89 km^2 are of building structures. On the other hand, the city of Gurabo has 73.22 km^2 of the surface area of which 3.33 km^2 are of buildings.

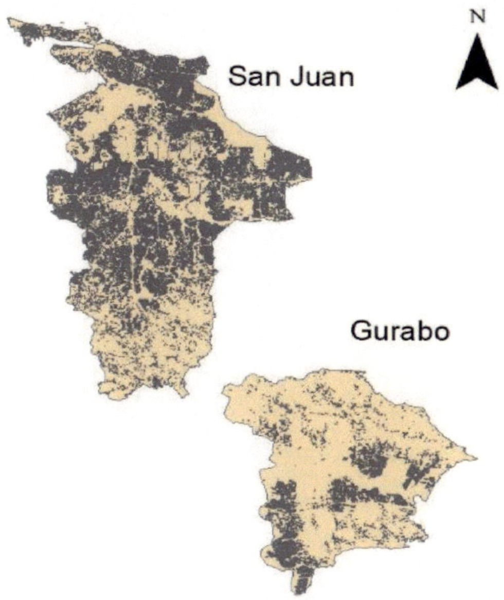

Figure 12.
Building structures San Juan vs. Gurabo, Puerto Rico.

The complex heat exchange between buildings' mass and adjacent air changes the intensity and patterns of airflow in urban canyons where wind patterns are also affected by the canyon-like structure of streetscapes surrounded by tall buildings. Urban surface materials' thermal characteristics (specific heat, mass, conductivity, and diffusivity), color, texture, and coverage alter heat exchange in urban settings and are important to determine UHI.

Some construction materials have properties that tend to exacerbate the heat island problem. Two material properties are important to heat storage: thermal conductivity and heat capacity. Materials with high thermal conductivity tend to conduct heat into their depths. Materials with high heat capacity can store more heat in their volume. A combination of these properties, called thermal diffusivity, is an important indicator of how easily heat can penetrate a material. Thermal diffusivity is calculated by dividing a material's thermal conductivity by its heat capacity. Rural areas tend to be composed of materials of lower thermal diffusivity, while urban areas have higher diffusivities. This enhances the storage of heat during the day and its slow release at night [12].

Thermal diffusivity (mm^2/seg) is given by the following relationship:

$$\alpha = \frac{k}{\rho C} \tag{1}$$

where k is thermal conductivity (W/mK), ρ is density (kg/m^3), and C is specific heat (J/kgK).

Another significant material characteristic is the albedo. Urban areas are considered with a low albedo, while rural areas are considered with higher albedos. Most urban materials reflect less incoming solar energy than materials commonly found in rural areas. **Figure 13** shows differences of temperature between two widely used materials in Puerto Rico, such as asphalt and concrete, at 14:00 hours on a spring day.

The asphalt shows temperatures that reach 61°F, compared to the temperature of the concrete that decreases by almost 20°C. This demonstrates that under normal conditions, in a country like Puerto Rico where the solar radiation received

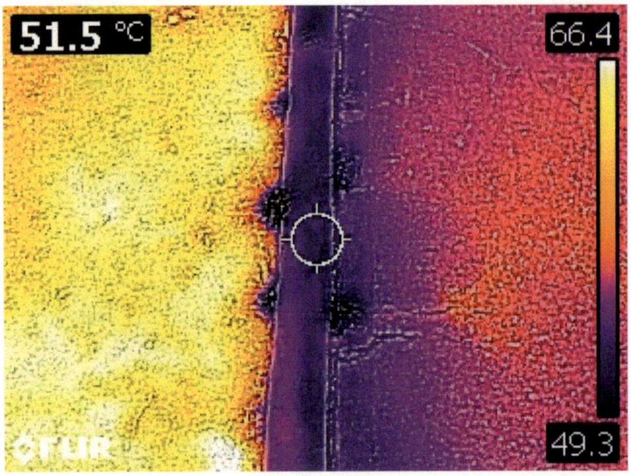

Figure 13.
Asphalt (left) vs. concrete (right) temperatures (°C).

throughout the year is constantly high, the roads built on asphalt should be changed to another material such as concrete, where heat emissions decrease considerably.

Commonly in urbanized places, two prominent materials have low values of solar reflectance: asphalt paving and built-up roofing (black roofs). The prevalent use of these materials lowers the overall solar reflectance of communities. Some of the characteristics of the materials commonly found in urban and rural areas are presented in **Table 1**, where the specific heat capacity and albedo are important features when reference is made to UHI.

In addition to the materials used in structures, road, and roofs, detailed computational and wind tunnel study shows that building packing density or how packed buildings are erected in a unit area also affects the movements of polluted and heated air from the pedestrians' levels, which eventually affects UHI. Reda et al. [13] performed computational fluid dynamics (CFD) simulations using OpenFOAM as well as ran experiments using blocks in wind tunnel using the Kuala Lumpur City Center, at five different locations representing different building densities. The results show that the more packed buildings are spaced, the harder for the incoming winds to clear the polluted and heated air trapped at low levels. Based on further details of the CFD simulations and atmospheric-scaled measurements, there are complex interactions such as vortices and large-scale features in the bulk movements of air and the more static air at near the grounds [14].

One of the characteristics of cities with UHI effects is the increase in impervious areas, where there is change in land use, i.e., grass or trees have been converted to parking lots and roads. These changes have increased flash floods in cities, leading

Material	Thermal diffusivity (mm^2/s)	Albedo
Asphalt	1.017	0.04
Concrete	0.677	0.22–0.55
Wood	0.26	0.15
Grass	—	0.25

Table 1.
Material characteristics.

to major economic disasters and human losses. A personal flood alert system is being proposed as a mitigation method for this type of event, and these are shown in the following topic.

6. Flood alert system

One of the effects of UHI is flash floods, which is a consequence of the increase of the impervious areas. Floods are the most frequent disaster type and cause more humanitarian needs than other natural disasters. The use of new technologies with higher accuracy, covering areas missed by radars, is important for flood warning system efforts and for studying and predicting atmospheric phenomena [15]. Almost 90% of all-natural disasters in the United States lead to flooding, and 20% of all flooding claims happen in low to moderate flood risk areas [16].

This information is important when judging where to live. Home insurance does not cover anything that is not attached to the house. Insurance policies for cars, for example, that are the most affected by flooding since most of them are outside must have specific insurance called comprehensive insurance. This system not only can save a family's pocket and vehicle, but it can also save lives by notifying them of the flash flood occurring in the area. In addition, it can help gather information about the flood in the area of the disaster.

Many people lose their vehicles due to flash floods at parking lots. A way to avoid such disasters is by creating a flood alarm system that detects flood levels from their own cars making this alert more specific depending on individual cases. The system must be able to do the following: detect and measure water elevation considering the type of vehicle and its size. The alarm will function via phone call or text. As an alternative method, if there is no signal in the area, the cars' alarm system will alarm the owners instead of a phone call or text. Giving security to any of these two methods to notify the affected person.

How the system works is indicated in **Figure 14**. Once the system detects the sudden rise in water level, it will alert the owner through phone call or text. As a final step, the owner of the vehicle will move the car to a safer location if it is possible.

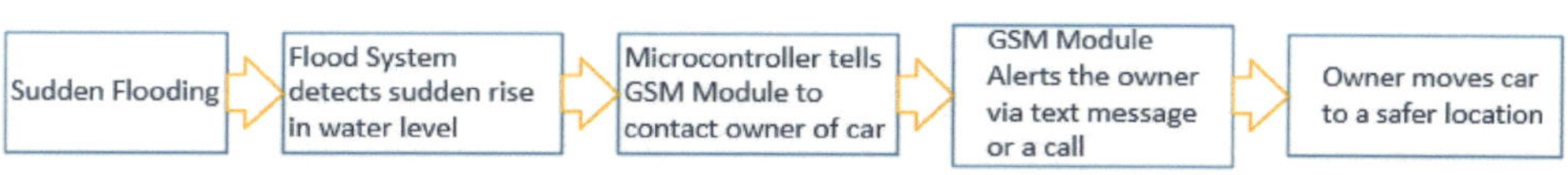

Figure 14.
Diagram of the process for a flood alert system.

7. Proposed system

The proposed system alerts the client when the water level is a threat to the vehicle. This system includes a set of sensors that will perceive different measurements in water level and vehicle acceleration. If the water level increases to a threatening level, it will alert the client via web service through a text message which notifies about a sudden flood every 10 seconds. When for any reason there is no signal, the system has the option to send a notification so that the vehicle horn starts to go off; this functions as another alternative. In case that the car is in motion, it will notify the client by an audible notification. The collected data will be used to report flooded areas in real time through a web service.

The system is located inside the car, close to each of its four tires. The Arduino will receive data from the accelerometer, ultrasonic, and the GPS module. Once the data has been processed, the device will detect if the water level has risen to a threatening level. There are two forms of operation. In the first one, when the vehicle is moving, the client will be alerted by an audible notification located inside the car. And in the second one, if the vehicle is stationary, the client will receive a text message sent by the web service. The web service stores the information received in our database and will provide information that the client may see from their devices. **Figure 15** presents interaction of the system, client, and webpage.

This system consists of the following components: ultrasonic sensors, the GSM module, and the accelerometer. Each of these components plays an essential part in the contraction of this prototype. A breadboard is used as an intermediary for the connection from the Arduino mega to each component. An ultrasonic sensor is used to measure the distance between the ground or water and the sensor. It contains four pins: VCC, trig, echo, and GND. In order to function, it needs a power of 5 V which can be found in the Arduino. They emit and receive sound waves in order to detect how far the object is. It does not measure the distance, but rather the time of emission and receiving [17].

A GSM module's operating voltage is between 3.4 and 4.4 V, and an external battery must be used to supply the power to the module. This module includes an individual antenna called helical antenna that is connected such as a button and a net pin. In addition, at the back of the module is a micro SIM card slot. The module is used to connect to the cellular network in order to notify the owner of the vehicle via email, website, or text message. The module has an LED light that indicates the status of the cellular network. When blinking every 1 second, the module is running and looking for a network, when blinking every 2 seconds the GPRS data connection requested is active, and for every 3 seconds, the module has successfully connected to the network. To make the module work, it is necessary to connect the pins accordingly such as in **Figure 16** since the module only handles 3.4–4.4 V.

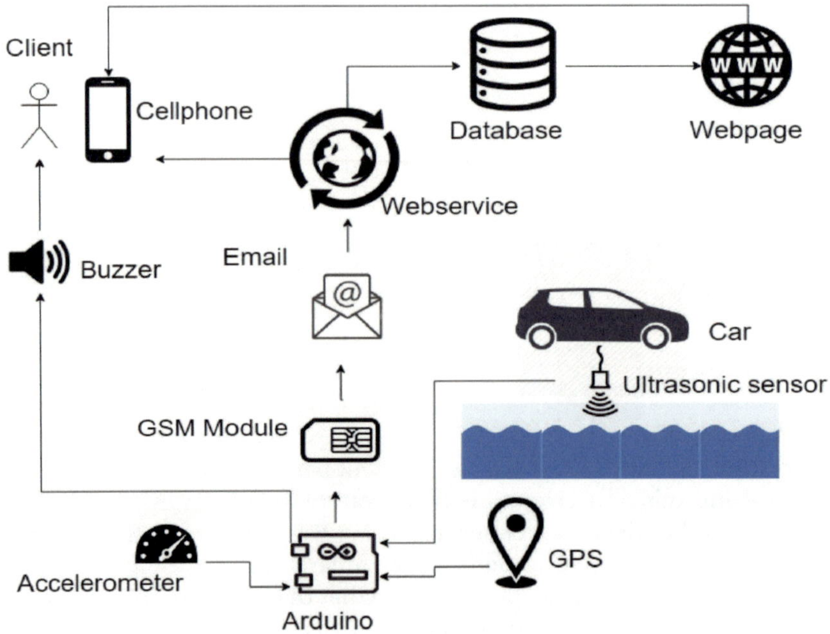

Figure 15.
Interaction of the system, client, and webpage.

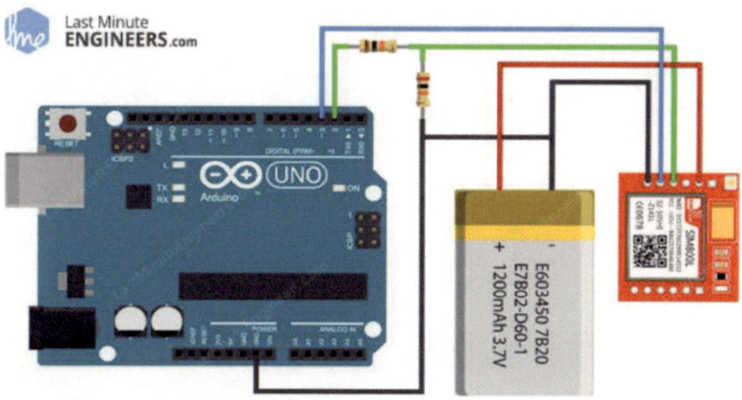

Figure 16.
Circuit assembly for GSM module.

An accelerometer is used to detect if the car is in motion. It works by sensing the acceleration of gravity [18]. This sensor has several pins. For this case four were used: the VCC, which needs 3.3 V to be powered, GND, serial clock line (SCL), and serial data line (SDA). **Figure 17** shows the interaction of the devices. The car battery shall be connected to a voltage regulator next to the Arduino in order to protect it. The Arduino is connected to the sensors that are responsible for obtaining the readings for when the water level is rising. It will also have an accelerometer connected in order to know when the car is in movement. There will also be a GPS module (to recognize the coordinates of the vehicle), an alarm (to notify the client when the car is moving), and a GSM module (to be able to send a message of the coordinates to an email for the web service to extract its information and process the data to store it in the database).

Once the accelerometer detects that the car is not in motion, the ultrasonic sensors start measuring the distance between the ground and itself. If there is a sudden rise in water level, using conditions established in the program, the ultrasonic sensors detect it, and the GSM module will send a warning message to notify of the sudden rise.

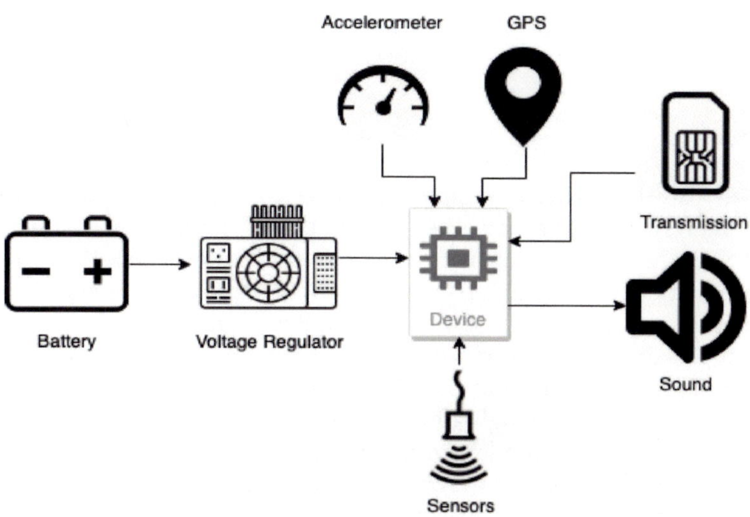

Figure 17.
Devices schematic diagram and interactions.

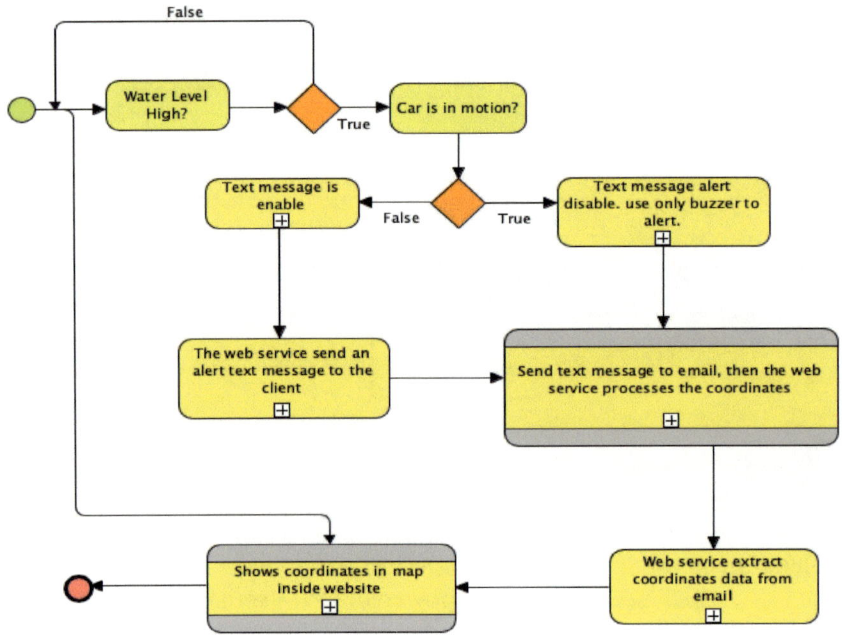

Figure 18.
System work diagram.

The software of the program consists of conditions based on thresholds, distances, and a counter. The first question is whether the vehicle is moving. This is where the accelerometer detects its movement using the Cartesian plane of x, y, and z. In the process diagram, in **Figure 18**, the first step is to detect if the water level is high. If the level is low, it shall keep checking if there is any change. If not, the vehicle shall proceed to check if the vehicle is in motion. If the car is in motion, then the text message alert will be disabled, and the client will be alerted through an audible notification. On the contrary, a text message will alert the client. Once the movement is detected, and the condition is established, a text message will be sent to the web server email containing the coordinates. The data is processed, and the web service extracts the coordinates from the email. After extracting the coordinates, a signal is made on the map shown on the web site where there are flooding areas.

8. Conclusion

Based on historical data and remote sensing technology, research on atmospheric characteristics and environmental effects of UHI has been conducted in the city of San Juan, Puerto Rico, providing a theoretical reference that evidenced the increment of UHI in the capital city. Based on analyses of temperature, solar radiation, evapotranspiration, and albedo data, it was found that the city of San Juan, Puerto Rico, is the most representative zone where the UHI is present in the last 20 years. The consequences of this effect have increased more rapidly in the last 10 years. In conclusion using high-albedo materials and pavements, having green vegetation and green roofs, implementing urban planning, and preserving pervious pavements, shade trees, and water bodies in city areas are the potential UHI mitigation strategies [19]. When urban planning is mentioned, a flood alert system is included. This paper represents the first time that water level technology has been used for hydrologic analyses, specifically for flash floods which are events caused by

UHI. The proposed system sends flood SMS alerts to the inhabitant of such zones for necessary action. The flood monitoring system can be expanded to cover a wider area than the one under study due to the protocol capabilities. In addition, this model can be used in flash floods, parking lots, streets prone to flooding, and other zones with a high probability of flood. Furthermore, UHI effects that develop in the city were also discussed, and through strategies such as the improvement of energy efficiency, urban landscape optimization, green roof construction, high reflectivity material utilization, and green land cultivation, UHI effects could be significantly mitigated.

Acknowledgment

The authors would like to thank CIESESE and Universidad Ana G Méndez (UAGM) for giving them the opportunity to participate in the PREC 2019.

Author details

Luz E. Torres Molina[1*], Sara Morales[2] and Luis F. Carrión[2]

1 Civil Engineering Department, Universidad Ana G Méndez, Gurabo, Puerto Rico

2 Engineering Department, Universidad Ana G Méndez, Gurabo, Puerto Rico

*Address all correspondence to: torresl6@uagm.edu

References

[1] EPA. Reducing Urban Heat Island; 2010

[2] Parry ML, Canziani OF, Palutikof JP, van der Linden PJ, Hanson CE, editors. Climate Change 2007: Impacts, Adaptation and Vulnerability. Contribution of Working Group II to the Fourth Assessment Report of the Intergovernmental Panel on Climate Change. United Kingdom: Working Group 3 Report: Society and Economy Cambridge University Press; 2007

[3] Gartland L. Heat Islands. 1st ed. London: Routledge; 2008. DOI: 10.4324/9781849771559

[4] Taylor M, Alfaro E. Climate of Central America and the Caribbean. In: Oliver JE, editor. Encyclopedia of World Climatology. Netherlands: Springer; 2005. pp. 183-189

[5] The United States National Weather Service. 2019. Available from: http://www.weather.gov

[6] Geological maps, Department of Geology, Puerto Rico. Available from: http://www.gis.pr.gov

[7] González JE, Comarazamy DE. Climate Impacts of LCLU Changes on a Tropical Coastal Region under a Changing Climate. 2nd International Conference on Countermeasures to Urban Heat Island. Berkeley, CA: Environmental Energy Technologies Department, E.O. Lawrence Berkeley National Laboratory; 2009

[8] Rosenzweig C, Solecki W, Slosberg R. Mitigating New York City's Heat Island with Urban Forestry, Living Roofs, and Light Surfaces. A Report to the New York State Energy Research and Development Authority; 2006

[9] Sailor DJ. Mitigation of Urban Heat Islands—Recent Progress and Future Prospects. Paper Presented at the Paper presented on American Meteorological Society 6th Symposium on the Urban Environment and Forum on Managing Our Physical and Natural Resources; 2006

[10] Ramos-González OM. The green areas of San Juan, Puerto Rico. Ecology and Society. 2014;**19**(3):21. DOI: 10.5751/ES-06598-190321

[11] Puerto Rico Agricultural Water Management. Available from: https://pragwater.com/

[12] Subhrajit G. A review of Heat Islands: Understanding and mitigating heat in urban areas. Journal of the American Planning Association. 2013;**79**(3):256-257. DOI: 10.1080/01944363.2013.811377

[13] Reda E, Zulkifli R, Harun Z. Large Eddy simulation of wind flow through an urban environment in its full-scale wind tunnel models. Journal of Mechanical Engineering Science. 2017;**11**(2):2665-2678

[14] Lotfy ER, Abbas AA, Zaki SA, Harun Z. Characteristics of turbulent coherent structures in atmospheric flow under different shear–buoyancy conditions. Boundary-Layer Meteorology. 2019;**173**:115-141

[15] Torres M. Flood alert system using sensor networks in flood prone zone. In: 16th LACCEI International Multi-Conference for Engineering, Education, and Technology: Innovation in Education and Inclusion, 19-21 July 2018. Lima, Peru; 2018

[16] Consumer Reports. Why flooded-out cars are likely total losses. Available from: https://www.consumerreports.org/car-insurance/why-flooded-out-cars-are-likely-total-losses/

[17] Latha A. Distance sensing with ultrasonic sensor and Arduino.

International Journal of Advance Research, Ideas and Innovations in Technology. 2016;**2**(5)

[18] Kosobitsky P, Ferens R. Statistical analysis of noise measurement system based on accelerometer-gyroscope GY-521 and Arduino platform. In: 14th International Conference the Experience of Designing and Application of CAD Systems in Microelectronics (CADSM). Ukraine: Lviv-Polyana; 2017. pp. 405-407

[19] Nuruzzaman M. Urban Heat Island: Causes, effects and mitigation measures - a review. International Journal of Environmental Monitoring and Analysis. 2015;**2**(3):67-73. DOI: 10.11648/j.ijema.20150302.15

Chapter 10

The Effect of Liquid Viscosity on the Rise Velocity of Taylor Bubbles in Small Diameter Bubble Column

Olumayowa T. Kajero, Mukhtar Abdulkadir, Lokman Abdulkareem and Barry James Azzopardi

Abstract

The rise velocity of Taylor bubbles in small diameter bubble column was measured via cross-correlation between two planes of time-averaged void fraction data obtained from the electrical capacitance tomography (ECT). This was subsequently compared with the rise velocity obtained from the high-speed camera, manual time series analysis and likewise empirical models. The inertia, viscous and gravitational forces were identified as forces, which could influence the rise velocity. Fluid flow analysis was carried out using slug Reynolds number, Froude number and inverse dimensionless viscosity, which are important dimensionless parameters influencing the rise velocity of Taylor bubbles in different liquid viscosities, with the parameters being functions of the fluid properties and column diameter. It was found that the Froude number decreases with an increase in viscosity with a variation in flow as superficial gas velocity increases with reduction in rise velocity. A dominant effect of viscous and gravitational forces over inertia forces was obtained, which showed an agreement with Stokes law, where drag force is directly proportional to viscosity. Hence, the drag force increases as viscosity increases ($5 < 100 < 1000 < 5000$ mPa s), leading to a decrease in the rise velocity of Taylor bubbles. It was concluded that the rise velocity of Taylor bubbles decreases with an increase in liquid viscosity and, on the other hand, increases with an increase in superficial gas velocity.

Keywords: structure velocity, cross-correlation, drag force, dimensionless number, electrical capacitance tomography

1. Introduction

Slug flow is characterized by Taylor bubbles, which has large pockets of bullet shaped bubbles occupying almost the entire cross-section of the column. The Taylor bubble is surrounded by a thin film of liquid, and below, it is the liquid slugs, which are agglomerate of small bubbles. Zukoski [1], Tomiyama et al. [2] and Mandal and Das [3] described the Taylor bubble length to exceed 1.5 times of the tube diameter or its diameter is greater than 60% of the tube diameter. The rise velocity of a single isolated Taylor bubble is dependent on inertia and drag forces [4]. A number of parameters affect the rise velocity of Taylor bubbles through a stagnant liquid; such

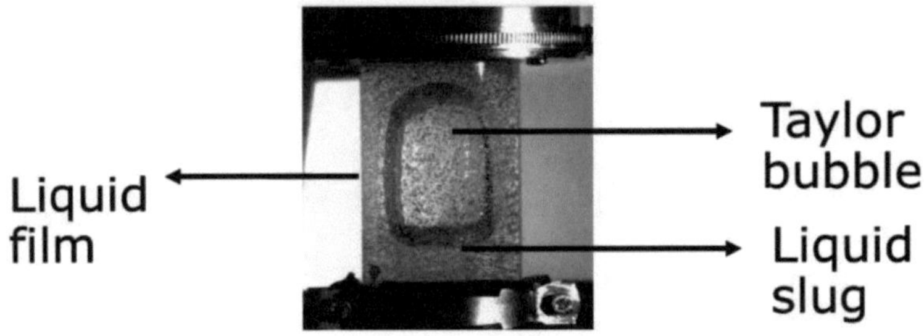

Figure 1.
A single Taylor bubble rising through a stagnant silicone oil liquid (viscosity, 1000 mPa s).

parameters include density of liquid, surface tension of liquid, liquid viscosity, acceleration due to gravity, diameter of bubbles etc. [5]. Mao and Dukler [6] explained that in a situation whereby the liquid is flowing, the rise velocity of a Taylor bubble must depend on the velocity of the liquid flowing upstream as well as the rise due to buoyancy. A typical example of bubble rising through a stagnant liquid as taken from a high-speed video camera (from the current study) is shown in **Figure 1**.

1.1 Background

The rise velocity of Taylor bubbles is otherwise known as structure velocity. It can also be defined as the velocity of periodic structures in the slug [7, 8]. Some researchers have carried out studies on the rise velocity of Taylor bubbles through stagnant liquid. Mao and Dukler [9] from their experimental results explained that for a wide range of viscosity and surface tension, the rise velocity can be expressed in terms of a constant Froude number.

$$Fr = \frac{U_N}{\sqrt{gD}} = constant \tag{1}$$

Hence, the rise velocity of Taylor bubble is given as:

$$U_N = Fr\sqrt{gD} \tag{2}$$

where g is the acceleration due to gravity and D is the diameter of the tube.

The rise velocity of Taylor bubbles in stagnant liquids was first studied by Dumitrescu [10] and Davies and Taylor [11] in which observed bubbles were of characteristics shape referred to as Dumitrescu or Taylor bubbles. Griffith and Wallis [12] eventually proposed the name as Taylor bubbles. Dumitrescu [10] carried out a study on the rise velocity of bubbles using water in a vertical tube and it was established both theoretically and experimentally that the bubble velocity was:

$$U_N = 0.35\sqrt{gD} \tag{3}$$

This is synonymous to the proposition of Mao and Dukler [9], where Froude's number is given as 0.35.

Dumitrescu [10] assumed that the bubble would have a spherical nose, solving simultaneously the flow around the bubble and the asymptotic film which eventually led to the bubble velocity [13] given as Eq. (3).

Davies and Taylor [11] gave the bubble velocity as:

$$U_N = 0.328\sqrt{gD} \tag{4}$$

after solving the problem using different assumptions.

Nicklin et al. [7] later postulated that the Davies and Taylor [11] solution was not unique but should tend to the limiting value given as:

$$U_N = 0.346\sqrt{gD} \tag{5}$$

Eqs. (3) and (4) proposed by Dumitrescu [10] and Davies and Taylor [11], respectively, assume that the Taylor bubble was obtained from a gas of zero density [13]. Neal [14] proposed that if the bubble density is significant, the bubble velocity is given as:

$$U_N = c\sqrt{gD\left(\frac{\Delta\rho}{\rho_L}\right)} \tag{6}$$

where c is approximately 0.35, $\Delta\rho = \rho_L - \rho_G$, ρ_L and ρ_G are the liquid and gas densities respectively.

Brown [15] from his experimental studies found that the solutions of Dumitrescu [10] and Davies and Taylor [11] were not suitable for high viscosity liquids, that they only describe the behaviour of gas bubbles in low viscosity liquids [13]. So, Brown [15] gave the bubble velocity as:

$$U_N = 0.35\sqrt{g(D - 2\delta_o)} \tag{7}$$

$$\text{where } \delta_o = \frac{D\sqrt{1+N_{LB}} - D}{N_{LB}} \tag{8}$$

and

$$N_{LB} = \left(\frac{14.5\rho_L^2 D^3 g}{\mu_L^2}\right) \tag{9}$$

where N_{LB} is the liquid viscosity number and μ_L is the liquid viscosity.

Zukoski [1] proposed an expression for velocity of large bubbles in a closed horizontal pipe with large diameter (neglecting surface tension effects) given as:

$$U_N = 0.54\sqrt{gD} \tag{10}$$

where the Froude number is 0.54.

A correlation for the bubble rise velocity was proposed by Griffith and Wallis [12] based on the studies on vertical slug flow given as:

$$U_N = (U_{SG} + U_{SL}) + K_1.K_2\sqrt{gD} \tag{11}$$

where U_{SG} and U_{SL} are the superficial gas and liquid velocities respectively, and K_1 = 0.35.

They investigated the effect of different velocity profiles in the liquid slug by varying K_2 [16].

Nicklin et al. [7] from their vertical slug experiments proposed the rise velocity of a Taylor bubble in the liquid in a vertical tube as:

$$U_N = C_o(U_{SG} + U_{SL}) + U_o \tag{12}$$

where U_o is the translational velocity in a stagnant liquid or velocity of bubble propagating into stagnant liquid, given as:

$$U_o = 0.35\sqrt{gD} \tag{13}$$

where C_o is the distribution coefficient which is close to 1.2 for fully developed turbulent flow (low viscous liquid e.g. water) and close to 2 for laminar flow (high viscous liquid e.g. 1000 mPa s silicone oil) [7, 17–19].

From the experimental work reported by Sylvester [20], the rise velocity of Taylor bubble was presented as:

$$U_N = C_o(U_{SG} + U_{SL}) + C_1\left[\frac{gD(\rho_L - \rho_G)}{\rho_L}\right]^{1/2} \tag{14}$$

where D is the pipe diameter. They proposed C_o and C_1 to be 1.2 and 0.35, respectively.

From the experiment carried out by Bendiksen [21] in a vertical tube with flowing liquid, the distribution coefficient, C_o was obtained to be 1.2 for Reynold's number in the range 5000–110,000, i.e. low viscous liquid.

Nicklin et al. [7] interpreted their equation as:

Rise velocity of Taylor bubble (structure velocity) is equal to the velocity of the liquid at the tip of the bubble nose plus the rise velocity of bubble in a stagnant liquid (translational velocity).

Mao and Dukler [6] called this effective upstream velocity (rise velocity of Taylor bubble), the centreline velocity of the liquid.

White and Beadmore [22] carried out an experimental investigation on the rise velocity of Taylor bubbles through liquids in a vertical tube using three dimensionless parameters: Froude number, Fr, Eotvos number, E_o and Morton number, M_o. A recent review on vertical gas–liquid slug flow which highlights previous works on the rise velocity of Taylor bubbles is provided by Morgado et al. [23]. They discussed experimental, theoretical and numerical methods of investigating the rise velocity of Taylor bubbles, where the so-called numerical methods involve the use of empirical correlations. The limitations of these studies that have been addressed in the current study include (1) low range of liquid viscosities and more emphasis on low viscosities rather than high viscosities, (2) consideration of column or pipe diameter greater than 50 mm, (3) limited exploration of the effect of forces such as surface tension, inertia, gravitational and viscous forces acting on Taylor bubble, (4) limited exploration of the relationship between fluid dimensionless parameters and the Taylor bubble rise velocity, and (5) detailed comparison between different methods for obtaining the Taylor bubble rise velocity.

The rising Taylor bubble in a stagnant liquid as observed from the high-speed camera and ECT instrument 3D image (from current study) can be seen in **Figure 2**.

1.2 Fluid flow studies using dimensionless numbers

A significant number of dimensionless parameters have been identified to be of relevance in fluid flow studies. Examples of such include bond number, capillary number, drag coefficient, Froude number, inverse dimensionless viscosity, Reynolds number and Weber number to mention a few. In this study, the Reynolds

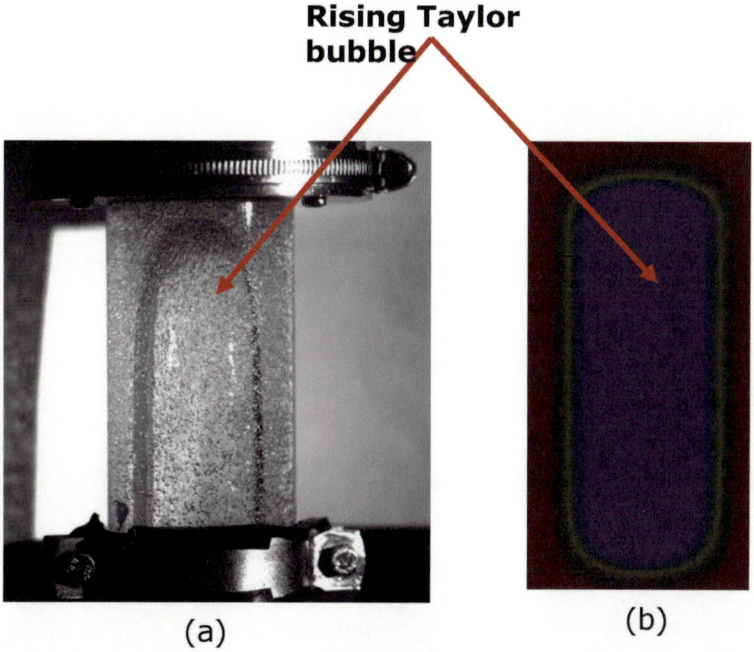

Figure 2.
Rising Taylor bubble from (a) high speed camera and (b) ECT instrument for 1000 mPa.s silicone oil at 0.361 m/s gas superficial velocity.

number, inverse dimensionless viscosity and Froude number were used to analyse the experimental results to clearly explain the effect of liquid viscosity on structure velocity.

1.2.1 Reynolds number, Re

The Reynolds number gives a measure of the ratio of inertia forces to viscous forces. Hence, it can be used to depict the competitive interplay between the effect of inertia forces and viscous forces [24].

$$\textbf{\textit{Reynolds number, Re}} = \frac{\textbf{\textit{Inertia forces}}}{\textbf{\textit{Viscous forces}}} \tag{15}$$

Reynolds number can also be used to characterize flow regimes into laminar or turbulent flow. The occurrence of laminar flow is at low Reynolds number in which viscous forces dominate. This is characterized by smooth, constant fluid motion. Turbulent flow on the other hand is at high Reynolds numbers which is associated with chaotic eddies, vortices and other flow instabilities [25].

The slug Reynolds number which is the Reynolds number of the rising slug in the gas–liquid mixture [26–28] is expressed as:

$$\boldsymbol{Re} = \frac{\boldsymbol{\rho U_M D}}{\boldsymbol{\mu}} \tag{16}$$

where $\boldsymbol{\mu}$ = dynamic viscosity of the fluid, $\boldsymbol{\rho}$ = density of the fluid, $\boldsymbol{D}$ = diameter of column and $\boldsymbol{U_M}$ is the mixture velocity [26, 29].

1.2.2 Inverse dimensionless viscosity, N_f

According to Lu and Prosperetti [30], the inverse dimensionless viscosity is directly proportional to the fourth root of the Eotvos number raised to a power of three and inversely proportional to the fourth root of the Morton number. Morton number, $\boldsymbol{M_o}$ is used alongside with Eotvos number, $\boldsymbol{E_o}$ to characterize the shape of bubbles or drops moving in a surrounding fluid or continuous phase.

The inverse dimensionless viscosity is given as:

$$N_f = \left(\frac{E_o^3}{M_o}\right)^{1/4} \tag{17}$$

where Morton number is given as:

$$M_o = \frac{g\mu_L^4}{\rho_g\sigma_L^3} \tag{18}$$

Eotvos number is given as:

$$E_o = \frac{\rho g D}{\sigma_L} \tag{19}$$

1.2.3 Froude number

Apart from the effect of viscous force, gravitational force also affects the rise velocity of Taylor bubble through the liquid. Froude number is a dimensionless parameter which gives a relationship between inertia and gravitational forces. It describes different flow regimes of open channel flow as in the case of the bubble column in this study.

The slug Froude number [31] is given as Eq. (1). Llewellin et al. [5] called the Froude number a dimensionless velocity.

2. Experimental arrangements

The bubble column experimental set-up consists of a 50 mm internal diameter and 1.6 m long perspex column in a vertical orientation. At the bottom of the column is a single nozzle gas distributor through which gas is introduced into the column. A phantom high-speed camera was used to obtain the video of the gas–liquid flow in the column. A frame rate of 1000 pictures per second (pps) and exposure time of 100 μs was used. The geometry specified from the high-speed camera setting gives the image width by image height as 512 by 512 pixel.

Fitted midway to the column is the twin-planes electrical capacitance tomography (ECT) sensor with an interplanar spacing of 30 mm. The 8-electrode system consists of measurement and driven guard electrodes, which is connected to the electrical capacitance tomography processor box, TFLR 5000–20. This sensor electronics gives 28 measurements which are relayed to the computer where image reconstruction occurs, and the data are acquired and processed to obtain the liquid holdup (which is the fraction of liquid in the gas-liquid mixture). The void fraction otherwise known as gas holdup is hence obtained from this.

The ECT is located about 0.7 m above the nozzle, while the liquid level is located about 0.095 m above the ECT sensor. On injecting the gas, the gas flows into the

Figure 3.
The twin-plane ECT sensor used.

bubble column through the single nozzle gas injector with an orifice diameter of 6.8 mm. A range of silicone oil with viscosities 5, 100, 1000 and 5000 mPa s was used.

The liquid holdup obtained from the ECT was used to obtain the structure velocity (rise velocity of Taylor bubbles) via cross-correlation between two planes—plane 1 and plane 2 putting into consideration the distance between the two planes, 30 mm. The ECT sensor used is shown in **Figure 3**.

Further details of the experimental arrangements are given in Kajero et al. [32, 33].

3. Rise velocity of Taylor bubbles

In this study, the rise velocity of Taylor bubbles was obtained from ECT (via cross-correlation between signals from planes 1 and 2 as shown in **Figure 3**), manual time series analysis and the high-speed camera. This rise velocity of Taylor bubbles is also referred to as structure velocity which is from the Taylor bubble periodic structures velocity at real-time measurements.

3.1 Cross-correlation

Correlation is the measure of the degree of linear relationship between two variables. Cross-correlation is a statistical method of estimating the degree to which two variables (in this case, time series data sets) are correlated.

The structure velocity was computed from the cross-sectional time averaged void fraction data measured by the ECT for both planes 1 and 2. The cross-correlation between the signals obtained from the two planes gave the structure velocity.

Given two functions $x(t)$ and $y(t)$, the cross-correlation function, $R_{xy}(\tau)$ between them is given as:

$$R_{xy}(\tau) = \lim_{T\to\infty} \frac{1}{T}\int_0^T x(t)y(t+\tau)d\tau \tag{20}$$

The correlation coefficient function is expressed as:

$$\rho_{xy}(\tau) = \frac{C_{xy}(\tau)}{\sqrt{C_{xx}(0)C_{yy}(0)}} = \frac{R_{xy}(\tau) - \mu_x\mu_y}{\sqrt{\left(R_{xx}(0) - \mu_x^2\right)\left(R_{yy}(0) - \mu_y^2\right)}} \tag{21}$$

where τ is the time delay, T is the record length (period), $C_{xy}(\tau)$ is the cross-covariance function, $C_{xx}(0)$ and $C_{yy}(0)$ are auto-covariance functions for x and y, respectively when time delay is zero, μ_x and μ_y are mean of the corresponding series, and $R_{xx}(0)$ and $R_{yy}(0)$ are the auto-correlation functions at a time delay of zero [34].

In this experimental work, the two functions $x(t)$ and $y(t)$ are time series data of planes 1 and 2, respectively.

The important parameters required for the computation of structure velocity using cross-correlation include:

a. Void fraction data for planes 1 and 2.

b. Number of data points.

c. Sampling frequency of data.

d. Distance between two planes (planes 1 and 2).

The time taken for the bubbles to travel between the two planes is calculated which then leads to the calculation of the structure velocity. This is done via an Excel Visual Basic Macro program used for the analysis of the time series data [35].

The time series, upstream and downstream with the corresponding correlation are shown in **Figure 4**. The time delay which is the time taken for the signal to travel between the two planes 1 and 2 is in the interval $-1 \leq \tau \leq 1$, where τ is the time delay.

Figure 4 indicates periodic structures of short slugs defined as advanced form of spherical cap bubbles gradually developing into clearly distinct slugs. These periodic structures are identified to be void waves in Taylor bubbles and liquid slugs in slug flow. Cross correlation of the time series data from the two axial locations can give

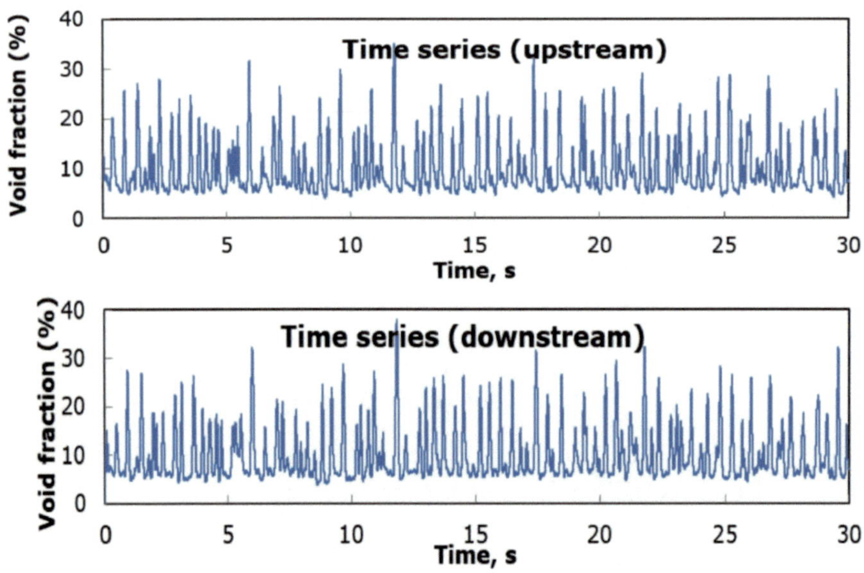

Figure 4.
Time series upstream and downstream for 5 mPa s silicone oil at a superficial gas velocity of 0.02 m/s.

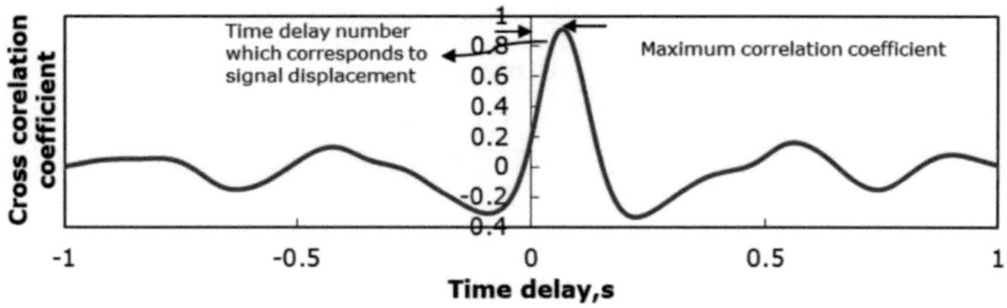

Figure 5.
Cross-correlation results for 5 mPa s silicone oil at a superficial gas velocity of 0.02 m/s.

the transit time (or time lag) which when used with the distance between the two planes leads to the calculation of the mean velocity for the slug structure.

From **Figure 5**, the y-axis on the cross-correlation plot is the correlation coefficient. This is in the range of −0.3 to +0.9 (though generally falls between −1 and +1). There could be perfect positive correlation (correlation coefficient of +1) or perfect negative correlation (correlation coefficient of −1). A positive correlation indicates that if a signal moves either up or down, the other signal will move in the same direction, while for a negative correlation, if a signal moves either up or down, the other signal will move by an equal amount in the opposite direction. When the correlation is 0, the movement of the signals gives no correlation and is completely random.

The mean velocity for the slug structure can hence be defined as distance between centres of measurement electrodes for two planes divided by the time delay [36–38].

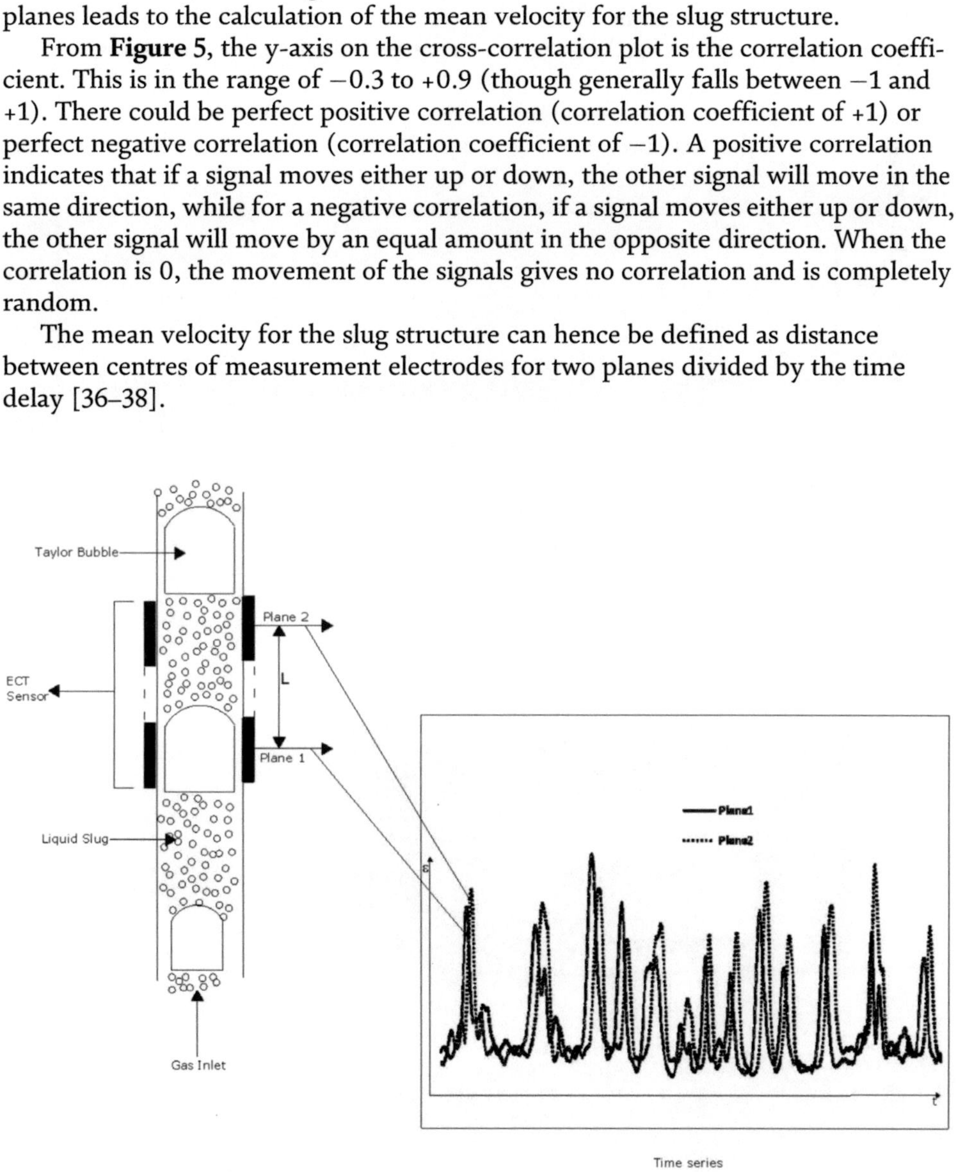

Figure 6.
ECT sensor signals generation for measurement of velocity in two-phase flows.

$$\textbf{Structure velocity }(\mathbf{m/s}) = \frac{\textbf{Distance between centres of measurement electrodes for two planes }(\mathbf{m})}{\textbf{Time delay }(\mathbf{s})} \tag{22}$$

The schematics of the ECT Sensor signals generation for measurement of velocity in two-phase flows is shown in **Figure 6**.

3.2 Manual time series analysis

The structure velocity can be obtained manually by estimating the time displacement of output signals from both planes 1 and 2 via the peaks of the time series plots for both planes. Since the distance between the centres of the measurement electrodes for both planes are known, the structure velocity is hence computed using Eq. (22).

The time displacement from the time series data of both planes 1 and 2 used in the calculation of the structure velocity alongside with distance between centres of

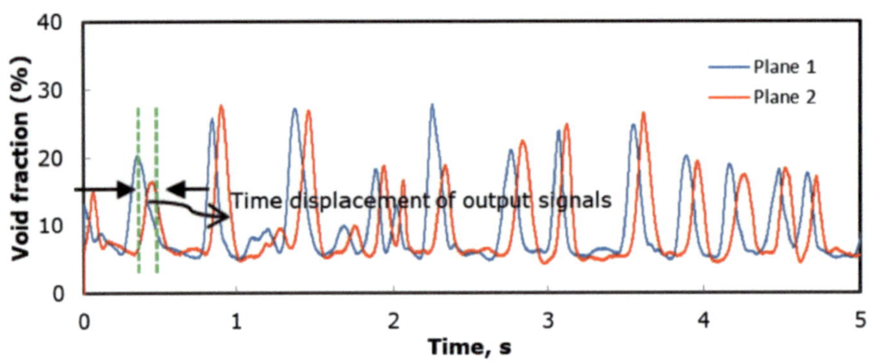

Figure 7.
Cross-sectionally averaged void fraction time series for plane 1 and 2 at a superficial gas velocity of 0.02 m/s.

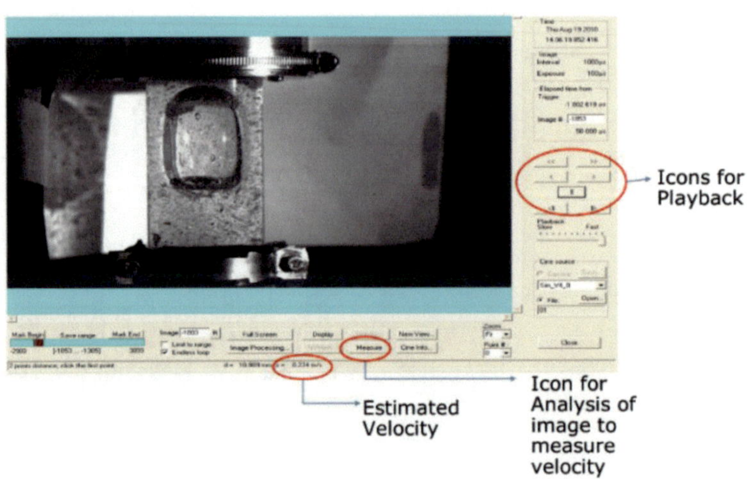

Figure 8.
Camera control software window used to estimate velocity.

measurement electrode is indicated in **Figure 7**. This is compared with that obtained via cross correlation later in this paper.

3.3 High-speed camera estimate

The camera control software of the Phantom High-Speed Camera can be used to obtain an estimate of the structure velocity. This is done via a playback in which the distance covered by the bubble at a given time is computed which gives a corresponding value for the velocity. A typical window of the camera control software is shown in **Figure 8**.

4. Results and discussion

4.1 Effect of viscosity on the rise velocity (structure velocity)

The effect of liquid viscosity on the rise velocity (structure velocity) has been studied by making a comparison between the respective structure velocities obtained from the ECT for the range of viscosities considered (i.e. 5, 100, 1000 and 5000 mPa s). The physical properties of the liquids used are given in **Table 1**.

A plot of structure velocity versus superficial gas velocity for all the viscosities considered is given in **Figure 9** which shows that structure velocity increases with an increase in superficial gas velocity which is in agreement with the observations of Abdulkaldir et al. [39] and decreases with increase in viscosity as shown in **Figure 10** (obtained from ECT Plot3d Image reconstruction software). The structure velocity of 5 and 100 mPa s is found to be approximately the same due to similar void fraction data values. The variation from small to bigger spherical cap and developing slug in 5 and 100 mPa s, and the slug flow in 1000 and 5000 mPa s (as shown in **Figure 9**) has been discussed by Kajero et al. [33].

This can be explained using the slug Reynolds number, a dimensionless parameter (Eq. 16).

A plot of slug Reynolds number versus superficial gas velocity is made at various viscosities as shown in **Figure 11**, with an indication of laminar flow.

Figure 11 reveals that as viscosity increases, slug Reynolds number decreases tending towards zero. This can be explained as follow:

i. Occurrence and prevalence of laminar flow as viscosity increases: According to Bendiksen [21], Reynolds number in the range 5000–110,000 (for low viscous fluids) give turbulent flow. As the Reynolds numbers of the viscosities considered are less than 5000, laminar flow prevails. For large slug Reynolds number, viscous effect will be negligible, while for small slug Reynolds number, viscous effect will be dominant [40]. So, since

Liquid	Viscosity, mPa s	Density, kg/m^3	Surface tension, mN/m	Relative permittivity
Silicone oil	5	915	19.7	2.60
	100	965	20.9	2.74
	1000	970	21.2	2.76
	5000	970	21.4	2.76

Table 1.
Physical properties of silicone oil viscosities used.

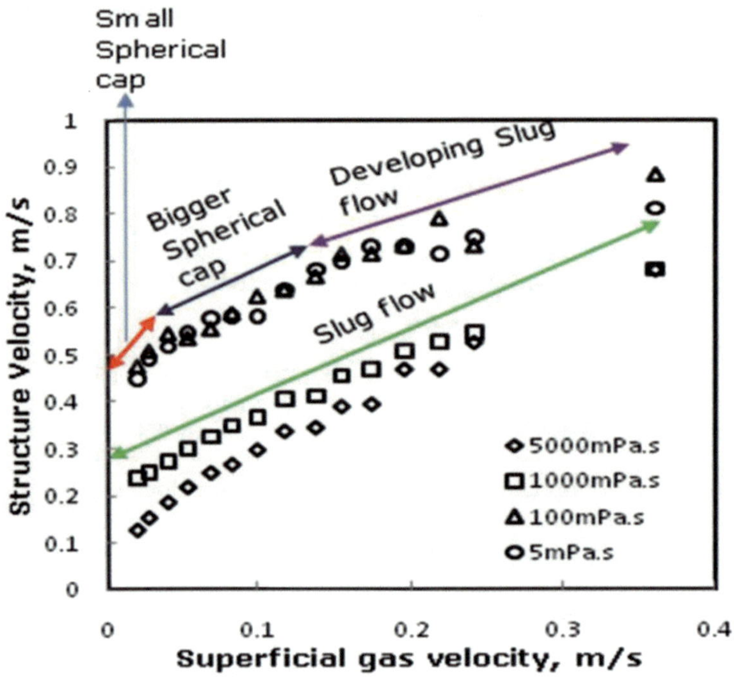

Figure 9.
Variation of structure velocity with superficial gas velocity at various viscosities.

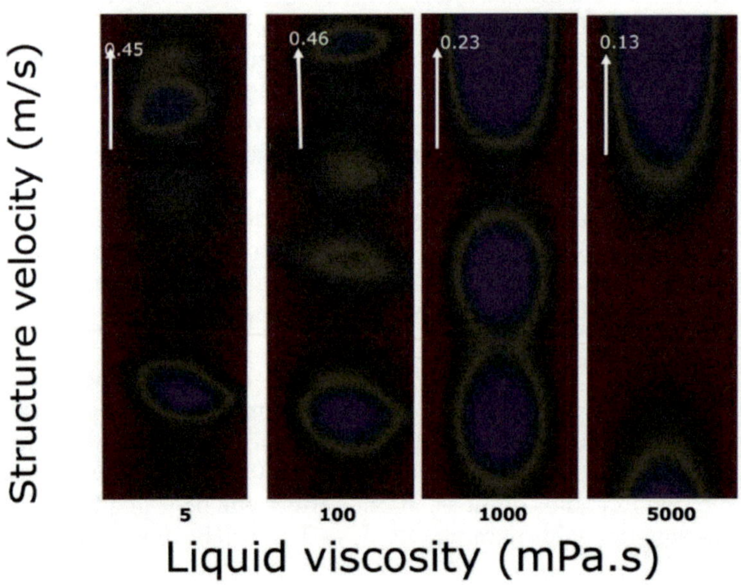

Figure 10.
Periodic structures in the slug as obtained from ECT Plot3d image reconstruction software (image display in single axial slice mode) at a superficial gas velocity of 0.02 m/s.

the slug Reynolds number for all the viscosities are small, viscous effect will be dominant.

ii. Dominating effect of viscous forces over inertia forces: Inertia forces are forces acting due to motion of bubbles through the liquid. It opposes any force that could resist motion. Viscous forces are forces acting due to the viscous nature of the liquid. Hence, from the plot, it can be inferred that viscous forces have a domineering effect over inertia forces [41].

The forces acting on the Taylor bubble are shown in **Figure 12**. These forces have an influence on its rise velocity. Surface tension force helps to hold the bubbles together due to the cohesive force existing between them. This accounts for why 'surface tension force' was indicated at the centre of the bubble in **Figure 12**. Based on the proposition of White and Beardmore [22], the effect of surface tension force can be neglected when Eotvos number is greater than 70. So, since for all the viscosities considered, Eotvos number is greater than 70, its effect on the rise velocity of Taylor bubbles can be neglected.

The dominating effect of viscous forces over inertia forces can be further confirmed using the Inverse dimensionless viscosity according to White and Beardmore [22].

From the various viscosities considered, the dimensionless property numbers are given as follows (**Table 2**).

It will be observed that as viscosity increases, Morton number, M_o increases while the Eotvos number, E_o decreases, which culminates in the decrease of the dimensionless inverse viscosity, N_f. This confirms the proposition of Fabre and Line [42]. As the dimensionless inverse viscosity decreases, viscous effect dominates [31].

White and Beardmore [22] proposed that viscous effects come into play when the square of dimensionless inverse viscosity, N_f is less than 3×10^5. i.e.

$$(N_f)^2 = \frac{(\rho_L^2)gD^3}{\mu_L^2} < 3 \times 10^5 \tag{23}$$

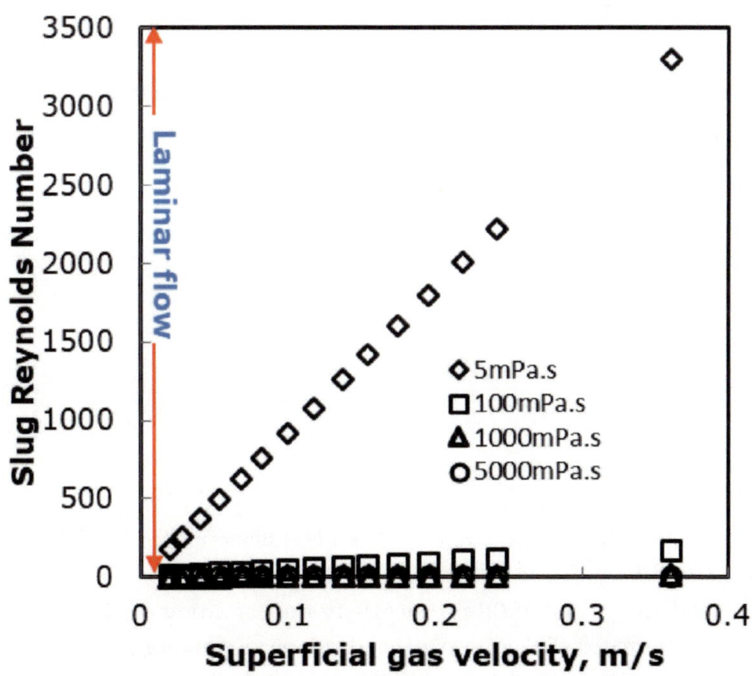

Figure 11.
Variation of slug Reynolds number with superficial gas velocity for various liquid viscosities.

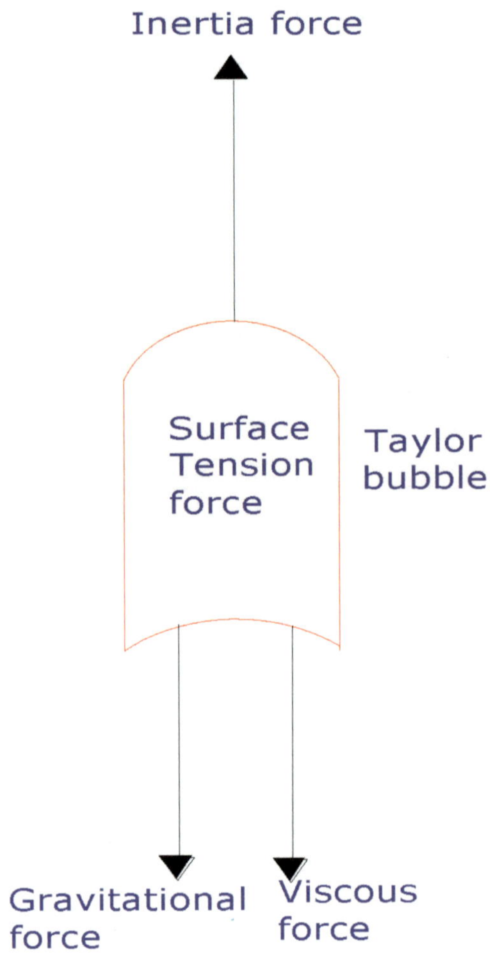

Figure 12.
Forces acting on Taylor bubble (surface tension force helps to hold the bubbles together).

Viscosity, mPa s	Dimensionless numbers		
	M_o	E_o	N_f
5	8.7645×10^{-7}	1139.105	6408.267
100	1.11353×10^{-5}	1132.374	337.9223
1000	1061.426	1122.134	33.96732
5000	644964.7	1111.647	6.793463

Table 2.
Morton, Eotvos and inverse dimensionless numbers.

The square of the inverse square dimensionless viscosity for 5, 100, 1000 and 5000 mPa s are 41,012,747, 114051.4, 1152.37 and 46.09481, respectively. Since 100 mPa s, 1000 mPa s and 5000 mPa s satisfy the condition of $(N_f)^2 < 3 \times 10^5$, viscous effect dominates. This dominating effect of viscous force over inertia force possibly causes a decrease in structure velocity with an increase in viscosity. This can be further confirmed by obtaining a relationship between drag force and viscosity taking superficial gas velocity as a parameter.

4.1.1 Drag force, F_D

From Stokes law, drag force is given as:

$$F_D = 3\pi\mu_L V d \tag{24}$$

based on the fact that the flow is laminar.

Drag force is the force due to the resistance provided by the fluid to the motion of a body through it.

From Eq. (24), Drag force is directly proportional to viscosity of the fluid.

Drag coefficient, C_D is given as:

$$C_D = \frac{24}{Re} \tag{25}$$

On substituting Eq. (25) into (24) given that $Re = \rho_L V d/\mu_L$,

$$F_D = \frac{72\pi\mu_L^2}{C_D\rho_L} \tag{26}$$

Eq. (26) is used to compute drag force, F_D which is plotted against superficial gas velocity for all viscosities considered as shown in **Figure 13(a)**. This can be interpreted as follows:

i. Drag force is directly proportional to viscosity, which confirms Eq. (26) according to Stokes law. Drag force is the force due to the resistance provided by the viscous nature of the silicone oil fluids to the motion of the large bubble through them. So, as viscosity increases, there is a linear increase in drag force.

ii. The increase in drag force shows an opposition to the motion of the bubbles through the liquid acting parallel to direction of relative motion, hence causing the bubbles to rise at much slower rate. This obviously leads to a decrease in the rise velocity [42].

To further confirm the effect of drag force, drag force was plotted against void fraction, which is the volume fraction of gas in the gas–liquid mixture as shown in **Figure 13(b)**.

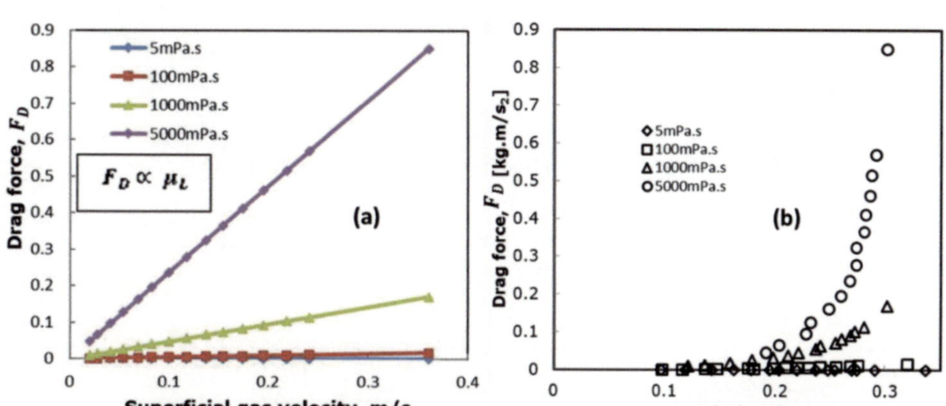

Figure 13.
Drag force versus (a) superficial gas velocity, (b) void fraction.

As void fraction increases, the drag force increases. This is probably due to the increase in the volume fraction of gas (void fraction); hence, an increase in the bubble size relative to the column diameter. This increase in bubble size is due to coalescence and it subsequently requires an increase in drag force to be effective enough to oppose the motion of the bubbles through the liquid. Hence, it can be said that the drag force is exponentially proportional to the void fraction. The drag force was found to be low at low viscosities (5 and 100 mPa s) and increases as viscosity increases. This is due to viscous effect as explained earlier on. As viscosity increases, void fraction increases; hence, an increase in drag force subsequently leads to hindered rise velocity.

The relationship between drag force and drag coefficient is shown in **Figure 14**. As viscosity decreases, the curves of the respective viscosities are tending towards zero. As drag coefficient increases, the drag force decreases tending towards zero. This gives an inverse relationship, which fits well into a power law expression, with the inverse proportionality constant increasing with increase in viscosity as shown in **Table 3**.

A plot of drag coefficient against Reynolds number on a log–log plot agrees with Stokes law, which gives an inverse relationship between the drag coefficient and Reynolds number as shown in **Figure 15**, with purple, green, blue and red for 5, 100, 1000 and 5000 mPa s respectively. This further confirms the fact that the drag force is directly proportional to the liquid viscosity. At low Reynolds number, drag coefficient is high, while at high Reynolds number, drag coefficient is low. Hence, the structure velocity of the latter is greater than the former.

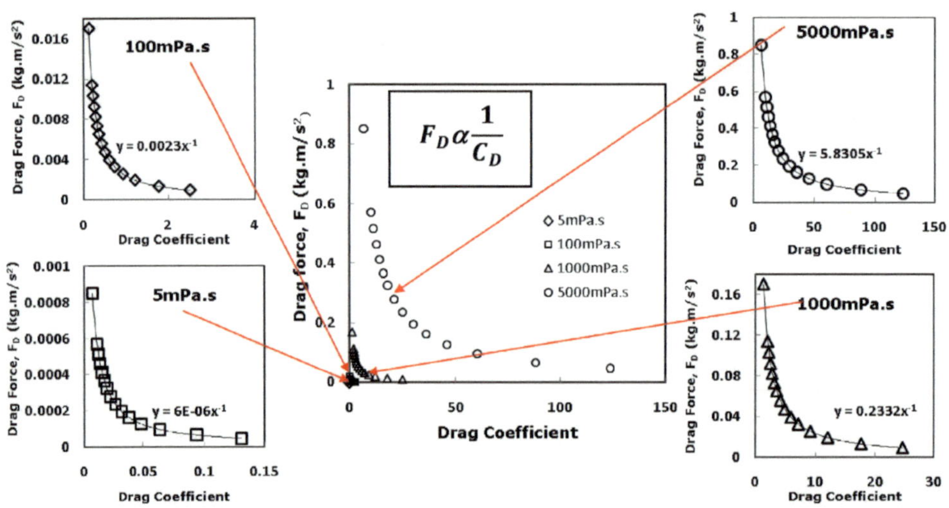

Figure 14.
Inverse relationship between drag force and drag coefficient for all the liquid viscosities considered.

Viscosity (mPa s)	Inverse proportionality constant
5	6×10^{-6}
100	0.0023
1000	0.2332
5000	5.8305

Table 3.
Inverse proportionality constants from the power law relationship between drag force and drag coefficient.

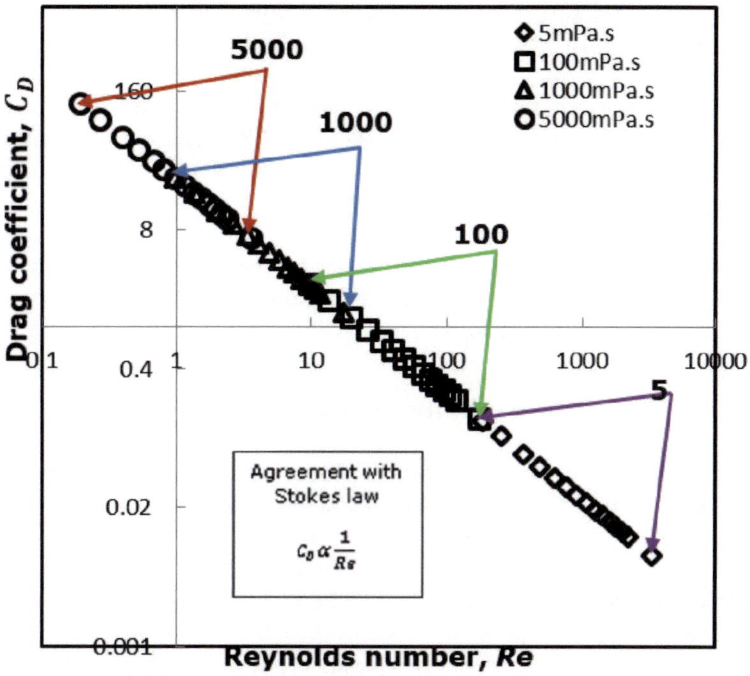

Figure 15.
Drag coefficients for the rising bubbles.

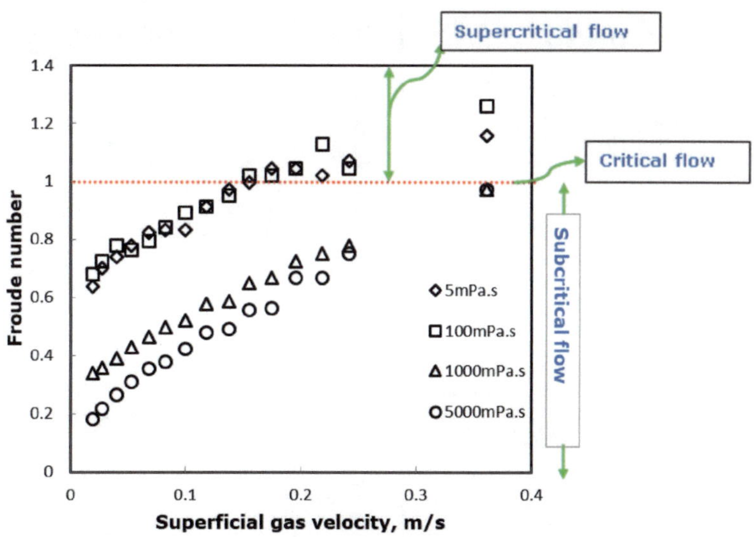

Figure 16.
Effect of Froude number.

A plot of Froude number versus the superficial gas velocity is shown in **Figure 16**. The following can be inferred from the plot:

1. At a superficial gas velocity of 0.01–0.17 m/s for both 5 and 100 mPa s viscosity, $Fr < 1$ which implies subcritical flow (slow/tranquil flow) due to low superficial gas velocity; while between 0.17 and 0.361 m/s, $Fr > 1$ which implies supercritical flow (fast rapid flow). At 0.2 m/s, $Fr = 1$ which implies critical flow [42]. In the case of both 1000 and 5000 mPa s, at all superficial

gas velocities, $Fr < 1$, this implies subcritical flow. The viscous effect makes the flow very slow, with low rise velocity of Taylor bubbles.

2. As viscosity increases, Froude number decreases which indicates the dominant effect of gravitational force over inertia force. This also has a retarding effect on the rise velocity of Taylor bubbles. Hence, this shows that the combined effect of the viscous and gravitational force causes a decrease in the rise velocity of bubbles as viscosity increases. For the high viscous liquids, 1000 and 5000 mPa s, as superficial gas velocity increases, the Froude number tends to be the same. This can be seen at 0.242 and 0.361 m/s. This implies that the dominant effect of gravitational force over inertia force tends to be the same.

4.2 Comparison of structure velocity computation methods

The structure velocity obtained from ECT was compared with the manual estimate from the time series [34], high-speed camera and empirical models such as modified Viana et al. [43], modified De Cachard and Delhaye [44], model and Nicklin et al. [7] model. These are shown in **Figure 17**.

The modified form of De Cachard and Delhaye [44] model is given as:

$$U_N = \left\{ 2.29\left[1 - \frac{20}{E_o}\left(1 - e^{-0.0125E_o}\right)\right]\right\} U_m + \Gamma\,(gD)^{1/2} \tag{27}$$

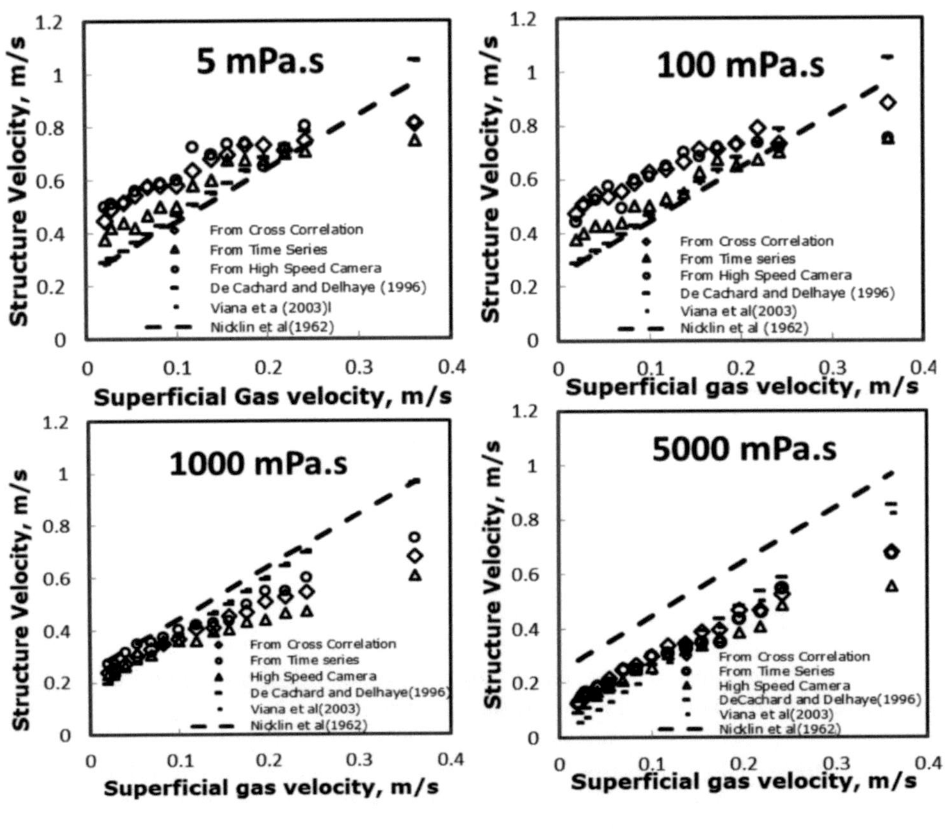

Figure 17.
Structure velocity comparison.

where $\boldsymbol{\Gamma}$ is given as:

$$\boldsymbol{\Gamma} = 0.345\left(1 - e^{-\frac{0.01N_f}{0.345}}\right)\left[1 - e^{(3.37 - B_o)/m}\right] \tag{28}$$

N_f is the inverse dimensionless viscosity.
Bond number, B_o is given as:

$$B_o = \frac{(\rho_L - \rho_G)gD^2}{\sigma} \tag{29}$$

Condition for m:

$$m = 10 \text{ when } N_f > 250 \tag{30}$$

$$m = 69(N_f)^{-0.35} \text{ when } 18 < N_f < 250 \tag{31}$$

$$m = 25 \text{ when } N_f < 18.$$

The modified form of Viana et al. [43] model is given as:

$$U_N = \left\{2.29\left[1 - \frac{20}{E_o}\left(1 - e^{-0.0125E_o}\right)\right]\right\}U_m + \sqrt[Fr]{gD} \tag{32}$$

where Fr is obtained from Eqs. (27) to (32).

$$Fr = L[R; A, B, C, G] \equiv \frac{A}{\left(1 + \left(\frac{R}{B}\right)^C\right)^G} \tag{33}$$

$$A = L[E_o; a, b, c, d] = \frac{a}{\left[1 + \left(\frac{E_o}{b}\right)^c\right]^d} \tag{34}$$

$$B = L[E_o; e, f, g, h] = \frac{e}{\left[1 + \left(\frac{E_o}{f}\right)^g\right]^h} \tag{35}$$

$$C = L[E_o; i, j, k, l] = \frac{i}{\left[1 + \left(\frac{E_o}{j}\right)^k\right]^l} \tag{36}$$

$$G = m/C \tag{37}$$

and the parameters $(a, b, \ldots\ldots\ldots, l)$ are:
$a = 0.34; b = 14.793; c = -3.06; d = 0.58; e = 31.08; f = 29.868; g = -1.96; h = -0.49; i = -1.45; j = 24.867; k = -9.93; l = -0.094; m = -1.0295.$

On the average, a good agreement exists between the structure velocity from the ECT and that measured from the high-speed camera. A reasonably fair agreement exists between the former and that estimated from the time series. The modified Viana et al. [43], modified De Cachard and Delhaye [44], and Nicklin et al. [7] model gave roughly similar pattern with structure velocity from cross-correlation, manual time series analysis and high-speed camera. Modified Viana et al. [43] and modified De Cachard and Delhaye [44] models showed good agreement with the velocity obtained from cross-correlation, manual time series analysis and high-speed camera at low superficial gas velocity for 1000 and 5000 mPa s, while Nicklin et al. [7] over predicts it.

The variations observed in the agreement could be due to the viscosities of the liquids used. Viana et al. [43] used silicone oil of viscosity range 1–3900 mPa s. De Cachard and Delhaye [44] and Nicklin et al. [7] used water. Also, from the structure velocity plots using cross-correlation, the distribution coefficient is in the range 1.07 to 1.6, while that of modified Viana et al. [43] and De Cachard and Delhaye [44] is approximately 2.25. A distribution coefficient of 2.0 was used for Nicklin et al. [7] as proposed for laminar flow.

The video technique of determining the rise velocity of bubbles gave errors of 4.3, 4.6, 7.3 and 11.5% for 5, 100, 1000 and 5000 mPa s, respectively when compared with cross-correlation technique.

5. Conclusions

From the foregoing, the following conclusions can be drawn:

1. The forces acting on a Taylor bubble in a 50 mm column diameter include inertia force, surface tension force, viscous force and gravitational force. Surface tension force can be neglected based on Eotvos number greater than 70. The remaining forces have an influence on the rise velocity of Taylor bubble. Furthermore, viscous and gravitational forces were observed to have dominant effect over inertia forces, hence causing the rise velocity of Taylor bubbles to decrease as viscosity increases.

2. The dimensionless parameters: Froude number, Reynolds number and inverse dimensionless viscosity all played vital roles in affecting the rise velocity of Taylor bubbles in various viscosities for a 50 mm diameter column. The dimensionless parameters being functions of the fluid properties and column diameter. Froude number helped to categorize the flow in the four viscosities considered into subcritical (slow and tranquil flow, $Fr < 1$), with lower rise velocity of Taylor bubbles, critical ($Fr = 1$) and supercritical flow (fast rapid flow, $Fr > 1$), with higher rise velocity of Taylor bubbles.

3. The rise of large bubbles through the liquid in the column agrees with Stokes law where drag force is directly proportional to viscosity and an inverse relationship exists between drag coefficient and Reynolds number, as superficial gas velocity increases. The drag force is also exponentially proportional to the void fraction and it retards the motion of the Taylor bubbles through the liquid. An inverse relationship in the form of Power law expression also exists between the drag force and drag coefficient. The drag coefficient was high at low Reynolds number but low at high Reynolds number, which contributed to the rise velocity of Taylor bubbles decreasing from low to higher viscosity liquids.

4. The rise velocity of Taylor bubbles increases with an increase in superficial gas velocity for each viscosity considered (i.e. 5, 100, 1000, 5000 mPa s).

5. The comparison between the rise velocity of Taylor bubbles obtained from the ECT, high-speed camera, cross correlation, manual time series, Nickel et al. [7] model, modified models of De Cachard and Delhaye [44] and Viana et al. [43] gave reasonably fair agreement.

Dedication

This publication is in loving memory of Late Prof. Barry Azzopardi.

Nomenclature

U_N	rise velocity of Taylor bubble, m/s
μ_L	liquid viscosity, kg/m.s
D	column diameter, m
σ_L	surface tension (N/m)
ρ_L	liquid density, kg/m^3
g	acceleration due to gravity, m/s^2
U_m	mixture velocity, m/s
U_{SG}	superficial gas velocity, m/s
U_{SL}	superficial liquid velocity, m/s
Re	Reynolds number, dimensionless
E_o	Eotovos number, dimensionless
B_o	Bond number, dimensionless
Fr	Froude's number, dimensionless
M_o	Morton number, dimensionless
N_f	inverse dimensionless viscosity, dimensionless
ρ_G	gas density, kg/m^3
C_D	drag coefficient, dimensionless
F_D	drag force, N

Author details

Olumayowa T. Kajero[1]*, Mukhtar Abdulkadir[2], Lokman Abdulkareem[3] and Barry James Azzopardi[4]

1 Department of Chemical and Process Engineering, University of Surrey, Guildford, United Kingdom

2 Department of Chemical Engineering, Federal University of Technology, Minna, Nigeria

3 Department of Petroleum Engineering, University of Zakho, Zakho City, Northern Iraq

4 Department of Chemical and Environmental Engineering, University of Nottingham, United Kingdom

*Address all correspondence to: ot.kajero@gmail.com

References

[1] Zukoski EE. Influence of viscosity, surface tension, and inclination angle on motion of long bubbles in closed tubes. Journal of Fluid Mechanics. 1966;**25**:821

[2] Tomiyama A, Nakahara Y, Adachi Y, Hosokawa S. Shapes and rising velocities of single bubbles rising through an inner subchannel. Journal of Nuclear Science Technology. 2003;**40**:136

[3] Mandal TK, Das G, Das PK. Prediction of rise velocity of a liquid Taylor bubble in a vertical tube. Physics of Fluids. 2007;**19**:128109

[4] Mario ART. Terminal velocity of a bubble rise in a liquid column. World Academy of Science, Engineering and Technology. 2007;**28**:264-268

[5] Llewellin EW, Bello ED, Taddeucci J, Scarlato P, Lane SJ. The thickness of the falling film of liquid around a Taylor bubble. Proceedings of the Royal Society A. 2011;**2012**(468):1041-1064

[6] Mao ZS, Dukler AE. The motion of Taylor bubbles in vertical tubes—II. Experimental data and simulations for laminar and turbulent flow. Chemical Engineering Science. 1991;**46**(8): 2055-2064

[7] Nicklin DJ, Wilkes JO, Davidson JF. Two-phase flow in vertical tubes. Transactions of the Institution of Chemical Engineers. 1962;**40**:61-68

[8] Omebere-Iyari NK, Azzopardi BJ. A study of flow patterns for gas-liquid flow in small diameter tubes. Chemical Engineering Research and Design, Institution of Chemical Engineers. 2007; **85**(A2):180-192

[9] Mao ZS, Dukler AE. The motion of Taylor bubbles in vertical tubes. I. A numerical simulation for the shape and rise velocity of Taylor bubbles in stagnant and flowing liquid. Journal of Computational Physics. November 1990;**91**(1):132-160

[10] Dumitrescu DT. Stromung an einer Luftblase im senkrechten Rohr. Zeitschrift für Angewandte Mathematik und Mechanik. 1943;**23**:139-149

[11] Davies RM, Taylor GI. The mechanics of large bubbles rising through extended liquids and through liquids in tubes. Proceedings of Royal Society London. 1949;**200**(A):375-392

[12] Griffith P, Wallis GB. Two phase slug flow. Journal of Heat Transfer. 1961;**83**:307-318

[13] Govier GW, Aziz K. The Flow of Complex Mixtures in Pipes. New York: Van Nostrand Reinhold Company; 1972

[14] Neal LG. Analysis of slip in gas-liquid flow applicable to the bubble and slug flow regimes. KR-62, Kjeller Research Establishment. Kjeller, Norway. 1963

[15] Brown RAS. Mechanics of large gas bubbles in tubes. I. Bubble velocities in stagnant liquids. Canadian Journal of Chemical Engineering. 1965;**43**:217-223

[16] Bendiksen KH, Malnes D, Nydal OJ. On the modelling of slug flow. Chemical Engineering Communication. 1996; **141–142**:71-102

[17] Collins R, De Moraes FF, Davidson JF, Harrison D. The motion of a large gas bubble rising through liquid flowing in a tube. Journal of Fluid Mechanics. 1978;**89**:497-514

[18] Bendiksen KH. On the motion of long bubbles in vertical tubes. International Journal of Multiphase Flow. 1985;**11**:797-812

[19] Douglas JR. A theoretical and experimental study of airlift pumping

and aeration with reference to agricultural applications [PhD thesis]. Cornell University; 1987

[20] Sylvester ND. A mechanistic model for two-phase vertical slug flow in pipes. ASME Journal of Energy Resources Technology. 1987;**109**:206-213

[21] Bendiksen KH. An experimental investigation of the motion of long bubbles in inclined tubes. International Journal of Multiphase Flow. 1984;**10**: 467-483

[22] White ET, Beardmore RH. The velocity of rise of single cylindrical air bubbles through liquids contained in vertical tubes. Chemical Engineering Science. 1962;**17**:351-361

[23] Morgado O, Miranda JM, Araujo JDP, Campos JBLM. Review on vertical gas-liquid slug flow. International Journal of Multiphase Flow. 2016;**85**:348-368

[24] Batchelor GK. An Introduction to Fluid Dynamics. Cambridge: Cambridge University Press; 1967. pp. 211-215

[25] Rott N. Note on the history of the Reynolds number. Annual Review of Fluid Mechanics. 1990;**22**(1):1-11

[26] Shoham O, Dukler AE. Heat transfer during intermittent/slug flow in horizontal tubes. Industrial and Engineering Chemistry Fundamentals. 1982;**21**:312-319

[27] Campos JBLM, Guedes De Carvalho JRF. An experimental study of the wake of gas slugs rising in liquids. Journal of Fluid Mechanics. 1988;**196**: 27-37

[28] Chukwu GA. Study of transportation of GTL products from Alaskan North Slope (ANS) to markets. In: Final Report. The US Department of Energy, National Energy Technology Laboratory; 2002

[29] Dukler AE, Hubbard MG. A model for gas-liquid slug flow in horizontal and near horizontal tubes. Industrial and Engineering Chemistry Fundamentals. 1975;**14**(4):337-347

[30] Lu X, Prosperetti A. A numerical study of Taylor bubbles. Industrial and Engineering Chemistry Research. 2009; **48**(1):242-252

[31] Zheng D, He X, Che D. CFD simulations of hydrodynamic characteristics in a gas-liquid vertical upward slug flow. International Journal of Heat and Mass Transfer. 2007;**50**: 4151-4165

[32] Kajero OT, Abdulkareem L, Azzopardi BJ. Effect of liquid viscosity on slug flow in a small diameter bubble column. In: Proceedings of the International Conference, Experimental Fluid Mechanics, Jicin, Czech Republic. 2011

[33] Kajero OT, Abdulkadir M, Abdulkareem L, Azzopardi BJ. Experimental study of viscous effects on flow pattern and bubble behaviour in small diameter bubble column. Physics of Fluids. 2018;**30**(9):093101

[34] Bendat J, Piersol A. Engineering Application of Correlation and Spectral Analysis. New York, USA: John Wiley and Sons; 1980

[35] Kaji R, Hills JH, Azzopardi BJ. Extracting information from the time series data in vertical upflow. Multiphase Science and Technology. 2009;**21**(1–2):185

[36] Sekoguchi K, Takeishi M, Hironaga K, Nishiura T. Velocity measurement with electrical double sensing devices in two-phase flow. In: Measuring Techniques in Gas-Liquid Two Phase Flows: Symposium. 1984. pp. 455-477

[37] Abdulkareem LA. Tomographic investigation of gas-oil flow in inclined

risers [PhD thesis]. Department of Chemical and Environmental Engineering, University of Nottingham; 2011

[38] Crowe CT. Multiphase Flow Handbook. USA: CRC Press/Taylor and Francis Group; 2006

[39] Abdulkadir M, Hernandez-Perez V, Sharaf S, Lowndes IS, Azzopardi BJ. Experimental investigation of phase distributions of an air-silicone oil flow in a vertical pipe. World Academy of Science, Engineering and Technology (WASET). 2010;**61**:52-59

[40] Robert WF, Philip JP, Alan TM. Introduction to Fluid Mechanics. 7th ed. India: John Wiley and Sons; 2010

[41] Thizon P, Contanceau M. Wall effect on the bubble behaviour in highly viscous liquids. Journal of Fluid Mechanics. 1981;**107**:339-373

[42] Fabre J, Line A. Modeling of two-phase slug flow. Annual Review of Fluid Mechanics. 1992;**24**:21-46

[43] Viana F, Pardo R, Yanez R, Trallero JL. Universal correlation for the rise velocity of long gas bubbles in round pipes. Journal of Fluid Mechanics. 2002;**494**:379-398

[44] Cachard F, Delhaye J. A slug-churn flow model for small-diameter airlift pumps. International Journal of Multiphase Flow. 1996;**22**:627-649